THE COLLECTION BOULLIAU

(BN, FF. 13019–13059)

An Inventory

ISMAËL BOULLIAU (1605–1694)

Figure 1. Engraving by Pieter van Schuppen (1623?–1702), "le petit Nanteuil," after a portrait by his son Jacob van Schuppen (1670–1751). From Charles Perrault, *Les hommes illustres* (Paris, 1696–1700).

THE COLLECTION BOULLIAU

(BN, FF. 13019–13059)

An Inventory

by
Robert A. Hatch

Foreword
by
Harcourt Brown

THE AMERICAN PHILOSOPHICAL SOCIETY
INDEPENDENCE SQUARE
PHILADELPHIA
1982

Library of Congress Catalog Card No. 82-72157
International Standard Book Number: 0-87169-984-2

To Margaret

Busie old foole, unruly Sunne,
Why dost thou thus,
Through windowes, and through curtaines call on us?
Must to thy motions lovers seasons run?
Sawcy pedantique wretch, goe chide
Late schoole boyes, and sowre prentices,
Goe tell Court-huntsmen, that the King will ride,
Call countrey ants to harvest offices;
Love, all alike, no season knowes, nor clyme,
Nor houres, dayes, moneths, which are the rags of time.

* * *

—Donne, "The Sunne Rising"

Table of Contents

Illustrations

Foreword

Perspectives & Background: The Place of Ismaël Boulliau

For most students of classical French culture, the erudition and science of the seventeenth century have been represented rather by the Trissotins and Vadiuses of Molière or the astrologer of La Fontaine than by the colleagues of Descartes and Pascal. The presence of the mathematician and physicist has not effaced the satirical and comic figures which the fringe of *femmes savantes* and their fashionable protégés offered the witty satirist whose function was to amuse the bourgeois and his wife. Yet a little critical thought makes it apparent that the scientific revolution was already at work, that the outlook of the "new sciences" made the satire sharper, more apt, and that there were points of contact in such a comedy as *Le Malade imaginaire,* that Doctor Diafoirus and his absurd son appeared the more ridiculous as the intelligent reader or spectator thought of the innovations of Harvey and Pecquet. So it was also with astronomy; the comet of 1665 was discussed in the *Journal des Savants,* and it took little perception to distinguish between the innocent speculations of the learned Jesuits and the precise mathematical argument of Adrien Auzout, both set out at some length in successive pages in the spring of that fruitful year. Delight in the easy fun of the *comédiens du Roy* and the pages of books from the shop of Barbin was enhanced by the undercurrent of science already making itself felt in conversation and the circulation of news.

Those who have taught in the field of seventeenth century French literature know how few of the standard books pay any attention to this substantial body of thought. What has been described as the dissociation of sensibility, the developing breach between the sciences and the humanities, particularly apparent perhaps in the educational system of France, has made it hard for the teacher of French to speak with interest, not to say authority, of scientific developments, and to create understanding of the world as the classical author saw it. Generations of students of French literature have been deprived of knowledge of the solid work done in science, French and also European generally, contemporary with the literary works under review. The perspective of the French imagination has thus been limited, even in such an obvious case as the common interpretation of *Micromégas,* a hilarious farce remotely based on the notable expedition to Lapland of 1736–1737, led by Maupertuis and Anders Celsius to determine the curvature of the earth on a meridian crossing the Arctic circle. What could lead to instructive discussion of the way in which a great comic author distorts the facts for the purposes of the immediate amusement of Frederick the Great and his court has been neglected for comment on the nascent antagonism of Voltaire and Maupertuis; the actual perspective in which the arduous efforts of the French academicians should be seen is lost in the comic debate about ultimate philosophy in which the *volée de philosophes* indulges for the edification of the giant from *l'étoile Sirius*. This chasm exists generally through literary courses; the extensive and deep influence that science and scientists have exerted in much of French literature is simply left aside, even in places and times where it appears as fancy or the fantastic. One thinks of André Chénier and Paul Valery among the poets who touch the subject with intellectual insight, of Victor Hugo who exploits the vein with exuberant irresponsibility, of Rimbaud who absorbs the imagery with delight and vision, and in prose, of Balzac, Zola, and Jules Romains, who handle the arguments and language of science with individual degrees of precision, but still with a sense of the larger values involved.

Invitation to write a kind of foreword to a major work of seventeenth century scientific documentation came to me by a series of very odd circumstances. Some fifty years ago I happened to light on the relatively unknown Ismaël Boulliau in the course of doing something else; I believe that this is called serendipity from a character in Samuel Johnson's *Rasselas*. Nobody had ever mentioned Boulliau to me, nor did I discover his name in the works we usually read concerning the *grand siècle* in which he was in fact a not inconsiderable figure. At a certain point in a scholar's development one looks for an area in which there is room for study and possible discoveries and publication. I found, in a period that interested me in a general way, that there was one niche offering space and materials for work, a period inadequately explored and analyzed, occasionally recognized as such in the books but whose significance was quite unknown. The *solution de continuité,* as Rabelais would call it, lay between the death of Descartes in 1650 and the founding of the Académie des Sciences in 1666, and extended in some senses to the period of the acceptance of the Newtonian theory in France about

1732. This was the age of the salons, of *préciosité,* of the high intelligence of Pascal and Molière, of Racine and Boileau, authors usually taught for their aesthetic qualities and style rather than for the sheer intelligence which made their style exceptional.

My interest in this epoch increased as I tried to account for what has been called the *crise de conscience* preceding the age of the Enlightenment, a time known at least since H. T. Buckle's *History* as the Scientific Revolution, a phrase which has hardly impinged on the historians of literature. I had had smatterings of science from my early years from a father who was an amateur polymath interested in field botany, practical chemistry, the simpler parts of mathematics, a carpenter and joiner, worker in metals, and possessing a degree in pharmacy. The result was that as I read Voltaire in college, in particular *Micromégas* and the *Lettres philosophiques,* along with *pages choisies* of Montesquieu and Diderot, I was ready to find the accepted outlook on French literature of the eighteenth century highly unsatisfactory. Even Gustave Lanson's two or three excursions into the transition from what was described as "classicism" to the "mouvement philosophique" seemed less than adequate.

At this stage in a student's life, the great quest was, and perhaps still is, for a topic suitable for a doctoral dissertation, the *sine qua non* for a would-be academic. Taking the advice of an elder and wise scholar, Fortunat Strowski, I set about reading widely in literature of and about my period, and noticing phenomena not much regarded by previous workers. A first discovery was the importance of the learned journal, begun by Denis de Sallo in 1665, quickly followed by Henry Oldenburg; a second was the fact that the journals were closely related to the meetings of learned or at least cultivated men, lawyers, amateurs of the arts and sciences, travelers—sessions held in various private houses at more or less regular intervals. On these occasions news of noteworthy developments was exchanged; often the presence of an active library and museum of curiosities lent a background of erudition and even charm to the proceedings. Typical of all these *cabinets* or *conférences,* as they were called, was the effort to remain free of ministerial intervention. The *académie de chez Conrart* had been infiltrated by an agent of Richelieu and changed by pressure into the *Académie française,* its membership restricted, its program determined, and its sessions watched with political suspicion. Even under Louis XIV there were Frenchmen who took their liberties seriously.

A result was of course that these groups in Paris remained as obscure as possible, and hence it has been difficult to determine what they did, what was said in their sessions, and even who took part in their activities. The average professor of French has been therefore quite unaware of these activities—in which the "new sciences" were often discussed—and this whole aspect of seventeenth century life in Paris and the provinces has remained concealed from the general view of students, in particular because the historians who write the textbooks have rarely looked beyond the printed masterpieces to examine the milieu of circulating ideas in which the *chef-d'oeuvres* were composed. The correspondence of the mid-seventeenth century is copious and rich in texture; the subjects discussed range widely beyond what could be printed, the style is allusive and often in a kind of code which conveys meanings inaccessible to the eye of a potential censor, sometimes in words which have to be read with an implied negative. When, for instance, Henry Justel writes that a certain book is *dangereux* the initiate knows that it serves a common cause and should be read and circulated. And the sheer bulk of the letters preserved, bound in folio volumes that stand in rows sometimes under the eyes of the reading public, is an ominous warning that brief Thursday afternoons will not suffice for justice to be done to the decades of correspondence the books contain.

After three visits to Paris in successive summers, I was told that there was unplumbed wealth for a French scholar in the libraries of London, and we went there in 1932 to spend an autumn that turned into a good part of the winter. The Royal Society had much material, either direct from French writers, Justel, Auzout, and others, but also from English and Scots who corresponded with Oldenburg, and told of what was going forward in Paris. In the present context then was the first letter from Boulliau that we had seen, a contact set up by the services of Justel, and offering a new lead in the jungle of material of the Bibliothèque Nationale. Catalogs of that library were then incomplete, but secondary sources, Haag et Bordier, Chauffepied, and others guided us around a new area of research, and soon we undertook to look a little at the Fonds Dupuy, which offers a background for the Collection Boulliau, so thoroughly explored and cataloged by Robert Hatch.

In the 1930s it was something to point the way; the History of Science was not yet a recognized discipline, and the nascent Ph.D.

in French literature could hardly justify spending more time on the witty, erudite, and sometimes esoteric Boulliau, genial though he might have been in both French and English senses of the word. The present writer waves an aging hand to the waiting reader, now about to enter a formidable field of knowledge, under expert guidance, with adequate charts and abscissae.

Harcourt Brown

Preface

The life and career of Ismaël Boulliau virtually spanned the century of genius—the age of the *homme de science* and the *libertin érudit.* An accomplished astronomer, mathematician, and classical scholar, Boulliau was generously proclaimed, shortly after his death, "l'un des génies le plus universel de son temps." On balance, he was an activist in the cause of Scientific Revolution, a respected citizen in the Republic of Letters. His voluminous correspondence (rivaling the combined efforts of Mersenne and Oldenburg) placed him at the heart of one of the largest intelligence networks of the century. His communications with Hevelius, Gassendi, Huygens, Leopold of Tuscany, and numerous others wove the rich and often opposing strands of his age into a vast but telling tapestry, a portrait of the learned world in transition. Elected unanimously to the Royal Society, his major contributions were to an earlier age, to the halcyon days of the Cabinet Dupuy and similar assemblages foreshadowing the Académie des Sciences. But above all he was a man of science. An outspoken Copernican and defender of Galileo, an influential yet critical student of Kepler, he was the foremost mathematical astronomer in seventeenth-century France.

The most complete published account of Boulliau's life came at his death. His éloge in the *Journal des Sçavans* (XXIII, 1695, pp. 110–127) began: "Notre siecle tout fertile qu'il est en gens de letres, en a peu produit qui ayent reüni autant de qualitez diferentes qu'en avoit M. Boulliau." The first sketch of Boulliau's career consists of some four-thousand words. It indicates that he was born in Loudun in 1605 to Calvinist parents, at age twenty-one converted to Catholicism, at twenty-five was ordained a priest. It hints at his scientific interests, his contacts with learned Europe, and concludes with his retirement to the Abbaye Saint-Victor in Paris where he died in 1694.

Although his name plays prominently in seventeenth-century studies—from science and mathematics to philology and libertinage—until recently Boulliau has been undeservedly neglected. While the present study ventures to redress this neglect, it is far from complete; at best it is a beginning. It is increasingly clear that the definitive biography of Boulliau will not admit to hasty conclusion.

The proper materials are too diverse in kind, too dispersed in space, too numerous to be assembled with an eye on construction cost and production schedule. As the larger project continues, an attempt has been made to clear the ground, to scratch a crude outline in *terra firma*.

But the purpose of this work is not merely to introduce Boulliau; it is designed primarily as a guide to a correspondence network that helped to unite the learned world of his time. Accordingly, the volume includes a short sketch of Boulliau's life and career, followed by an Inventory (and Name Index) of the forty-one volumes of the Collection Boulliau (BN, FF. 13019–13059). Consisting of over 22,000 pages of manuscript materials, the Collection ranks as one of the largest sources of its kind in the seventeenth century.

Interest in scholarly correspondence is not new. Since the turn of the century, well-orchestrated efforts have produced monumental editions containing the correspondence of Kepler, Galileo, Descartes, Huygens, Newton, Mersenne, and Oldenburg. The published correspondence of these prominent scientists and "intelligencers" has enriched our understanding of science—its first inklings, second guesses, and second thoughts—as well as the complex communities that gave science shape. The purpose of the present Inventory is to continue this tradition, however modestly, by providing a *via media* between the published editions of eminent scientists and the virtual eclipse of other illuminati, their communities, and their communication networks. If the Inventory succeeds, it will shed light on the crucial decades preceding the *Philosophical Transactions* and the *Journal des Sçavans,* and not the least, promote scholarship "among the manuscripts of a dusty past."

In the extended course of this study I have incurred many debts both personal and professional. It is a pleasure to take this opportunity to acknowledge them. To begin, I must record my continued debt to David Lindberg who introduced me to the history of science as an undergraduate, who guided my graduate studies, and who has since persisted in his kindnesses. An inestimable debt is owed to my friend and colleague, John Leavey. His amiable skepticism, honesty, and criticism (often unsolicited) is deeply appreciated and highly prized. Cordial thanks also to Mark Smith for his reflections and advice in the early stages of the work. For their assistance and encouragement, I would like to thank John Neu, Robert Westman, Wilbur Applebaum, Albert Van Helden, Curtis Wilson, and Maureen Hyland Carver. My list of debts would be incomplete without mentioning my colleagues, Todd Savitt, Fred Gregory, Herb Kanetz, and other members of the I.C.B.S. Would that space afforded a *compte courant* of my students past and present. Their zest and curiosity have been a constant source of inspiration.

I am equally pleased to record my debt to other scholars—many I have never met—who have graciously offered information, assistance, and advice. I especially wish to thank Professor Harcourt Brown for writing the Foreword to this volume. Professor Brown's professional insight and personal generosity have loomed large since first I saw the name Boulliau in his classic study, *Scientific Organizations in Seventeenth Century France*. To Elsa Zilberbogen, who shares a keen interest in Boulliau, I express continued enthusiasm for an epistolary friendship not unknown to Boulliau and his contemporaries. I have also benefited from communications with other specialists, among them H. J. M. Nellen, Guy Picolet, and Professor René Taton. Finally, cordial thanks to Carole Le Faivre of the American Philosophical Society. Her editorial skill and patience have eased my task.

Lest the obvious remain unstated, the rationale of my research is rooted in the efforts of others. I would like to take this opportunity to acknowledge the life-long labors of Alexandre-Guy Pingré, Guillaume Bigourdan, Pierre Humbert, René Pintard, and again Harcourt Brown. Theirs was a prodigious and provocative beginning, one worthy of articulation and synthesis.

In preparing this volume I have been privileged to use the facilities of several major research libraries. First and foremost, I would like to acknowledge the assistance of the staff in the Salle des Manuscrits at the Bibliothèque Nationale, particularly the able efforts of M. François Avril. Although the focus of this study is the Collection Boulliau, I must also acknowledge the assistance of Mme Alexandre of the Bibliothèque de l'Observatoire, and M. La Fourniere of the Archives du Ministère des Affaires Etrangères. Finally, cordial thanks to the courteous and capable staff at the Library of the Royal Society, the Bodleian Library, and the British Library. I only regret that Boulliau had so few English correspondents.

Access to these and other facilities would have been impossible without assistance of a different sort. My maiden voyage to Europe was made possible by a Henry Vilas Travel grant and an Amos B. Lytton fellowship from the University of Wisconsin; I could not have begun my work without these timely fellowships. My initiation to other manuscript collections in the French capital was in turn made possible by an Iowa State University Faculty Research Grant—this too was welcome and timely. Finally, I must record a special thanks to the Division of Sponsored Research at the University of Florida for supporting my research in Europe in 1980 and 1981; and to the American Philosophical Society (Penrose Fund) and the Gilbert Chinard foundation (Institute français de Washington) for their generous support last Spring. This volume represents the first installment in a larger debt.

My greatest debt is untold. For her skill and good humor in typing the manuscript, for her continued selflessness as companion, I have dedicated this volume to Margaret.

University of Florida
Vernal equinox 1982

Robert A. Hatch

Biographical Introduction

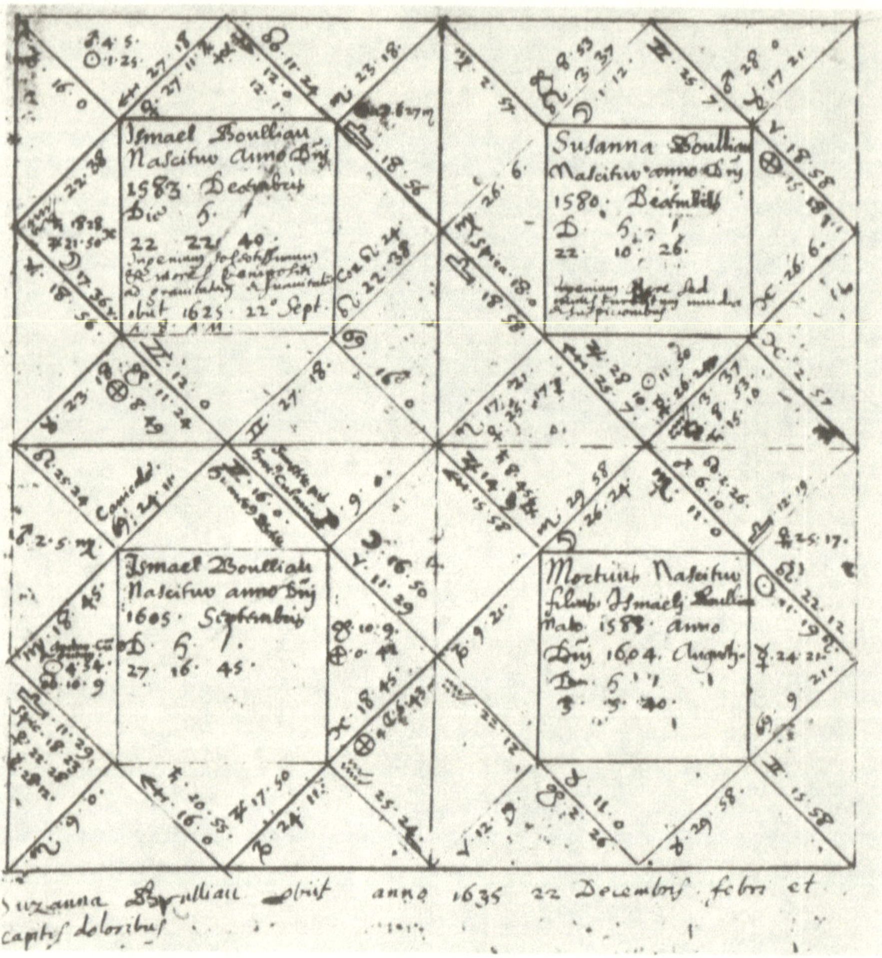

Figure 2. Horoscopes of Ismaël Boulliau (1605–1694), lower left; his father (1583–1625), top left; and presumably his paternal aunt, Susanna (1580–1635), top right. FF. 13028, fol. 155v (Courtesy of Phot. Bibl. Nat. Paris).

Biographical Introduction

Early Life and Career, 1605–1632

Situated southwest of Paris in the province of Poitou, the Huguenot town of Loudun was dominated by a medieval fortress and the spire of Saint-Pierre du Marché. Seventeenth-century Loudun could claim little else save some 14,000 inhabitants.[1] Until the late sixteenth century the fate of this quiet agrarian center in the diocese of Poitiers was determined by the natural rhythms of planting and harvest, and the unpredictable cycles of famine and plague.[2] But the early decades of the seventeenth century brought a radical transformation. As its Protestant population grew, the town was torn by religious controversy, witchcraft, plague, and the policies of Richelieu. Caught, as it were, between the forces of heaven and earth, it ceased to be a tranquil town in provincial Poitou.

The celebrated astonomer Ismaël Boulliau was born in Loudun on Wednesday, 28 September 1605.[3] In all likelihood he was the eldest child of his Protestant parents, Ismael, a young notary and procurer for Loudun,[4] and Susanna Motet, his wife.[5] A volume of autograph

[1]For information on Loudun see Auguste-Louis Leorsey, *Loudun, histoire civile et religieuse*; Pierre Delaroche, *Une épidémie de peste à Loudun en 1632*, pp. 33–46; and Howard M. Solomon, *Public Welfare, Science, and Propaganda in Seventeenth Century France, the Innovations of Théophraste Renaudot*, Chap. 1.

[2]Between 1350 and 1632 Loudun was struck by nine major outbreaks of plague; an estimated 3,700 died in 1632. Delaroche, pp. 39–40.

[3]See Boulliau to Viviani, 16 Mar 1678, FF. 13026, fol. 313v, and MS Gal. 256, c. 184v (birth date 28 Sep 1605, H. 4 '40); cf. Boulliau to Viviani, 7 June 1683, FF. 13026, 311v, and MS Gal. 256, c. 252v, Biblioteca Nazionale Centrale (the latter gives a horoscope date of 27 Sep 1605, H. 16 '40 calculated from the Rudolphine Tables). Boulliau's original horoscope (FF. 13028, fol. 155v) gives 27 Sep 1605, H. 16 '45. Cf. FF. 13028, fol. 168v. There is no precise record of Boulliau's date of birth in the Archives Communales at Loudun.

[4]A nativity for Ismael Boulliau, Sr., is found in FF. 13028, fol. 155v, giving the dates 22 Dec 1583 and 22 Sep 1625; he is characterized as having "a very clever mind and a character suited to gravity and pleasantness." Susanna apparently gave birth to a son on 3 Aug 1604 that did not survive (155v); there was a third surviving Ismael Boulliau, Susanna's grandson (son of Daniel) born at Loudun 2 Mar 1644, died at Paris 20 Apr 1674 (fol. 323v).

[5]The horoscope for Susanna (Motet) Boulliau is found in FF. 13028, fol. 160v; it gives the dates 2 Apr 1582 and 15 Aug 1634. Boulliau to Gassendi, 7 Sep 1634, MS 1810, fols. 47–50, Bibliothèque d'Inguimbert, confirms the year and approximate date of her death. Her birth date (April 1582) is confirmed by copies from the Protestant register (made between 1692 and 1693 by a priest of the parish of Saint-Pierre du Martray) found in the Archives Communales at Loudun. These very incomplete copies show that Ismaël was born in September 1605, his sister, Marie, in January 1607. There is no record of Boulliau's baptism or his parent's mar-

horoscopes found in the Collection Boulliau suggests that their marriage produced over a dozen children, most dying at birth or of childhood diseases. Four of the Boulliau children appear to have married and raised families of their own.[6]

One of the most prolific correspondents of his century, Boulliau left no record of his childhood or early education. Most of our information is therefore derivative or conjectural.[7] Standard biographical sources agree, however, that his earliest studies were at Loudun where, as was the custom, he initially studied the humanities, perhaps in public classes. Having begun formal studies in his native city, Boulliau presumably traveled to Paris to study philosophy and theology, and possibly Poitiers to study law.[8] The least we can suppose—without firm evidence—is that Boulliau concentrated his formal studies in theology, philosophy, history, and mathematics.

But Boulliau's major intellectual interest was inspired by his father, an avid astronomer. Archival evidence indicates the older

---

riage. Further information on the Motet family is found in FF. 13028, fols. 160r–161v. Boulliau appears to have had a sister Susanna (1612–1617), fol. 157v; and, apparently, a paternal aunt, Susanna Boulliau (1580–1635), fol. 155v.

[6]Ismaël Boulliau (1605–1694) appears to have had six sisters and five brothers, though the relationships are not always clear. See FF. 13028, fols. 4r, 6rv, 53r, 101rv, 155rv, 157r–161r, 170v–171r, 323v, and 327v.

[7]Standard biographical dictionaries are of little help. The best traditional source of information is the "Eloge de Monsieur Boulliau," *Journal des Sçavans* (7 Feb 1695), pp. 110–126. Other standard sources include Louis Moréri, *Le grand dictionnaire historique ou le Mélange curieux de l'histoire sacrée et profane*; Eugène and Emile Haag, *La France protestante, ou Vies des protestants français*; J.-F. Dreux Du Radier, *Bibliothèque historique et critique du Poitou*; Louis-Ellies Du Pin, *Nouvelle Bibliothèque des auteurs ecclésiastiques* (Paris, 1731); Charles Perrault, *Les hommes illustres qui ont paru en France pendant ce siècle*; Jean Pierre Nicéron, *Mémoire pour servir à l'histoire des hommes illustres dans la République des Lettres, avec un catalogue raisonné de leurs ouvrages*; and H. & P. Beauchet-Filleau, *Dictionnaire historique et généalogique des familles du Poitou*. Of recent works, the best treatment of Boulliau's life is the article of Hans Bots, "Ismaël Boulliau, ses voyages en Hollande," *Lias* 1 (1974): 29–62, which consists of twelve annotated letters of Boulliau to Jacques Dupuy (Sep–Dec 1651) and an excellent seven-page sketch of Boulliau's career emphasizing the years 1651–1662. H. J. M. Nellen has recently completed a Dutch dissertation, "Ismael Boulliau (1605–1694), Nieuwsjager en correspondent," Nijmegen, 1980.

[8]Dreux Du Radier relates an anecdote from Urbain Chevreau's *Chevraeana*: "Son père lui ayant demandé compte de l'argent qu'il lui donné pour sa dépense à Paris, le jeune Boulliau ne se trouvant pas fort en état de compter *de Clerc à Maître* mais voulant pourtant satisfaire aux ordres qu'il avait reçus, exagéra sa dépense par plus de soixante articles qu'il commençait toujours par le mot d'usage, *item*. Malgré cette ressource ne pouvant encore faire cadrer la dépense avec la recette, il donna cette forme à son dernier article: *item, mon père il faut vivre*." Dreux Du Radier, pp. 274–275.

Boulliau observed planetary conjunctions and occultations, as well as the comets of 1607 (Halley's Comet) and 1618.[9] His son seems to have taken special care to preserve and identify these observations, and he published several in his *Astronomia Philolaica*. To be sure, the elder Boulliau was neither a famous nor a fastidious observer, but his son would become one of the outstanding astronomers of his age. As Jacques de Valois later wrote Mersenne, Boulliau was "un tres docte homme, curieux et diligent, et qui a l'astronomie . . . la portée en sa tete depuis le berceau."[10]

Some evidence suggests that Boulliau's father died in September 1625 at the age of forty-one[11] (his mother, Susanna, died nine years later in the summer of 1634). At this time the young Boulliau was only nineteen years old, and though precocious in his studies, his social position and lack of funds presented formidable obstacles. Boulliau had been born to Calvinist parents, and in seventeenth-century France many avenues traditionally open to Catholics were closed to him. He had already sampled the excitement and stimulation of Paris, and like Renaudot, a Protestant native of Loudun, he elected to leave the reformed religion for Catholicism, and eventually, Loudun for Paris.[12]

Five years after his conversion to Catholicism Boulliau was ordained a priest.[13] His first position—one not free from controversy—was that of vicar to the ill-fated Urbain Grandier, the

[9]See, for example, MS B 12, pp. 91–115 and pp. 275–276, BO; Boulliau writes: "Observatio Ismaelis Bullialdi Ismaelis patris" (94); "Hic cometa apparuit/anno 1607. mense Septembri/ Ab Ismaele Bullialdo parente meo./visus" (95); "Ab Ismaele Bullialdo parente meo." (102); "Descripto Meteori quod una cum parente meo Ismaele Bullialdo . . ." (115). Boulliau sent several of his father's observations to Wilhelm Schickard in his letter of XVII Kal April 1633 [16 Mar 1633 NS]; see Briefsammlung Cod. 10889, fol. 278rv (observation of 22 Feb 1608) Österreichische Nationalbibliothek.

[10]MC XII: p. 339. Jacques de Valois (1582–1654) of Grenoble was a friend and correspondent of Gassendi, Mersenne, and Boulliau (for letters to Boulliau, see particularly FF. 13030). An avid astronomer, Valois would eventually send Boulliau an astronomical quadrant; see de Valois to Boulliau, FF. 13030: 25 May 1653 (75r–76v), 5 Oct 1653 (106r–107v), and 16 Nov 1653 (118r–119v).

[11]The evidence is uncertain; see FF. 13028, fol. 155v; cf. MC III: p. 400; and documents in Série BB, 30, Archives Communales (dated May and August 1632), purportedly signed by Ismael Boulliau, Sr.: "Extrait des registres du Greffe des Affaires Communes de la Ville de Loudun, fait et délivré par moi I. BOULLIAU, Greffier de la ville."

[12]For a discussion of the problems faced by Théophraste Renaudot, see Soloman, Chap. 1.

[13]M. Dumoustier de la Fond, in his *Essais sur l'histoire de la ville de Loudun*, p. 127, suggests: "Il fut d'abord Avocat; mais ses parens s'étant opposes à un mariage qu'il vouloit contracter, il abjura le Calvinisme à l'âge de 21 ans, & fut prétre à l'âge de 25."

central figure in one of the most scandalous trials of the century.[14] Grandier, who was eventually burned at the stake for witchcraft, was suspended from his duties in 1630; during his absence Boulliau, François Grandier (Urbain's brother), and Gervais Meschin directed the parish of Saint-Pierre du Marché (Loudun).[15]

The trial and execution of Grandier made a lasting impression on Boulliau. The affair began on 3 January 1630, when Grandier was charged, and ended on 18 August 1634, when he was tortured and executed.[16] Outraged by the ordeal, in a letter of 7 September 1634, Boulliau related the essential elements of the affair to his friend Pierre Gassendi, a fellow astronomer and priest. That Boulliau respected Grandier and maintained his innocence is beyond doubt; but with characteristic candor, Boulliau's hostility was reserved for exorcists and their dogma:

> D'ailleurs Je deplore la Condition en laquelle on veult mettre les Chrestiens de les faire mourir sur la deposition des diables, doctrine dangereuse, Impie, erronée, execrable et abominable, qui rend les Chrestiens Idolatres, ruine la Relligion Chrestienne dans ses fondements, ouure la porte à la Calomnie, et fera, si dieu par sa prouidence, ne remedie à ce mal, que le diable se fera Immoler par les homes des victimes humaines non plus soubs le nom de Moloch, mais à la faueur d'un dogme Diabolique & Infernal. . . .[17]

Though he would later return, Boulliau left Loudun before the Grandier affair ran its course.[18] The precise date and circumstances of Boulliau's arrival in Paris are uncertain, although he appears to have taken residence in 1632. By this time he had visited the French

[14]In his much-quoted study of the Grandier affair, Gabriel Legué remarks: "Le vieux Barot, président aux Élus de la ville, dans son impatience n'attendit même pas que le procès fut terminé et fit prendre un dévolu sur les bénéfices de Grandier au profit d'Ismaël Boulliau, vicaire de Saint-Pierre. Il est juste de dire que Boulliau, qui devint plus tard une des gloires de Loudun, ne voulut être mêlé en aucune façon à toutes ces misérables intrigues. D'après le témoignage de Ménage, qui a pu apprécier ses belles et nobles qualités, le vicaire de Saint-Pierre conserva toujours une grande estime et de la reconnaissance pour son curé." Gabriel Legué, *Urbain Grandier et les possédées de Loudun*, p. 79. The best sources concerning the Grandier affair, in addition to Legué, include Edmund Goldsmid, trans. and ed., *The History of the Devils of Loudun; Le Procès de Urbain Grandier*, Les Grand Procès de l'histoire de France XVIII: pp. 9–156. Aldous Huxley's *The Devils of Loudun* is a lively and generally accurate account.

[15]Lerosey, p. 221. I have found no archival evidence for this claim.

[16]See Legué, Chaps. VII–XV.

[17]Boulliau to Gassendi, MS 1810, fol. 49r (copy), Bibliothèque d'Inguimbert.

[18]He returned in September 1634, to attend to the estate of his mother who died two months earlier. See also CD 18, fol. 42r.

capital, made astronomical observations there, and perhaps met members of the Parisian circle.[19] It is possible (though speculative) that Boulliau initiated and maintained Parisian contacts while still in Loudun. Scévole de Sainte-Marthe, a noted scholar and grammarian, was the focus of Loudun's intellectual circle,[20] and the group that met in his Hôtel included Théophraste Renaudot, a native of Loudun and eventual founder of the Bureau d'Adresse and *Gazette de France*, Louis Trincant, Urbain Grandier, and numerous visiting scholars and dignitaries.[21] Between 1619, when Sainte-Marthe returned to Loudun, and 1623, this provincial circle enjoyed visits of Jacques-Auguste de Thou I, the celebrated historian, and no less a figure than Charles I of England.[22] Through this circle Boulliau may have become acquainted with the family de Thou and, eventually, the brothers Dupuy, friends and correspondents of Peiresc.[23] In any case, Boulliau seems to have met Gassendi in Paris before the latter left for Digne in October 1632.[24]

The Early Parisian Period, 1632–1645

By spring 1633 Boulliau was comfortably settled in Paris. As his earliest letters suggest, he was by this time acquainted with François Luillier, a close friend of Gassendi,[25] and Marin Mer-

[19] Boulliau seems to have observed an occultation in Paris on 5 July 1623; see MS B5 12, p. 277, BO; and Boulliau's *Astronomia Philolaica*, lib. 3 c. 6 [p. 159].

[20] See Delaroche, p. 36.

[21] Legué, pp. 22–23.

[22] Soloman, p. 17.

[23] For the relations of Peiresc and the Cabinet Dupuy see Harcourt Brown, *Scientific Organizations in Seventeenth Century France (1620–1680)*, Chap. 1. I have found no evidence that Boulliau studied with Joseph Gaultier (1564–1647), the astronomer from Aix. Gaultier was known to Peiresc and is the reputed teacher of J. B. Morin, Boulliau, and Gassendi. An extended study of provincial schools and their relation to the "Parisian Circle" is long overdue; see Seymour L. Chapin's excellent article, "The Astronomical Activities of Nicolas Claude Fabri de Peiresc," *Isis* 48 (1957): 13–29.

[24] On 12 Dec 1632 Gassendi (at Digne) wrote Luillier at Paris: "A propos envoiez s'il vous plaist quelqu'un de vos gens vers le Sr Renodot affin qu'il vous indique où loge Mr Boliaud de Loudun par ce qu'il est ce me semble son parent afin de le faire prier de vous aller trouver, et quad il vous verra priez-le de ma part d'envoyer à Mr Schickard toutes les observations qu'il a de Mercure." Pierre Gassendi, *Lettres familières*, pp. 30–31; cf. MC III: p. 400.

[25] François Luillier (1583–1652) became a close friend and correspondent of Boulliau; see FF. 13042, fols. 1–19 for letters exchanged between 1642–1650. For Gassendi's relations with Luillier see Pierre Gassendi, *Lettres familières*.

senne.[26] In one of his earliest letters to Gassendi, dated 21 June 1633,[27] Boulliau shared his first impressions of the minim priest and his thoughts on Kepler's astronomy. Throughout the 1630s Boulliau sought to reconcile Gassendi's observations with the best astronomical tables available, particularly Gassendi's data for the transit of Mercury (6 November 1631). The technical issue at hand was the accuracy of the tables of Adrien Vlac, Philippe van Landsberge, Longomontanus, and Kepler. The overriding concern, however, was Kepler's celestial physics, of which Boulliau was skeptical:

> I am forced to marvel at the meditations of Kepler concerning Physics . . . he owned that the motions [of the moon] were real and Physical and conceded they occurred through true and real spaces, whose dimensions are derived from Geometry; however, the magnetic virtue of the Sun that Kepler introduced, by which the Sun attracts Planets in proportion to the measure of their distance, is not demonstrable. . . . Reasonably Astronomy and Geometry reject as mere fancy those hypotheses and physical equations of motion which exclude Geometrical proof. To this should be added the axiom that celestial motions are simple circular motions, or compounded circular motions, and for this reason I reject and repudiate his Ellipses, unless they are described by a Copernican or Tychonic Epicycle. For it is easily demonstrated that an ellipse is described by a circle whose diameter is

[26]Marin Mersenne (1588–1648) requires no introduction. It is interesting to compare Boulliau's first reaction (quoted in note 27 below) to a later assessment. In a letter to Huygens twenty-six years later, Boulliau compared Pierre Petit (1594–1677), a mathematician and friend of Pascal, to the late Mersenne. In his letter of 4 July 1959 Boulliau wrote: "Monsieur Petit est comme le feu Pere Mersenne, il propose tousjours plusieurs questions & cependant il en resoult peu; & si vous vous laissez aller a ses demandes & questions & que vous vouliez entreprendre d'y respondre vous n'en verrez jamais le bout, je vous donne cet aduis, que je vous prie de tenir secret." HO II: pp. 430–431.

[27]"Inchoans autem Laetiora auspicia habere non possum quàm ab inuocatione nominis R. P. Marini Mersenni sub vexillis S. francisci de Paula stipendia spiritualia facientis. huius sanè Aquilae pullus generosissimus est; siquidem Parens ille dissolutis corporis vinculis, intuitu fixo nec vacillanti diuinam essentiam rerum omnium in se rationes et χαρακτῆρας ἀκραιφνεῖς haben[tem] contemplatur, Non degener filius inconniuentibus oculis illas rationes quas ἀρχετύπους oculo corporeo cernere nequit, scrutat[ur] et rimat[ur] ἐγγλυφεῖσας καὶ ἐγγραφεῖσας in omnibus rebus quae earum signaculo informatae esse suum a deo conditore participant. Piget me tandiu huic viro ignotum Parisiis vixisse, cum etiam anno superiore per te ad illum aditus pateret. Inexhaustae scientiae et sapientiae thesaurum in pectore gestat, fucatam illam Scholarum et corymbis quarumdam distinctiuncularum cincinnatam auersatur, quae sane non veritatem jnuestigat et perquirit sed quibusdam argutiis veluti praestigiis auscultantium animos percellere conatur et contendit, dum antiquas et sane antiquandas opiniones ne dicam errores tuetur barbaris verbis sui auditores perturbat dumque intelligi non vult vel sese explicare nequit, abstractas non solum a rebus singularibus sed ab ipsa etiam veritate formas curiose et sollicitè confingit." Boulliau to Gassendi, 21 June 1633, FF. 13050 fol. 258r.

equal to the difference of the axes of the Ellipse, provided the Planet is understood to make a double motion on the epicycle, while the center of the epicycle revolves once on the deferent.[28]

Except for a brief interruption in August,[29] Boulliau continued his astronomical researches through the summer of 1633. By this time he had been a confirmed Copernican for perhaps a decade.[30] No archival evidence exists from this period to indicate the precise direction of his thought, but if his later concerns are any measure, Boulliau must have quickly seized upon the fundamental simplicity of the Copernican model.[31] In any event, though he decried Galileo's

[28]"Non possum non mirari Keppleri meditationes de Physicis . . . fatet[ur] esse motus reales et Physicos, concedat etiam per vera et realia spacia fieri, quorum dimensio a Geometrica dependent; non autem demonstrabilis est illa virtus magnetica Solaris, qua Planetas trahit [Sol] pro modulo suae distantiae ab illis, quam induxit Kepplerus. . . . Astronomia sane et Geometri semper has hypotheses et physicas motuum aequationes, quae demostrationes Geometricas excludunt, vt mera somnia reiicient. Standum est huic axiomati, coelestes motus circulares esse simplices, et ex circularibus compositos, & hanc ob ration[ent] Ellipses illius respuo et repudio nisi per Epicyclium Coperniceum aut Tychonicum describi velit. facilè enim demonstrat[ur] descripta Ellipsis, assumpto circello, cuius diameter aequalis sit differentiae axium Ellipseos, & Planeta duplam in illo reuolutionem facere intelligatur interim dum Epicyclij centrum in orbe magno semel reuoluitur." Boulliau to Gassendi, 21 June 1633, FF. 13050, fols. 258v–259r.

[29]Boulliau accompanied an expedition to Nancy located in northeastern France. Heavily fortified by Charles III (1604–1675), duke of Lorraine, Nancy underwent siege on 28 August and fell on 24 September.

[30]In his *Astronomia Philolaica* of 1645, Boulliau claims to have been a Copernican since 1624: ". . . between 1624 and 1631 I made a number of observations of Mercury, and attempted to correct the errors of the Prutenic Tables, retaining the Copernican hypothesis; but I could never get my observations to agree with the calculus, unless in the superior part of the orbit of Mercury [near aphelion] I retarded its motion, and in the inferior part [near perihelion] I speeded it up. But my attempt did not have a happy outcome, since I made this inequality of motion commensurable with the Earth's motion, while in fact it has to be referred to the motion of Mercury in its own orbit. Not yet had Kepler's tables and his *Commentaries on the Motions of Mars* fallen into my hands, and I had scruples about departing from the circular hypothesis, nor had it occurred to me that the orbit of the planet could be an ellipse: lacking which, if anyone should attempt to twist the motions of the planets into other circuits, . . . he would but lose time and trouble, and revolve the stone of Sisyphus." Ismaël Boulliau, *Astronomia Philolaica*, p. 355 (quoted and trans. in Curtis Wilson, "The Inner Planets and the Keplerian Revolution," *Centaurus* 17 (1972): 205–248, p. 243.

[31]Appropriately characterized as a "hyper-Copernican," Boulliau repeatedly underscored his commitment to geometrical simplicity. In a fascinating letter to Leopold dated 23 July 1666, Boulliau evaluated Borelli's *Theoricae mediceorum planetarum*: "All these things [Borelli's physical hypotheses] subsist as a supposition of mind, certainly one which has been thought out by the effort of a superior intellect, [but] they do not reveal the truth. The opinion is simpler and closer to nature, it seems to me, which works by the simplest means, which admits that the planets are moved by their own forms . . . nevertheless, I allow anyone to choose as he pleases, since nothing about Astronomical Hypotheses can be demonstrated from the causes or reasonings of physics." FF. 13027, fol. 10rv.

fate, Boulliau was committed to publication, and in an early letter to Wilhelm Schickard he described his activities and his plan:

> In the previous autumn, when we were returning from the expedition to Nancy, I thought of treating the mobility of the Earth as leisure permitted. Therefore, having returned to Paris, I approached the matter of demonstrating the motion of the earth with a new argument which was geometrical and optical. I know that this was hitherto untried . . . I do not dare publish the tract lest I invoke ill-will from certain individuals. I sent a copy to M. Gassendi, whom I shall entreat to have it transcribed and sent to you; if leisure allowed I would transcribe it myself. Its title is: *Philolaus sive de terrae mobilitate* . . .[32]

By spring of 1634, Boulliau had completed his *Philolaus*, and true to form, he circulated the manuscript privately before publication. By June of that year, Luillier would send Peiresc "un discours en quatre caiers, fait par Mr Boulliaud, que j'ai fait coppier pour l'usage de Mr Gassendi et pour vous demeurer en propre quand il vous quittera."[33]

Boulliau's decision to publish the *Philolaus* came at an awkward time, particularly for a young Catholic priest. In a well-known letter to Mersenne (end of November 1633), Descartes claimed he would rather suppress his *Le Monde* (based on Copernican principles) than gain the disapproval of the church.[34] Galileo's condemnation was no small matter, and Descartes's reticence should in no way be viewed as a lack of courage or conviction. Nonetheless, Descartes was astonished to hear of a treatise on the motion of the earth; in April 1634 he wrote Mersenne, "Pour le Mouuement de la Terre, ie m'estonne qu'vn homme d'Eglise en ose escrire, en quelque façon qu'il s'excuse."[35]

In August 1634, Boulliau's mother died, and he was obliged to return to Loudun at the end of September. In his four-month ab-

[32]"Dum Autumno superiore ab expeditione Nanceiana redibamus Cogitaui de Terrae mobilitate tractare, vbi per otium Liceret, ad Parisios igitur redux factus rem aggressus, novo argumento eoque Geometrico et optico exantlavi Terrae mobilita[tem] demonstrans; Scio intentatum huc usque fuisse . . . tractatum publicare non audeo, ne in me inuideam quorundam vertam. Exemplar transmisi ad Dnm̃ Gassendum, quem orabo vt describi curet et ad te transmittat, si per otium vacaret, transcriberem ipse. Titulus illius est, Philolaus, seu de Terrae mobilitate authore Ismaele Boullialdo." Boulliau to Schickard, 19 June 1634, MS Cod. hist. 2°, 563ab, Wurttembergische Landesbibliothek, Stuttgart.

[33]MC IV: p. 185; FF. 9539, pièce 80.

[34]Descartes to Mersenne, end of Nov 1633, DO I: pp. 270–271.

[35]Descartes to Mersenne, Apr 1634, DO I: p. 288.

sence from Paris he worked on his manuscript of the *Philolaus* and recorded several astronomical observations.[36] In September he wrote Gassendi:

> Apres mon retour de Loudun ou je m'achemine bien tost pour recueillir la succession de ma mere, decedeé depuis deux mois, je faicts estat de mettre la derniere main à mon traicte & l'augmenter de la Theorie des planetes dont je vous ay parlé. J'attendray neantmoins v[ot]re abbouchem[en]t pour en conferer ensemble sur l'edition . . . [37]

Returning to Paris in early 1635, Boulliau continued to lodge at the house of the Marquis de Sourdis near the Louvre.[38] Though he would make a number of extended journeys—to Italy, Constantinople, Germany, Holland, and Poland—he henceforth considered himself a Parisian. Throughout his early years in Paris he was in regular contact with Gassendi,[39] Mersenne, Luillier, Jacques and Pierre Dupuy,[40] Théophraste Renaudot, François-Auguste de Thou (and his brother Jacques-Auguste),[41] Jean-Baptiste Morin,[42] the pugnacious Roberval,[43] and other habitués of the Mersenne circle and the Cabinet Dupuy. During this period he also came into contact

[36]Among his observations at Loudun Boulliau mistakenly concluded that he had observed an unknown star the evening of 25 Oct 1634, FF. 13058, fol. 103r.

[37]Boulliau to Gassendi, 7 Sep 1634, MS 1810, fols. 49v., Bibliothèque d'Inguimbert.

[38]Charles d'Escoubleau, marquis de Sourdis (1588?–1666); an address is found on letters in FF. 13037, e.g., fol. 8v and 12r. During this period (1633–1636) Gassendi addressed his letters to Boulliau as "prof^{r} en Ste Theologie" (1633), "docteur en droit" (1634), and "Estudiant en Theologii/au College de navarre" (1636), FF. 13037, fols. 14v, 17v, 28v; the Collège de Navarre was located on the current site of the Ecole Polytechnique (Paris). In his letter of 18 June 1639 Florimond Debeaune wrote: "A Monsieur/Monsieur Boulliau/docteur en medicine. . . ." FF. 13037, fol. 217v.

[39]Although Gassendi did not take residence in Paris until 1641, he made a number of visits; in the interim he and Boulliau maintained a steady correspondence. The Collection Boulliau alone contains over 30 letters from Gassendi to Boulliau for the years 1633–1654, most unpublished.

[40]The brothers Dupuy, bibliophile-savants, directed the famous Cabinet Dupuy; Boulliau was their associate until their deaths.

[41]François-Auguste de Thou (1604–1642), who know Boulliau during his early years in Paris, was tragically executed in the infamous Cinq-Mars affair in 1642. Jacques-Auguste de Thou (1609–1677), François' brother, would eventually be appointed French Ambassador to Holland; Boulliau served as his first secretary.

[42]Jean-Baptiste Morin (1583–1656), astronomer-astrologer, was one of Boulliau's few foes.

[43]Giles Personne de Roberval (1602–1675) came to Paris in 1627 and lodged for a time with Mersenne. Boulliau judged this creative mathematician and opponent of Descartes to be boorish, ill-mannered, and consumed with self-esteem. Pierre Des Noyers (1607–1693) was one of Roberval's major correspondents.

with Hevelius, who was to become one of his closest friends.[44]

Boulliau's scientific pursuits during these early years were not restricted to astronomy. Though the list of his astronomical observations during the 1630s is remarkable in itself,[45] his other concerns ran the gamut from music theory to the optics of fish eyes.[46] Next to astronomy, his major interest was optics, and his *De Natura Lucis* of 1638 was foreshadowed in his correspondence with Gassendi. In addition, Boulliau was not to be outdone by Jean-Baptiste Morin who, like many of his contemporaries, labored over the problem of determining longitude at sea. Boulliau was less than sanguine, and his thoughts on the question remain unpublished.[47]

A crucial change occurred in 1636 when Boulliau moved to the Hôtel de Thou on rue des Poitevins.[48] The Hôtel de Thou had become one of the most exclusive centers of scholarly activity in all of Europe.[49] Except for the Bibliothèque du Roi, the library housed at the Hôtel was the most extensive in Paris.[50] The historian

[44]Boulliau and Johannes Hevelius (1611–1687) met in Paris when the latter was on tour to Leiden, England, and France during the years 1630–1634; a student of Peter Krüger, Hevelius was an excellent naked-eye observer. The Collection Boulliau contains over 160 letters and drafts exchanged between Boulliau and Hevelius, not to mention copies and extracts of astronomical data.

[45]See Alexandre-Guy Pingré, *Annales célestes du Dix-Septième Siècle*, pp. 76–142, for a convenient chronological listing; Pingré drew his information from manuscripts at the Bibliothèque de l'Observatoire (B3, 11; B5 11–12) and the Collection Boulliau (FF. 13058). By this time Boulliau had constructed an extensive set of astronomical tables, but they were not published in the *Philolaus*. As Mersenne wrote: "Monsieur Boulliau, l'un des plus excellens astronomes de nostre siecle, auquel nous aurons l'obligation de tres exact observations celestes et de nouvelles tables astronomiques. . . ." See MC IV: p. 436.

[46]Boulliau wrote a short unpublished tract on music, MSS B5 11, pp. 349–368, BO. His interest in optics is evident in the correspondence with Gassendi.

[47]For Boulliau's "Des Moyens Astronomiques pour Cognoistre/La distance des Meridiens et des Longitudes/sur la proposition de Naviger en longitude," see MSS B5 11, pp. 439–442, BO. Boulliau was forced into several disputes with Morin; nearly a decade later, on 16 Dec 1644, Boulliau wrote Mersenne that: "non obstant mon charactere et ma profession il me menace de voye de faict, s'il ne me prend en assassin je me deffendray de luy le plus vigoureusement qu'il me sera possible." N.a.f. 6205, fol. 112.

[48]The site of the Hôtel de Thou is now occupied by the Hôtel des Sociétés Savantes on rue Danton, near Place St. Michel.

[49]For details see Harcourt Brown, *Scientific Organizations in Seventeenth Century France* (1620–1680), Chap. I, and Isaac Uri, *Un cercle savant au XVIIe siècle: François Guyet (1575–1655): d'après des documents inédits*, passim.

[50]A useful history of the library is Henry Harrisse, *Le Président de Thou et ses descendants, leur célèbre bibliothèque, leurs armoiries et les traductions françaises de J.-A. Thuani Historiarum sui Temporis*; see also Jack A. Clarke, "A Book Buying Tour in 1645: A Note on Ismael Boulliau in Italy," *The Journal of Library History* 4 (1969): 330–336.

Jacques-Auguste de Thou[51] had carefully selected the nucleus of the Biblioteca Thuana, and after his death the direction of his library was turned over to his nephews, Jacques and Pierre Dupuy.

The intellectual interests of "les Frères Dupuy" had no limit.[52] During their early years they tended to the Biblioteca Thuana and developed skills as scholars, antiquarians, and historians. Both Jacques and Pierre were regular correspondents of Peiresc and, like their provincial counterpart, enjoyed learned company. In the gallery of the Hôtel de Thou, they established and directed the famous Cabinet Dupuy, one of the most exclusive and enduring assemblies of its kind. Initially known as the Cabinet du Président de Thou, this assemblage of savants, poets, philosophers, and amateurs of science would include as many as fifty carefully-selected members. The reputation of the Académie Putéane spread throughout Europe, and its importance as a center of learning and communication cannot be overestimated. In the opinion of Peiresc, the Cabinet was "le vray domicile de la vertu, et le rendez vous de touts les gents de lettres tant de la nation que estrangers."[53]

The commodious Hôtel de Thou was much to Boulliau's liking. His new lodgings granted him ready access to some 8,000 books and 1,500 manuscripts, and his duties, though not always to his taste, seemed to suit his scholarly disposition.[54] More importantly, his daily obligations did not impinge on his personal research; the repose and relative stability of his position afforded ample time for astronomy and mathematics. In his letter to l'Empereur[55] of 25 June 1641, Boulliau described his first five years at the Hôtel de Thou:

> Je vous diray donc que depuis cinq ans i'ay l'honneur de demeurer chez Monsieur de Thou, avec son frere Monsieur l'Abbé de Bonneval, leur faueur et bienveuillance m'ayants donné le moyen d'estudier,

[51]On Jacques-Auguste de Thou I, see Samuel Kinser, *The Works of Jacques-Auguste de Thou*.

[52]Some of these interests are described by Hans Bots, *Correspondance de Jacques Dupuy et de Nicolas Heinsius (1646–1656)*, Introduction.

[53]Peiresc to Pierre Dupuy, 12 May 1628. Nicolas-Claude Fabri de Peiresc, *Lettres de Peiresc* 1: p. 607.

[54]On the growth of the Biblioteca Thuana see Harrisse, pp. 73–74. One of Boulliau's duties was to develop a catalog of the library, an obligation he completed "not without distaste and boredom." Boulliau to Storanus, 11 July 1653, FF. 13043, fol. 147r.

[55]L'Empereur was secretary of the French Ambassador at Constantinople; his letters are found in FF. 13042, fols. 214–240.

auec grand loisir et grand repos, et d'apprendre auec facilité, et commodité beaucoup de choses, qui ne fussent iamais venües a ma cognoissance. Vne partie de mes estudes s'est employee dans les Mathematiques, et i'ay trauaillé entre autres dans l'Astronomie, et a la contemplation des mouuemens du ciel, et desia i'ay mis au jour quelque ouurage sur cette science, et i'en ay encores vn tout prest a sortir, i'ay faict vn recueil le plus grand qu'il m'a esté possible des observations anciennes. . . .[56]

Early Publications

During his early years at the Hôtel de Thou, Boulliau read and wrote continuously; these were the most creative and productive years of his life. The first public evidence of his scientific efforts was his *De Natura Lucis* of 1638. His first thought upon publication was to send a copy to Galileo. Boulliau's letter of 30 October 1637 served several purposes:

> I am drawn by numerous and pressing reasons to send you this letter together with a short book, *De Natura Lucis*, which I recently completed; for in the future I hope to become known to you (whom I most greatly esteem) and . . . I would appreciate your judgment concerning this short work. . . . Furthermore, you will not be displeased to hear that the *Philolaus* is being published in Amsterdam; it seeks to demonstrate by necessary conclusion this system of the world by reasons previously unknown, deduced from geometry and optics. The delay in printing is bitterly regretted, since there are many who wish to hear your opinion once you have read it. Meanwhile your state of health and age keep me solicitous and anxious; you must excuse my liberty, far removed from good manners, and I hope you will grant forgiveness with a noble and open heart. I have commended this little packet to M. Diodati; he is well-known to you, and joined with me in friendship.[57]

[56]Boulliau to l'Empereur, 25 June 1641, FF. 13042, fol. 212r.

[57]"Multis vrgentibusq[ue] rationibus adductus sum ad hanc epistolam tibi mittendam, vnaque libellum quem nuperrimé scripsi de Natura Lucis, in posterum enim tibi, quem maximi facio, innotescam, et . . . de illo opusculo iudicium tuum intelligam. . . . graue tibi porro non erit audire Philolaum Amstelodami typis exornari, is systema mundi rationibus in hanc vsque diem ignotis, a Geometria et optica deductis necessaria conclusione demonstrare contendit; Typographi mora acriter reprehenditur, quia vbi illum legeris, quid de illo senties audire multi cupiunt, Valetudo interim tua, aetasque sollicitos, et anxios tenent: hanc libertatem meam a ciuili forsan comitate nimis detortam excusatam habebis, et ingenuo atque aperto animo veniam dabis, hunc D[omi]no Diodato fasciculum commendaui, is tibi notissimus est, et amicitia mecum iunctus." Boulliau to Galileo, 30 Oct 1637, FF. 13026, fol. 47r.

In the year following the publication of *De Natura Lucis*, Boulliau published his *Philolai, sive Dissertationis de vero systemate mundi*. Publication was achieved with difficulty;[58] the publisher Blaeu (of Amsterdam) delayed scheduled printing for more than three years, causing some concern and consternation on Boulliau's part. In his letter of 16 September 1639, Boulliau described the process to Galileo:

> At last, Illustrious Sir, the *Philolaus* has appeared, after it lay hidden through three years and three months among the Dutch, restrained by the fetters of long delay. The elegance of the type and the subtle engraving of the illustrations completely alleviated the annoyance of unwelcome delay and calmed whatever anger would have been expressed towards the printer. I am sending a copy to your very Illustrious Lordship. . . . Would that God, who binds their wounds, restore the light of your eyes, which is such a loss to us, so that you might read this little book and see clearly the succession of reasons without the labor of another's eyes. But if this wish cannot be granted, I ask one thing, should the state of your health permit, that you take care that some pages be recited to you, and that you indicate to me with freedom and mathematical skill your opinion.[59]

The natural outgrowth of Boulliau's *Philolaus* was the monumental *Astronomia Philolaica*. Published in 1645, the large folio volume was the most influential astronomical work published in the period from Kepler to Newton.[60] As we have seen, Boulliau was familiar

[58]Boulliau sent his manuscript to Amsterdam in 1636. Mersenne wrote Rivet (15 Sep 1636): "L'un de mes amis a envoyé à Amsterdam un tres beau traité, où il pretend demonstrer geometriquement que la Terre se meut autour du Soleil et qu'il est stable, comme vous sçavez qu'on tient dans l'opinion de Copernic." MC VI: p. 138. Hugo de Groot served as an intermediary in the publication, MC VI: p. 140; see also Boulliau to Blaeu, [10 Jul 1636], FF. 13026, fol. 2v; Blaeu to Boulliau, 6 Feb 1638, FF. 13027, fol. 98; Hortensius to Boulliau, 10 Mar 1637, FF. 13037, fol. 94r–95v; and Mersenne to André Rivet, 23 Jan 1638, MC VII: p. 32.

[59]"Tandem, Vir Illustrissime, prodiit Philolaus, postquam per triennium et trimestre inter Batavos, diuturnae morae veluti compedibus constrictus, latuit. Ingratae morae molestiam, typorum nitor ac schematum sculptura subtilis admodum levarunt, et quicquid bilis in typographum efferbuerat sedarunt. Unum exemplar Illustriss^{ae} Dominationi tuae mitto. . . . Utinam Deus, qui *alligat contritiones suorum*, restituat oculorum lumen tibi ademptum, nobisque tale damnum resarciat, ut ipse legas libellum, et rationum seriem sine alienorum oculorum opera dispicias. Sed si voto damnari non datur, unum interim, si per valetudinem Dominationis tuae licet, rogo, ut recitari tibi ex illo aliquot paginas cures, et quid sentias cum libertate et ingenuitate mathematica mihi significes." Boulliau to Galileo, XVI Kal Oct 1639 [16 Sep 1639], MS Gal. 94, fol. 169r, Biblioteca Nazionale Centrale.

[60]See for example, John L. Russell, "Kepler's Laws of Planetary Motion: 1609–1666," *British Journal for the History of Science* 2 (1964): p. 16: "The most important planetary treatise

with the basic elements of Keplerian astronomy in the early 1630s, and his *Philolaus* indicates that he had read several of Kepler's works.[61] Nonetheless, Boulliau did not come to accept planetary ellipses until the late 1630s,[62] and in his *Astronomia Philolaica* he continued to question the foundations of Keplerian astronomy. Although he praised Kepler's genius and the accuracy of his planetary tables, Boulliau was ruthless in singling out contradictions in Kepler's complex (and often contradictory) cosmology.[63] In essence, Boulliau preferred geometrical demonstration to physical speculation;[64] if planetary motion had a cause, Boulliau maintained, that motion (and the resulting path) was the expression of the planet's form.[65]

---

of the 1640s was, however, Ismaël Boulliau's *Astronomia Philolaica* . . ."; Isaac Newton, *The Correspondence of Isaac Newton* 4: 60, ed. n. 13: "Bullialdus . . . wrote [the] *Astronomia Philolaica* . . . perhaps the most important work on planetary systems between that of Kepler and the *Principia*."

[61]For example, Boulliau to Gassendi, 21 June 1633, quoted in note 28 above. In his *Philolaus* of 1639 (privately circulated five years before), Boulliau indicates his familiarity with Kepler's *Ephemerides*, *Epitome*, and *Astronomia pars optica* (*Philolaus*, pp. 24, 26). Boulliau's autograph copy of the *Philolaus* is found at the BN, MS Lat. 17866.

[62]Boulliau was troubled by the irregular orbit of Mercury as early as 1624; Gassendi's data for the transit of Mercury in 1631 played an important role in Boulliau's acceptance of elliptical orbits. See, for example, MS B5 11, pp. 657–660, BO.

[63]Boulliau's conical hypothesis, which assumes elliptical orbits, is developed in Bk. I, XV[XIV]–XVI; Kepler's celestial physics is assaulted in Chap. XII, "An Sol moueat planetas." An abbreviated English translation of Boulliau's punctilious critique (*Astronomia Philolaica*, pp. 21–24) of Kepler's theory of a solar motive force (*Commentaria de Motibus Stellae Martis*, Chap. XXXIII) is found in Alexandre Koyré, *The Astronomical Revolution, Copernicus, Kepler, Borelli*, Trans. R. E. W. Maddison (New York, 1973), Appendix III: pp. 371–375. It is an effective attack on Kepler's celestial physics.

[64]On 16 Dec 1644, with the *Philolaica* achieved, Boulliau wrote Mersenne: "J'ay veu l'*Aristarchus* de M^{r}. Roberval. Il devine a sa mode, comme M^{r}. des Cartes à la sienne, comme le monde a esté faict, le νοῦς d'Anaxagore, l'ἔρος d'Empedocle, et la chaleur du Soleil expliquent aussi bien l'un comme l'autre la coordination et agencement de toutes les parties du monde. Comme je suis autant grossier et d'un esprit bas que ces Messieurs sont subtils et relevés en spiritualité, j'estime avoir beaucoup faict lors que je me suis asseuré que ὅτι τόδέ ἐστὶ, et je me perds dans cet' abysme, ou ces fortes cervelles se plongent hardiment pour aller prendre sur le sable de διότι, et la demonstration a priori, qui est a la verité un tresor plus riche que la mine de Potassi, ou sa devanciere celle de Porca. Et tous les pescheurs de perles à la rade de Goa ne rapporteront jamais d'huistre qui porte un perle plus orientalizee, et de plus belle eau, que ces Messieurs, qui nageans dans la vaste estendue de leurs belles pensees, nous apportent de belles et claires verités. Ne me deceles pas pourtant, car ces esprits delicats croiroient que ces louanges que je leur donne, sont des illusions, et diroient qu'en les encensant, je leur donne de l'encensoir par le nez." N.a.f. 6205, fols. 112r–113v; MC XIII, 269–270.

[65]Boulliau's most succinct published statement comes in the introduction to a later work, the *Astronomia philolaica fundamenta clariùs explicata & asserta*: "In our *Astronomia Philolaica* we showed from the general circumstances of celestial motions that a revolving Planet completes an elliptical path. From particular circumstances Kepler demonstrated this same thing in a brilliant work, in which, by enormous labor and powerful exertion of mind, he

The *Astronomia Philolaica* was Boulliau's greatest scientific achievement, the culmination of his most fruitful years of research. Only recently have the nature and the extent of his influence been adequately addressed. The astronomer-historian Jean-Baptiste Delambre (writing in the nineteenth century) was the first to evaluate Boulliau's planetary hypothesis; observing from a rather narrow perspective, Delambre viewed Boulliau's work critically but not always correctly.[66] In recent years Boulliau's planetary hypothesis has received more sensitive and subtle analysis.[67] While his contribution continues to be called a "retrograde step," the linear view of scientific development has lost favor in contemporary circles. It is now clear that post-Keplerians (uncognizant of their pre-Newtonian status) were justifiably critical of Kepler's celestial physics, his "empirical laws," and his mathematical methods. But Boulliau's position in the so-called Keplerian Revolution remains ambiguous. In leading astronomers of the "elliptical way" through the middle decades of the century, he finally emerged as Kepler's most influential but critical student. Along the way Boulliau played the double role of conservative-progressive, carefully conserving and promoting the essential elements of Copernicus, Kepler, and Galileo. In the end he translated a cosmological disturbance and an astronomical revolution into an elegant—but short-lived—Copernican synthesis.

---

investigated the motions of Mars. But by removing the equal and mean motion and all innate power of self-movement from the body of the planet, he attributed all the efficacy of movement to the Solar body: and consequently, he stated that the mean motion was fictitious. Yet we showed that this mean or equal motion is not at all fictitious, but is a principle of fixed perpetuity, and that the planet from its own form, and from the innate force of being driven in a circle, should be supposed for urgent, valid reasons; thus, we approached the matter by a different method and showed *a priori* that planetary motion is elliptical." Boulliau, *Astronomia philolaica fundamenta*, p. 5.

[66]Sections dealing with Boulliau are found in Delambre's, *Histoire de l'astronomie moderne* 2: 142–144, 146–161, and 168–172. Delambre assessed Boulliau in the following terms: "Il avait de la science et de l'érudition, c'est à cela que se borne son mérite. Il fut observateur et calculateur; mais ses théories n'auraient été propres qu'à faire rétrograder la science." Delambre, 2: 172. In treating Boulliau's conical hypothesis of 1645, Delambre mistakenly assumes Boulliau's acceptance of the simple empty-focus model. Delambre 2: 151–152.

[67]The best treatment to date is Curtis Wilson, "From Kepler's Laws, So-Called, to Universal Gravitation: Empirical Factors," *Archive for the History of the Exact Sciences* 6 (1970): 89–170, particularly 106–136. Other sources include Russell, "Kepler's Laws"; D. T. Whiteside, "Before the *Principia*: The Maturing of Newton's Thoughts on Dynamical Astronomy, 1664–1684," *Journal for the History of Astronomy* 1 (1970): 5–19; D. T. Whiteside, "Newton's Early Thoughts on Planetary Motion: A Fresh Look," *The British Journal for the History of Science* 2 (1964): 117–137; and Victor E. Thoren, "Kepler's Second Law in England," *The British Journal for the History of Science* 7 (1974): 243–258.

Italy, the Levant, and Paris, 1645–1650

In spring 1645, Boulliau embarked on an extensive and exhausting journey far from the polite parlors of Paris. In May he left the Hôtel de Thou on a tour of Italy and the Levant.[68] Traveling with the wife of the French Ambassador to the Venetian Republic (Nicolas Bretel de Grémonville)[69] and several others, Boulliau proceeded by way of Lyon, Turin, Parma, Cremona, and Ferrara.[70] En route each town warmly welcomed the entourage, and few local curiosities escaped their attention—not the least among the rarities, a parchment Psalter, a Saintly shirt, and the head of Lazarus. At Parma, Boulliau and his group traveled the Po to Venice, where they arrived the morning of 2 July 1645.[71]

When not fulfilling his book-buying obligations,[72] Boulliau visited nearly every city of importance, learned the language, and made a number of scholarly contacts. From his temporary residence in Venice at the house of de Grémonville, he made excursions to Verona, Padua, and Florence, although he assiduously avoided Rome.[73] Visiting libraries and making scholarly contacts, he met in Padua his friend and countryman Gabriel Naudé,[74] and in Florence Boulliau and young Nicolaas Heinsius visited Italy's most illustrious patrons

[68]Boulliau's journey would last nearly two and one-half years; he left Paris on 19 May 1645 and returned in mid-October 1647. The first letter he received en route (at Lyon) was from Jacques de Valois, 21 May 1645, FF. 13042, fols. 20r–21v. Other letters from Luillier and Valois during this period are found in FF. 13042, fols. 12r–35v. Our major source of information is vol. 18 of the CD; Boulliau kept Jacques Dupuy constantly apprised of his activities. One final source is an extensive Latin mémoire written by Boulliau for Ferdinand (in 1651) relating aspects of his trip, FF. 13039, fols. 11r–34r.

[69]In addition to letters in the Collection Boulliau, de Grémonville's correspondence is located at the BN, N.a.f. 5107–5112, and at the Archives du Ministère des Affaires Etrangères, Correspondance Politique, Venise, vols. 54–56.

[70]See Boulliau to Jacques Dupuy, 8 July 1645, CD 18, fols. 8r–9v, and fols. 2–29, passim.

[71]Boulliau to Dupuy, 8 July 1645, CD 18, fol. 8r.

[72]For a brief survey of Boulliau's book buying activities see Clarke, "Book Buying Tour."

[73]Boulliau was in Padua in mid-December (Boulliau to Dupuy, 15 Dec 1645, CD 18, fol. 34) and at Florence from late September to early November 1646 (Boulliau to Dupuy, CD 18, fols. 15–162). On 22 September Christophe Dupuy wrote Boulliau from Rome: "Je ne connois point ce Pere Bonaventura de Cavalieri, et ne scai de quel ordre il est, et encore moins sc[ais-je] sur quoy il fonde l'opinion quil a que l'auteur de l'Astronomia Ph[ilolaica] recevroit quelque deplaisir, s'il venoit icy. Je n'ay point veu son Astronomia et ne puis dire par consequent sur quoi on la voudroit attaquer. Vous savez que le Galilei a esté en peine icy, et ie croi que vous en savez le sujet, c'est pourquoy ie voudrois que vous prissiez la peine de vous declarer plus particuli[erement] et plus distincte[ment] sur ce en quoy vous pouvez iuger à peu prez que l'on se pourroy icy attacher . . ." FF. 9778, fol. 82r.

of science, Prince Leopold of Tuscany and his brother Ferdinand II, the Grand Duke. A decade later, when the Accademia del Cimento began to flourish, Boulliau would become its major source of scientific information; his extensive correspondence with Prince Leopold opened lines of communication between France, Italy, Holland, and Poland.[75] As an expression of his respect for Boulliau, Ferdinand later offered Boulliau a professorship at Pisa, an opportunity the latter tactfully declined.[76]

By autumn of 1646 Boulliau's stay in Italy came to a close. In early September he left Venice for Florence,[77] remaining there until about the first of November; by the next week he traveled to Livorno and departed the following day for the Levant.[78] Boulliau's journey to the east lasted about a month, placing him in Smyrna in December 1646.[79] Touring the region of the Hellespont—near Pera and Constantinople—he remained in the east until spring 1647, when he began his long journey home.[80] By October he had reached Lyon[81] and arrived in Paris shortly thereafter. Boulliau's journey drew wide attention.[82]

[74]For details on Naudé see James V. Rice, *Gabriel Naudé, 1600–1653*. Letters from Naudé to Boulliau are found in the Österreichische Nationalbibliothek, Cod. 7049, fols 393r–408v.

[75]For an introduction to the activities of the Accademia del Cimento see W. E. Knowles Middleton, *The Experimenters*. Middleton writes ". . . there were a few outside Tuscany whose correspondence with Prince Leopold and some others was so voluminous that they ought probably to be considered corresponding members . . . . If volume of correspondence is any criterion, the Paris astronomer Ismael Boulliau ought to head the list, for there are about forty letters between him and Prince Leopold during the years 1657–1667 in the 'Galileo' manuscripts, and more in Paris." Middleton, p. 39. All told the Collection Boulliau contains over ninety letters and drafts exchanged between Boulliau and Leopold, not to mention mémoires. Boulliau also corresponded with Viviani, Rinaldini, and Burattini; Pierre Des Noyers (in Warsaw) was a constant source of information concerning the activities of Italian scientists in the Royal Court of Poland.

[76]See Nicolaas Heinsius to Christophe Dupuy, VIII Kal Feb 1648 [25 Jan 1648 NS], CD 663, fol. 91v; and Jacques Dupuy to N. Heinsius, 25 Jun 1649: ". . . et nostre ami Ismael Bullialdus avoit esté invité par le grand Duc pour succeder a ce dernier en la profession des belles Lettres a Pise, mais il s'en est excusé." Bots, *Correspondance*, p. 53.

[77]See the letters of 20 Sep to 1 Nov 1646, CD 18, fols. 150–162.

[78]CD 18, fol. 164r.

[79]CD 18, fol. 166r.

[80]See CD 18, fols. 174–186.

[81]Boulliau was in Marseille by late September; he wrote Dupuy on 11 Oct 1647 (from Lyon) before embarking for Paris, CD 18, fol. 188r.

[82]For example, Theodore Haak (1605–1690), the German scholar living in England, wrote Mersenne from London on 6 Aug 1647: "J'espere que quand Monsieur Bullialdus sera de retour du Constantinople vous nous en baillerez les nouvelles, et quels seront les fruit que la Republique litteraire s'en doibt promettre. Asçavoir si ce sera en la plus particuliere obser-

With his return to Paris in autumn, Boulliau took a needed rest; indeed, his extant correspondence from 1646 to 1650 is much less than expected, though it henceforth displays a decidedly Italian and eastern flavor. The Collection Boulliau contains perhaps a dozen letters of Boulliau during this period, a fraction of the total received; hence, our information is restricted, though artificially.[83] In any case, most of the correspondence received by Boulliau during this period comes from Italy, from Matharel, de Grémonville, Santini, La Boissière, and Storanus.[84] In addition to the letters he received from Italy and Constantinople,[85] Boulliau continued to correspond with such regulars as Jacques de Valois, Luillier, Leopold, Hevelius, Heinsius, Gassendi, and Mersenne.

Last Years with Dupuy, 1651–1656

In the spring of 1651 Boulliau embarked on yet another journey, this time to Germany and Holland.[86] Departing in early June, he and his companion, Antoine d'Aubray,[87] passed from Paris through Dijon, Basle, Strasbourg, and Frankfort-on-the-Main (where they remained for several weeks), arriving at Cologne on 17 August.[88] After this brief tour, Boulliau wrote from Danzig (on 12 September 1651): "Je partiray aujourdhuy de ce lieu Dieu aydant auec Mr Brauns Marchand Hollandois qui s'en retourne a Amsterdam, nous

---

vation qu'il y aura fait en Astronomie; ou mesme en la collection de plusieurs bon autheurs Arabiques avec intention d'en faire plus de profit au public, que n'ont pas fait plusieurs autres. On se promet des grandes choses de ce brave espirit." N.a.f. 6206, fol. 167; quoted in Brown, *Scientific Organizations*, p. 269.

[83] The Collection Boulliau, though extensive, reflects only part of Boulliau's epistolary efforts. For other manuscript locations, see *Sources*: Manuscript Sources of Ismaël Boulliau.

[84] Matharel, de Grémonville, and La Boissière wrote from Venice; their correspondence is concerned essentially with diplomatic and Papal news, with little emphasis on learned society. Several letters from Storanus and Santini deal with scientific issues.

[85] Most of Boulliau's letters from Constantinople are concerned with nonscientific issues. While in the environs of Constantinople, however, Boulliau made a number of astronomical observations, and the interest spawned there seems to have continued. Parenthetically, Boulliau's observations of the tides in the Hellespont led him to conclude that Galileo's tidal theory would be difficult to disprove.

[86] Boulliau left Paris in June and returned six months later in December, 1651.

[87] For later letters from Antoine D'Aubray (1633–1670) to Boulliau, see FF. 13042, fols. 62–105.

[88] See Boulliau to Dupuy, CD 18, fols. 192–200, letters from Dijon (15 Jun), Basle (26 Jun), Strasbourg (1 Jul), Frankfort (23 Jul & 6 Aug), and Cologne (17 Aug), where Boulliau mistakenly predicted he would be in Paris by November.

[89] Boulliau to [Blaeu?], 12 Sep 1651, FF. 13027, fol. 110r.

prenons le chemin de Berlin & de la Saxe pour aller a Francfort."[89] One week later Boulliau was in Amsterdam, and in succeeding months acquainted himself with Dutch politics and the Dutch countryside.[90] Although he would later return as secretary to the French Ambassador to Holland, Boulliau solidified a number of friendships in this first of four visits to Holland.[91] Holland at this time was a lively center of intellectual and political activity, a fact well-known to Boulliau, who had prior contacts with Johannes Blaeu,[92] Martin Hortensius,[93] Johann Friedrich Gronovius,[94] and Isaac Vossius,[95] to mention a few. During his visit Boulliau also met the polymath savante Anna-Maria van Schurman[96] and the mathematician François van Schooten.[97]

By December of that year, Boulliau was exhausted from his travels and eager to return home.[98] Upon his return to Paris, however, he was greeted with bad news; Pierre Dupuy, longtime friend and colleague, had died on 14 December. In the intervening years since 1645, the brothers Dupuy had entrusted the care of the Bibliotheca Thuana (on rue des Poitevins) to Boulliau,[99] and in June 1645 they

[90]See the article of Hans Bots, "ses voyages," which contains Boulliau's letters to Dupuy from 19 Sep 1651 to 6 Dec 1651; the originals are found in CD 18, fols. 202–220.

[91]In his second trip to Holland, as secretary to Jacques-Auguste de Thou, Boulliau arrived in mid-April 1657; he later returned for nearly four months (mid-Oct 1660 to early Feb 1661), and his final visit, following his trip to Poland, lasted from mid-Oct 1661 to mid-Jun 1662.

[92]The publishing family Blaeu printed Boulliau's *Philolaus* of 1639. See FF. 13026, fol. 2v.

[93]Boulliau corresponded with Martin Hortensius (1605–1639), a versatile scholar and astronomer, during his early years in Paris. See FF. 13026, fol. 2r and fol. 40rv; FF. 13037, fols. 94r–95r; 96r; 97r.

[94]Johann F. Gronovius (1611–1671) was a philologist and classicist. There is but one letter in the Collection Boulliau, FF. 13050, fol. 250r.

[95]The name Isaac Vossius (1618–1689) occurs repeatedly in Boulliau's correspondence, but there are no letters to or from him in the Collection Boulliau. A classicist of some renown, Vossius was well-known to Boulliau's other correspondents, e.g., Boecler, Edward Bernard, N. Heinsius, and others. Three letters of Vossius to Boulliau are found at the Österreichishe Nationalbibliothek, Cod. 7050, fols. 329r–334v.

[96]Anna-Maria van Schurman (1607–1678), a multi-talented savante; see Boulliau to Dupuy, 19 Sep 1651, CD 18, 202v. The Collection Boulliau contains a copy of a letter of Saumaise to van Schurman (Dec 1639), FF. 13040, fol. 154rv.

[97]Franciscus van Schooten (1615?–1661) was an accomplished mathematician and a correspondent of Huygens. See FF. 13050, fols. 261r–262v.

[98]On 6 Dec 1651 Boulliau wrote Dupuy: "Jamais je n'ay eu tant d'impatience de revoir Paris et mes amis que j'ay presentement. J'auray besoin de repos car je vous asseure que je suis extraordinairement fatigué. Je me reposeray en estudiant et regagnant le temps que j'ay consommé en ce voyage." CD 18, fol. 220r.

[99]Boulliau was later assisted in his duties by Jacques de La Rivière (1586–1663) until Boulliau departed for Holland ten years later; the correspondence of Boulliau and La Rivière is useful for news of the Cabinet Dupuy. See esp. FF. 13050, fols. 182–209.

left the Hôtel de Thou to administer the Bibliothèque du Roi, then located near the Sorbonne on the rue de la Harpe.[100] The Cabinet Dupuy continued to flourish under their direction, and it remained one of the most exclusive assemblies in Paris. With Pierre's death, however, Boulliau took new quarters on the rue de la Harpe, residing with Jacques Dupuy from late 1651 until the latter's death on 17 November 1656.[101]

In the meantime, Boulliau led an active scholarly life as mainstay of the Cabinet Dupuy. Among his many contacts during this period was the young and talented Pierre-Daniel Huet.[102] In his memoirs Huet described their early friendship:

> I was also very intimate with Ismaël Boulliau, who then lived with the Du Puys; nor when we were afterwards separated by my return to Caen and his continuance at Paris, did we intermit the offices of friendship; for we frequently interchanged letters, in which I acquainted him with the progress of my studies, and he gave me information of all that was going on among the learned. In this commerce we were by no means upon equal terms; for what entertainment could be derived from a provincial by one who was placed in the very seat of learning, and was himself a stranger to no kind of erudition? The rank, especially, that he held among the first astronomers of the age, may be estimated from his *Astronomia Philolaica* . . . and he abundantly proved his proficiency in polite literature, as well by his familiar conversation, as by his letters to numerous correspondents.[103]

In addition to his activities in the Cabinet Dupuy, Boulliau maintained an extensive correspondence tying together the corners of learned Europe. His principal correspondents during this period are worth mentioning, particularly Jacques de Valois (the astronomer

[100]See Jacques Dupuy to Saumaise, 1 July 1645, FF. 3934, fol. 70. The Bibliothèque du Roi was located on the site of rue Racine, near the Musée Dupuytren.

[101]On 9 Mar 1657 a young Dutch visitor wrote of Boulliau: "c'est une personne sçavante et fort estimée par tous les gens d'esprit; il a esté son bibliothequaire [de Thou] assez longtemps, et apres la mort du Sr Dupuy avoit esté mis dans la bibliotheque du Roy, pour y assister le frere du defunct." A. P. Faugère, *Journal du voyage de deux jeunes hollandais à Paris en 1656–1658*, pp. 92–93.

[102]Pierre-Daniel Huet (1630–1721), Bishop of Avranches, divided his time between Caen and Paris. An érudit of the first rank, he interested himself in theology, philosophy, and science. A small portion of their correspondence is found in the Collection Boulliau; see FF. 13037, fols. 182–191, and "Ismaelis Bullialdi adversaria literaria," MSS Add. 16912, fol. 1, British Library.

[103]Pierre-Daniel Huet, *Memoirs of the Life of Peter Daniel Huet, Bishop of Avranches*, pp. 59–60.

from Grenoble and friend of Gassendi),[104] J. A. Portner,[105] Hevelius, Leopold, Gassendi, J. H. Boecler at Strasbourg,[106] de Vair (the nephew of Valois),[107] Storanus at Ancona,[108] and de Thou. In addition, Boulliau also corresponded intermittently with Boüillon, a mathematician in the court of Gaston d'Orléans (and later with Bury also at Blois),[109] Chauvet at Chartreuse,[110] the Belgian astronomer and lunar cartographer Michael Van Langren,[111] Golius in Leiden,[112] Lenz at Ratisbon,[113] Bernegger at Strasbourg,[114] Langius at Copenhagen,[115] the two d'Aubrays (Dreux and Antoine) and, of course, Huet at Caen. Engrossed in his studies and frustrated by his publishers, Boulliau began to neglect his correspondents by the summer of 1656. In October he wrote Huet:

> Je responds bien tard à votre lettre du 21 Juin dernier . . . [Mais] vous avez l'esprit si bien fait que vous ne voudrez pas croire de moi que j'eusse tant tardé à vous faire réponse pour autre raison que celle que je vous ai dite d'abord [mes occupations d'esprit]. Il y a quatre mois que je combats avec les libraires et imprimeurs qui, depuis ce tems la m'ont fait 8 feuilles d'un traité de géométrie *de Lineis Spiralibus* accompagné de mes exercices géométriques et d'une réponse à Sethus Wardus qui a écrit contre mon Astronomie Philolaïque. Le tout pourra former 25 feuilles de sorte que j'en ai encore pour 8 mois ce qui m'a causé et me cause encore beaucoup de chagrin. J'ai aussi

[104]See particularly FF. 13030.

[105]Johann Albrecht Portner (1628–1687) held an administrative position at Ratisbon and was well known to literati and savants such as Ménage, Guyot, and Boecler; see FF. 13027, FF. 13034, and particularly FF. 13041.

[106]Johann Heinrich Boecler (1611–1672), a friend of Portner, was a classicist and philologist. See FF. 13026, fols. 21–35, and FF. 13027, fols. 63–76.

[107]After Jacques de Valois' death, his nephew de Vair continued to correspond with Boulliau. See FF. 13023, fols. 117–305.

[108]Christophe Storanus of Rome was interested in astronomy, cosmology, and the nature of comets. See particularly, FF. 13043, fols. 94–161. Storanus was in contact with Cassini.

[109]Gaston d'Orléans (1608–1660) was an avid amateur astronomer and sometime patron of science; Boulliau was unable to obtain support. For Boulliau's correspondence with Bury, see FF. 13043, fols. 66, 74, 75 and FF. 13040, fols. 134–136; for his correspondence with Boüillon, see FF. 13025, fols. 52–82, particularly fols. 75r–76r.

[110]See FF. 13042, fols. 36–47.

[111]Michel Florentz Van Langren (1600–1675); see FF. 13037, fols. 151–174. Van Langren's letter of 3 June 1645 (fol. 151rv) announced his intention of naming a lunar crater after Boulliau, which he did (lunar co-ordinates −20 longitude, −20 latitude).

[112]Jacob Golius (1596–1667); see FF. 13037, fols. 133–137.

[113]Most letters of Johann Kaspar Lenz (1630–1667) of Ratisbon were sent *via* Portner; see FF. 13041, fols. 30r–31r et seq.

[114]Matthias Bernegger (1582–1640); see FF. 13040, fols. 250–251.

[115]Wilhelm Langius (1624–1682); see FF. 13037, fols. 115–131.

travaillé à la version et à un commentaire sur un traité de Ptolémée qui n'a point été imprimé mais je ne sais quand il le sera.[116]

Boulliau's idyllic existence ended abruptly in autumn 1656. With the death of Jacques Dupuy on 17 November, Boulliau's future was suddenly unsettled. For sometime Jacques had served as director of the Bibliothèque du Roi,[117] and Boulliau was led to believe he would succeed Dupuy in that position—and perhaps direct the meetings of the Cabinet Dupuy.[118] Much to Boulliau's chagrin, Cardinal Mazarin saw fit to appoint someone else Royal Librarian, clearly a less likely candidate.[119] But bewilderment followed disappointment. Boulliau was all but overlooked in Dupuy's will; after Jacques' death Boulliau received a scant thousand livres.[120]

Years with de Thou, 1656–1666

Offered the opportunity to work with the new director at the Bibliothèque du Roi, Boulliau declined in favor of another option. Jacques-Auguste de Thou II, third son of the historian and brother of the ill-fated François, was to be appointed French Ambassador to the United Provinces, a position he would hold for five years.[121] Seeing his opportunity, Boulliau wrote Cardinal Mazarin; in his letter of 19 February 1657 he expressed his hopes and intentions:

[116]Boulliau to Huet, 1 Oct 1656, Correspondance adressée à Huet. MS in-4° 206, t. II, p. 179rv (copy), Bibliothèque de la Ville, Caen; cf. FF. 15189, fols. 86v–87v.

[117]See Brown, *Scientific Organizations*, p. 12.

[118]Concerning the future of the Cabinet Dupuy, Abraham Wicquefort wrote Boulliau on 30 Nov 1656: "C'est bien la plus mauvaise et la plus surprenante nouvelle que j'eusse pu recevoir que celle de la mort de feu M. du Puy . . . il ne faut pas pourtant permettre que nostre belle assemblée se dissipe, feu M^rs. Du puy vous ont considéré comme celuy qui estiez seul capable de la pouvoir continuer et j'espère que Mr le Cardinal vous rendra en cette occasion la justice qu'il vous doit, et qu'il vous fera succéder en l'employ et dans une partie des bénéfices du défunct." FF. 13042, fol. 108r; for Wicquefort's later reactions see FF. 13042, fols. 110–114.

[119]The appointee was Nicolas Colbert, brother of the minister. See Boulliau to Portner, 24 Nov 1656 and 22 Dec 1656, FF. 13041, fols. 310 and 316.

[120]For details see Boulliau to Portner, 24 Nov 1656, FF. 13041, fol. 310r; Boulliau to Heinsius, 23 Feb 1657, FF. 13027, fol. 117r; Wicquefort to Boulliau, 14 Dec 1656, FF. 13042, fol. 110rv; Des Noyers to Boulliau, 20 Jan 1657, FF. 13019, fol. 234r.

[121]Harrisse, p. 174. In his letter of 5 Jan 1657 to Christiaan Huygens, Claude Mylon described Boulliau as ". . . vn peu desorientè depuis la mort de Monsieur Dupuy, et demeure auec Monsieur le president de Thou qui est destinè Ambassadeur en vos Estats, vous aurez auec qui communiquer lorsque Monsieur Bouillaut sera en vostre pays dans l'employ de secretaire de l'Ambassade." HO II: pp. 1–2.

> Le Roy ayant faict l'honn[eu]r a M^r le President de Thou de le choisir pour son Ambassadeur ord[inai]re vers M^rs les Estats G[e]n[er]aux des Prouinces unies, & V.E. Monseig[neu]r ayant porté S.M. a faire ce choix, Mr de Thou m'a faict l'honneur de me receuoir pour luy servir de premier secretaire dans son Ambassade, et ce qui l'a principalement porté a jetter les yeux sur moy, c'est qu'il a creu que V.E. auroit pour aggreable que je rendisse seruice au Roy dans l'employ que S.M. luy donne & que V.E. luy procure; et je suis persuadé que V.E. croira, que m'estant attaché a ses interests, je seruiray le Roy comme un bon & fidele subject est obligé de faire. Ainsi Monseigneur je supplie tres humblement V.E. qu'elle ayt la bonté de me procurer un fauorablement traictement dans cette occurence, & qu'il luy plaise de me faire donner les appointemens que l'on a accoustumé de donner aux Secretaires de l'Ambassade.[122]

In the spring of 1657 Boulliau, together with de Thou and his friend, Eméric Bigot, departed for Holland. Their voyage began on 13 April from Dieppe, and they arrived in Rotterdam (after a difficult journey) on 22 April.[123] Spending his first night in The Hague on 25 April, Boulliau stayed in Holland this second time less than four months,[124] during which time he visited Christiaan Huygens, whom he had met several years earlier in Paris.[125] After his return home, Boulliau maintained an extensive correspondence with Huygens until the latter's visits to Paris (and eventual residence at the Bibliothèque du Roi) removed the necessity of written communication.[126]

[122]Boulliau to Mazarin, 19 Feb 1657, FF. 13027, fol. 243r. See also Boulliau to Portner, 12 Jan 1657, FF. 13041, fol. 314r.

[123]Leonard E. Doucette, *Emery Bigot, Seventeenth-Century French Humanist*, p. 9. See FF. 13026, fols. 17r; 8r–9r, and 329rv.

[124]For details see Boulliau to La Rivière, 9 Apr 1657, FF. 13026, fol. 17r; Boulliau to La Rivière, 3 May 1657, FF. 13026, fols. 8r–9r; Boulliau to Hevelius, 4 May 1657, FF. 13043, fol. 81r. See also the letters of La Rivière and Soëfve, FF. 13050, fols. 184–247.

[125]On the circumstances of Huygens' and Boulliau's first meeting, Huygens wrote to Constantyn Huygens, 6 Aug 1655: "Tassin m'a aussi promis de me mener chez Monsieur de Chamboniere, mais je veux differer les visites jusqu'àpres nostre voyage d'Angiers. Cependant je n'ay peu m'empescher d'aller chercher quelquesuns des fameux Mathematiciens, desquels j'ay veu Gassendi et Bullialdus. Cettuicy ma fait voir la bibliotheque du roy et mesne auprez Monsieur du Puis, qui m'a fait la faveur de me monster encore la siene et son cabinet. Il demeure dans la maison ou est la bibliotheque royale, et c'est la que s'assemblent tous les jours les illustres." HO I: p. 342.

[126]Huygens made three extended visits to Paris (July–September 1655; October 1660–March 1661; and April 1663–May 1664). An early member of the Académie Royale des Sciences, Huygens resided in Paris from May 1666 until 1681, making two trips to The Hague in 1670–1671 and 1676–1678 for reasons of health.

Following his four-month tour in Holland, Boulliau returned to Paris where he remained for the next three years.[127] In addition to his numerous duties as secretary to de Thou, he maintained a considerable correspondence with Huygens,[128] Hevelius, Prince Leopold of Tuscany, and Pierre Des Noyers. It was during this period of transition (as de Thou's secretary) that Boulliau became publically engaged in a celebrated controversy with Seth Ward, Professor of mathematics at Oxford,[129] on the foundations of Philolaic astronomy.[130]

Ward's first attack appeared in 1654 as *In Ismaelis Bullialdi astronomiae philolaicae fundamenta, inquisitio brevis.*[131] In effect, he claimed Boulliau's conical hypothesis of 1645 was equivalent to an "empty-focus model," that is, to an ellipse wherein the unoccupied focus served as an equant point.[132] Shortly thereafter, in his *Astronomia Geometrica* of 1656, Ward further maintained that his empty-focus hypothesis not only reduced the rigors of calculation, but was indeed true.[133] Boulliau's response appeared the following year as *Astronomia philolaica fundamenta clariùs explicata.* He be-

[127]Boulliau left The Hague on 5 August 1657 and arrived in Paris on 31 August. Boulliau to Portner, 12 Oct 1657, FF. 13041, fol. 346r.

[128]See, for example, FF. 13029, fols. 189–222; for sometime Boulliau served as an intermediary between Huygens and Leopold.

[129]There are five known letters exchanged between Boulliau and Seth Ward (1617–1689), four in the Collection Boulliau, (FF. 13037, fols. 177–181, fol. 208rv, and FF. 13050, fol. 263r), and Ward to Boulliau, 14 Feb 1656/7 OS, N.a.f. 6197, fol. 173rv. Boulliau's disposition during the controversy is evident in his letter to Ward of 9 Jul 1657: "Hic nunc Hagae Comitis in familia Ex^mi^ Dñi Legati dego, alii vitae instituto haud sponte mea implicitus, multumque diverso, ab eo, quod hactenus tenueram. Secretarius Dñi Legati ago, et ex otioso studiisque vacante negotiosus factus sum." FF. 13037, fol. 178r.

[130]See Boulliau to Wallis, 24 May 1655, FF. 13037, fol. 175v. On 28 May 1655 Pierre Des Noyers (temporarily in Paris) wrote Hevelius: "J'ay veu Monsr Boulliau qui se parle fort bien et qui vous baise les mains, il est apres a respondre a un anglais qui a escrit Contre ses Tables et qui le reprend d'avoir manqué a quelques prostapherese. En quoy il dit qu'il a raison, mais cet anglais ne desire pas le moyens de Coriger cette faute, et cest ce que Mr Boulliau donnera." Epistolae Hevelium, *AC* 1(3), No. 457v, BO. See also Hevelius to Ward, 17 Jul 1659, AC 1(4), No. 547. For an excellent survey of the issues see Wilson, "Kepler's Laws," pp. 115–122.

[131]The title page bears the date 1653, but is technically incorrect.

[132]Ward mistakenly assumed Kepler's hypothesis was equivalent to Boulliau's conical hypothesis of 1645, which, in turn, he assumed to be equivalent to the empty-focus (or simple elliptical) hypothesis. Ward was mistaken on all counts; see Wilson, "From Kepler's Laws," esp. p. 121. Boulliau's calculational procedure was itself internally inconsistent.

[133]The *Astronomia Geometrica* is constructed around the empty-focus model which Boulliau and others quickly dismissed. Boulliau rejected Ward's "new methods" for finding aphelions and nodes as impractical; see Boulliau's *Astronomia philolaica fundamenta,* "Monitvm," pp. 55–56.

gan by acknowledging an anomaly in his hypothesis, a "certain most small inequality" amounting to 2′ for the planet Mars.[134] In an early effort to eliminate the error[135]—detected when the *Astronomia Philolaica* was in press—Boulliau calculated a new eccentricity for Mars consistent with his equations;[136] in addition, he supplied the reader a second set of charts (derived from the *Rudolphine Tables*) that would also account for the discrepancy.[137] To be sure, Ward had noted the discrepancy, but he offered no correction or explanation. Boulliau's task was thus clear; he would correct the error and demonstrate that his "elegantly constructed and adorned hypothesis [was] supported by valid foundations."[138]

Boulliau's defense showed extraordinary ingenuity as well as respect for empirical accuracy. But having already answered Kepler's challenge in the *Astronomia Philolaica* (by demonstrating elliptical motion *a priori*), Boulliau responded to Ward by demonstrating the proposition "from particular and lower circumstances."[139] His results *a posteriori* were excellent; testing his hypothesis against 28 of Tycho's observations for Mars, Boulliau's calculations were in better agreement with the data than Kepler's.[140] An elegant and accurate alternative, Boulliau's modified hypothesis earned a number of adherents and imitators.[141] While still an undergraduate, Newton took note of the hypothesis in one of his small pocketbooks, copied virtually verbatim from Thomas Streete's *Astronomia Carolina*.[142]

[134]The inequality for other planets (excluding Mercury) was imperceptible by calculation. For references to the small inequality see *Astronomia philolaica fundamenta*, pp. 3; 32; 38–39.

[135]Two minutes of calculated error was at the limit of mid-century observational accuracy.

[136]Upon recalculation, Boulliau found his equations followed precisely from an ellipse with an eccentricity of 0.10192; see *Astronomia Philolaica*, Bk. I: p. 50 (first of two); and *Astronomia philolaica fundamenta*, p. 7.

[137]In book I of the *Astronomia Philolaica* Boulliau added two additional pages (numbered 49 and 50) between the pages originally numbered 48 and 49.

[138]Boulliau, *Astronomia philolaica fundamenta*, p. 4.

[139]Boulliau, *Astronomia philolaica fundamenta*, p. 6.

[140]Kepler's average error was 2′7″, the maximum 5′50″; Boulliau, with the same data, had an average discrepancy of 2′3″.2, the maximum 5′33″. The latter data is derived from the *Astronomia philolaica fundamenta*, p. 50; cf. FF. 13058, fols. 88r–89v.

[141]Boulliau's Philolaic Tables were adopted in whole or part by G. B. Riccioli, Jeremy Shakerley, and John Newton, and his influence is evident in the calculational procedures of Vincent Wing and Thomas Streete. Niklaus Mercator (rejecting the hypotheses of Comte de Pagan and Ward) acknowledged only his own and Boulliau's modified hypothesis. The last post-Keplerian—Newton himself—seems to have preferred the numbers of Kepler and Boulliau to others.

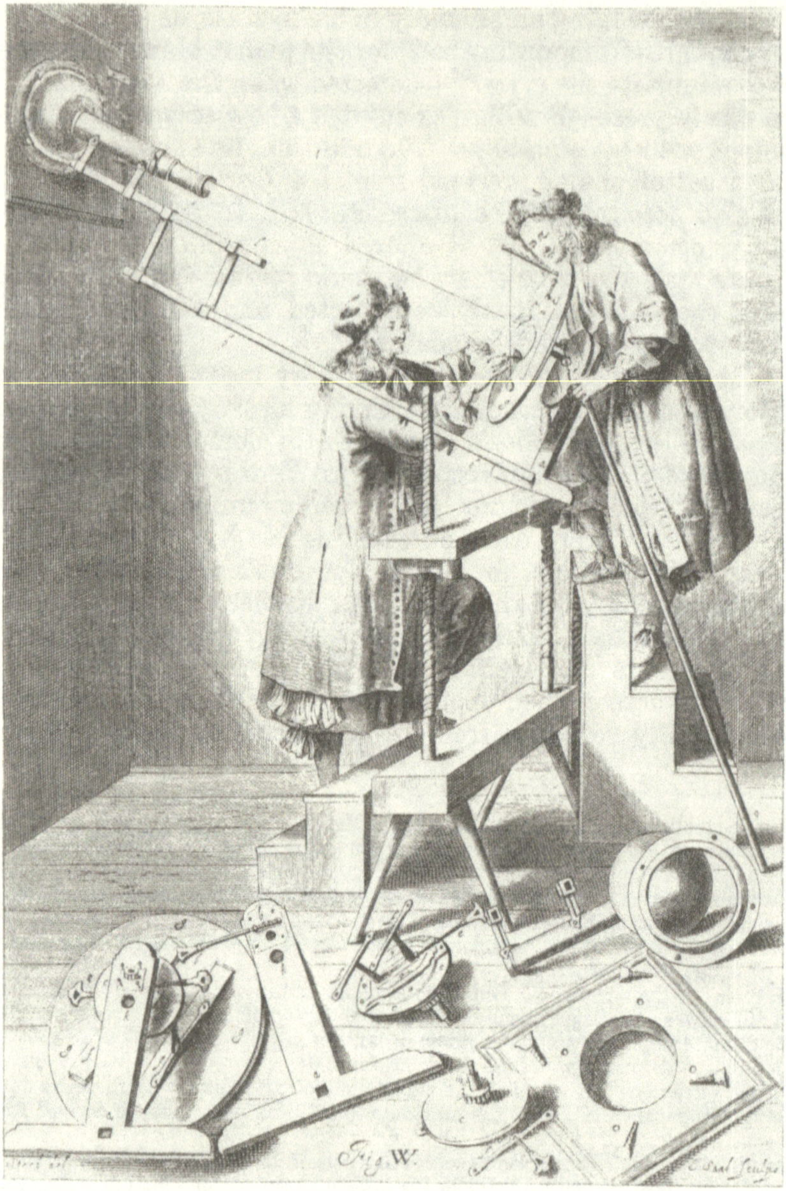

Figure 3. Boulliau and Hevelius sketching the solar eclipse of 29 March 1661 with the helioscope. En route to Warsaw on 3 May 1661, Boulliau was unable to observe the transit of Mercury. Illustration from Hevelius' *Machinae Coelestis Pars Prior*, Chap. XVIII.

At the end of September 1660, Boulliau set out for Warsaw and Danzig to visit his friends Des Noyers and Hevelius.[143] He first accompanied Madame de Thou to Holland, arriving in The Hague in mid-October;[144] he remained there nearly four months before departing for Poland.[145] Boulliau left Holland in early February, and after a brief stay in Hamburg arrived in Danzig in mid-March.[146] In Danzig he was the houseguest of Hevelius, an old friend and correspondent of nearly thirty years.[147]

Boulliau's six-week visit was anything but restful. The industrious Hevelius had spared no effort or expense in constructing his observatory, and he was eager to involve Boulliau in his astronomical activities.[148] For some time Hevelius had attempted to perfect a means of accurately observing and recording solar eclipses and sun spots.[149] The result of his labor was his helioscope, a device designed

[142]See Whiteside, "Newton's Early Thoughts," p. 123; and his "Before the *Principia*," p. 9.

[143]Boulliau left Paris on 30 Sep 1660; see his letter to J. A. Portner dated 6 Feb 1662, FF. 13027, fol. 228r. Boulliau's trip to Poland was long in planning and frought with delay. On 25 Jun 1659 Des Noyers wrote Boulliau: "Par votre lettre du 23 mai, j'apprends avec joie que vous persistez dans la volonté de faire un voyage en ces pays-ci. J'en ai donné avis a M. Hevelius; mais je n'ai reponse ni a cette lettre-là ni à deux autres, dont je ne m'étonne pas, car c'est sa coutume d'etre long a repondre." FF. 13020, fol. 219r.

[144]HO III: p. 66.

[145]Boulliau arrived in The Hague in mid-Oct 1660 and left 27 Jan 1661; he also spent time in Amsterdam, FF. 13050, fols. 145–146 and FF. 13026, fol. 103bis. Boulliau's departure for Danzig was delayed; in a letter of 9 Dec 1660 (Thursday) de Thou wrote [?Mazarin]: "Le S.ʳ Boulliau part Lundy pour aller voir le S.ʳ Hevelius a Danzic pour observeur avec luy au commencement de Mars deux eclypses de lune et de soleil. Touts les Messieurs les Astrologeurs ont une espèce d'association & de commerce entre eux qu'ils observent plus religieusement que s'ils y estoient obligez par des voeux." Correspondance Politique, Holland, no. 65, fol. 318v, Archives du Ministère des Affaires Etrangères. Boulliau wrote Huygens on 17 Jan 1659: "Jl est vray que j'ay escrit a Monsieur Heinsius que pourrois aller en Hollande auec Madame de Thou, & si sa santé luy permet d'y aller, je seray le voyage, & j'auray l'honneur de vous voir & de vous entretenir; & si entre cy & ce temps la je reçois les verres de Lunette que vous me faictes la grace de me donner, je vous rendraj compte de ce que j'auray veu et descouuert." HO II: p. 319.

[146]For details see FF. 13027, fol. 228r.

[147]Boulliau and Hevelius met in Paris during the latter's tour of Europe in 1632–1634. Their correspondence covers the years 1648 to the mid-1680s.

[148]On 16 Apr 1661 Boulliau wrote de Thou from Danzig: "Pour moy je philosophe auec Mʳ Hevelius, & je ne peux me lasser d'admirer son industrie, sa diligence, & la perspicacité de son esprit . . . Si dieu le conserue encore quatre ou cinq ans, il nous donnera des ouurages les plus beaux & les plus curieux & les plus exacts qui ayent encores paru dans l'Astronomie." FF. 13026, fol. 233v.

[149]His sharpened concern at this time was due to the impending solar eclipse of 29 Mar 1661 and, more importantly, the transit of Mercury of 1661. See, for example, Des Noyers to Hevelius, 23 Jan 1661, Epistolae Hevelium, AC 1 (5) No. 635, BO.

to project the solar image through a telescope at right angles onto a flat surface. With this projected image, sun spots, solar eclipses, and transit phenomena could easily be traced onto paper. The problem Hevelius faced—in the absence of an equatorial mounting—was to project a stationary image of the sun onto a flat surface. In his *Machina Coelestis* Hevelius explained:

> I, then, clearly seeing all these inconveniences, so early as the year 1661 applied to the eminent Boulliau, in those days one of my most valued friends, when he was with me on the occasion of an eclipse of the sun, as to how I could best obviate these difficulties, and cause the sun's reflection to remain steady on the object-table . . . in order to mark down spots and the different phases of eclipses in the most accurate manner, and without any vibratory movement. The illustrious Boulliau strongly advised me at that time to carry my project into execution as quickly as possible for the benefit of all observers. . . .[150]

After six weeks in Danzig (where he and Hevelius observed the solar eclipse of 29 March), Boulliau departed for Warsaw and arrived in early May.[151] Here he visited Pierre Des Noyers, secretary of Queen Marie-Louise de Gonzague, and by all accounts, was well received at the Polish court.[152] Boulliau remained there from May to early August before returning to Danzig (and another brief visit with Hevelius),[153] and in mid-September departed for Holland. His return journey lasted four weeks and brought him to The Hague by

[150]C. Leeson Prince. *The Illustrated Account given by Hevelius in his "Machina Celestes"* (London, 1882), pp. 33–34, quoted here with minor changes.

[151]See FF. 13026, fols. 233–236 and FF. 13044, fols. 5–6. Much to his dismay, Boulliau was unable to remain with Hevelius to observe the transit of Mercury on 3 May 1661. On 4 Jul 1661 Boulliau wrote de Thou from Warsaw: "J'estois en chemin pour me rendre de Dantzigt en cette ville le 3^e^. May lors que M^r^. Hevelius obserua la conjonction de Mercure auec le Soleil, & Mr Hugens l'obserua a Londre; Je regrette grandement d'auoir perdu un si beau spectacle, mais ma consolation est que mes Tables Philolaïques la representent mieux que toutes les autres, & ne s'en esloignent pas de 3. heures; ce qui est certainement de consideration, & qui donne vn tres-grand credit a mon ouurage, que je pretons de perfectionner d'avantage; pour quoi faire il ne me faut que 6. mois de temps, mais M. Heuelius m'a prié, que j'attendisse quil eust publié ses obseruations, qui pourront me seruir a cela comme vne reigle infallible." FF. 13026, fol. 237v. In London for the coronation of Charles II, Huygens observed the transit with Thomas Streete.

[152]Boulliau would later receive a handsome gift of 1,000 livres, see Boulliau to Marie-Louise de Gonzague, 8 May 1665, Correspondance Politique, Pologne, vol. 14, fol. 463r, Archives du Ministère des Affaires Etrangères.

[153]Boulliau wrote to Daniel Wachschlager, 30 Aug 1661, from Danzig, FF. 13027, fol. 189r.

mid-November;[154] Boulliau remained in Holland until mid-June 1662,[155] returning to Paris in early July.[156]

When he returned home in spring, Boulliau was in his fifty-seventh year and would never again leave the environs of Paris. The years 1662 to 1666 were uneventful; Boulliau spent them quietly fulfilling his duties as "Biblicaire en la Bibliothèque des Thuano," attending the meetings of the Cabinet and the early conferences of Henry Justel, and all the while maintaining his correspondence and his schedule of astronomical observation.

Collège de Laon, 1666–1689

In spring of 1666 a dramatic change occurred in Boulliau's career. At the end of May he left his quarters on rue des Poitevins after an argument with de Thou. It seems Boulliau was unjustly accused of meddling in a family matter between de Thou and his son, Louis-Auguste.[157] Shortly thereafter, Boulliau moved to the Collège de Laon, where he remained for the next twenty-three years.[158] In essence, he had lost a patron.

By this time Boulliau was regarded as the foremost mathematical astronomer in France.[159] In recognition of his rank, he was elected to the Royal Society of London in summer, 1667. Henry Oldenburg, who had met Boulliau in Paris in 1659–1660,[160] announced the news

[154]Boulliau to Hevelius, 14 Nov 1661, FF. 13026, fol. 109r; in addition to sketching his return journey Boulliau here relates Huygens' opinion of Hevelius' work. In a later letter to Leopold, Boulliau offered details of his trip: Boulliau to Leopold, 15 Dec 1661, FF. 13027, fol. 18rv; see also Boulliau to Portner, 6 Feb 1662, FF. 13027, fol. 228rv.

[155]In Boulliau to Fabricius, 8 May 1662, FF. 13027, fols. 132–133, Boulliau mistakenly predicts he will leave for Paris in two or three weeks; cf. Boulliau to Hevelius, 19 Jun 1662, FF. 13026, fol. 117v.

[156]Boulliau to Rautenstein, 28 Jul 1662, FF. 13027, fol. 154r; see also, Boulliau to Hevelius, 20 Jul 1662, FF. 13026, fol 118r.

[157]Boulliau left the rue des Poitevins about 28 May 1666; he related the incident to a number of correspondents, for example, Boulliau to Stephanus Gradius, 14 Jan 1667, FF. 13026, fol. 42r; Boulliau to Lubienietzki, 16 Jul 1667, FF. 13026, fol. 306r; Boulliau to Hevelius, 16 Jul 1666, FF. 13026, fol. 149v; and Boulliau to Leopold, 23 Jul 1666, FF. 13027, fol. 10r.

[158]The Collège de Laon was formerly under the auspices of the Benedictines; it was located between rue des Carmes and rue de la Montagne Saint-Geneviève, near the present day Marché de la Place Maubert.

[159]See, for example, OC III: p. 400, note.

[160]Acting as tutor of Richard Jones (Ranelagh), Oldenburg and the young nobleman ended their Grand Tour in Paris and stayed there from early spring 1659 to May 1660. During this period Oldenburg seems to have been the intermediary in an exchange of correspondence between Boulliau, Samuel Hartlib, and Niklaus Mercator.

in his letter of 22 August 1667 (OS):

> Ayant vû dans la lettre de Monsieur Justel le respect, que vous portez à la Societé Royale, et consideré au mesme temps vostre merite, et le rang, que vous tenez parmy les Scavans de ce Siecle; Je l'ay crû estre et du service de cet Illustre Corps, et du vostre, que Vous y fussiez receu pour Membre. Vous y ayant donc proposé comme Candidat en Assemblee publique, ie vous puis dire, Monsr, que ladite Societé vous receut d'un consentement universel. . . .[161]

Although he had been a leading member in several major Parisian societies and (for want of a better term) a corresponding member of the Accademia del Cimento, Boulliau was never appointed to the Académie Royale des Sciences.

In the same year he was elected to the Royal Society, Boulliau published a tract entitled *Ad Astronomos monita dvo*. . . .[162] The purpose of this work was to alert astronomers to two apparently variable celestial phenomena, the "new star" in the neck of the whale (the long period variable Mira Ceti), and the nebula in the northern part of the belt of Andromeda.[163] Boulliau sent a copy of his treatise to Prince Leopold of Tuscany, longtime correspondent and patron,[164] and in his letter of 7 January 1667 explained:

> I offer most courteously to your Most Serene Highness a small work I published in the last days of the declining year, and I desire that you accept it as a small gift. I advise astronomers in this treatise to direct their attention to the new star in the neck of Cetus next March, and furthermore, at present, to the nebula appearing after a period of two years in the northern part of the girdle of Andromeda; the academy of Astronomy at Paris and the Academy of all Mathematics urged me to place it in the public domain; although I am not a member of the Academy, I nevertheless will always strive to further the Mathematical sciences and the interests of the same Academy. Should this treatise please your Most Serene Highness, I shall receive ample re-

[161]Oldenburg to Boulliau, 22 Apr 1667 (OS), FF. 13034, fol. 91r.

[162]A summary of Boulliau's *Ad Astronomos monita dvo* . . . was published in the *Philosophical Transactions* 1: 381–383; see also pp. 459–460. Later contributions are found in 7, (No. 111): 238; 12, (No. 139): 969–975; and 13, 693. The *Journal des Sçavans* published summaries of Boulliau's *Ad Astronomos monita dvo* . . . (1667): pp. 11–14 and his *Arithmeticam infinitorum* (1682): pp. 388–390.

[163]Boulliau, *Ad astronomos*, pp. 5–14 and pp. 14–19 (includes illustrations).

[164]Leopold's patronage was not overwhelming, but he was responsible for sending Boulliau an eleven-foot telescope and a mercury thermometer. Boulliau dedicated his *Observatio secundi deliquii lunaris* (1653) to Leopold.

ward and extraordinary honor for my labor. If God grants life and health, I shall publish in the coming summer some work I hope will not be unappreciated by Mathematicians.[165]

Boulliau's first concern in this treatise was the "marvelous star" in the constellation Cetus; detailing its brief history and its periodic behavior,[166] he then showed that Mira's variation in luminosity (from about the third to the tenth magnitude) was probably due to the axial rotation of a semiluminous body with a period slightly under 333 days.[167] Although David Fabricius, Kepler, Johannes Phocylides Holwarda, Hevelius, and others had scrutinized this strange star, Boulliau was the first to establish its mean period with accuracy[168] and present a viable and enduring hypothesis. The observational skills of William Herschel over a century later added little to Boulliau's assessment.[169]

The second part of Boulliau's advisement concerned the variation in light of the Andromeda nebula (M31). As Boulliau related, this "vaporous convolution" was observed in 1612 by Simon Marius while investigating the satellites of Jupiter.[170] But as a skilled historian and astronomer, Boulliau was perplexed by the fact that Hipparchus, Tycho, and Bayer had not recorded the nebula in their star catalogs, especially when an anonymous fifteenth-century chart por-

[165]"Opusculum idcirco hisce postremis anni decursi diebus a me editum Celsit[udini] Tuae Ser[enissi]mae officiosissime offero, vtque gratum Tibi set munusculum valde cupio. Astronomos illa scriptiuncula moneo, vt ad nouam stellam in collo Ceti mense Martio proximo, praetereaque ad Nebulosam in Cinguli Andromedae parte Borea abiennio apparentem impraesens attendant. publici juris vt eam facerem, jnstituta Parisiis Astronomiae, totiusque Matheseos Academia me impulit; Cujus quamuis pars non sim, commodis tamen ipsius, & Mathematicis scientiis promouendis semper studebo. Si placuerit Ser[enissi]mae Celsit[udini] Tuae hoc opusculum meum, laboris praemium amplissimum gloriamque non mediocrem sum relaturus. Si Deus vitam & sanitatem mihi largitetur, aliquod opus sum aestate proxima editurus, quod non ingratum fore Mathematicis spero." Boulliau to Leopold, 7 Jan 1667, FF. 13027, fol. 6r.

[166]Boulliau, *Ad Astronomos*, pp. 5–8; Mira was first observed in 1596 by Fabricius, a major correspondent of Kepler.

[167]Boulliau, *Ad Astronomos*, p. 7; pp. 12–14, "Causae verisimilioris vicissitudinum ortus & occasus stellae nouae explicatio."

[168]Within 1/2 percent.

[169]One of Herschel's first publications dealt with Mira Ceti; he retained the hypothesis of axial rotation. In Bk. III of the *Principia*, Newton obliquely endorsed Boulliau's hypothesis: "But those fixed stars that appear and disappear by turns, and increase slowly by degrees, and scarcely ever exceed stars of the third magnitude, they appear to be of another kind, and in rotating on their axes, appear to show a light and dark side by turns." Isaac Newton, *Isaac Newton's Philosophiae Naturalis Principia Mathematica*, ed., A. Koyré and I. B. Cohen (2 vols., 3rd ed., Massachusetts, 1972) 2: 758.

[170]Boulliau, *Ad Astronomos*, p. 15.

trayed the nebula in perfect detail.[171] Other factors heightened his curiosity; in November of 1666 the nebula was completely invisible though, inexplicably, it had shone brightly two years before.[172] Offering no explanation for this variation, Boulliau concluded that periodic phenomena demand continuous and long-term scrutiny.

As he mentioned in his letter to Leopold, Boulliau was never elected to the Parisian Académie des Sciences.[173] He was an active member in the major assemblies and informal circles of astronomers and savants (whose membership overlapped), but had never been a whole-hearted partisan of formally structured academies, even when financially supported;[174] his opinion of the Académie française is a well-known case in point.[175] Yet, several of Boulliau's closest

[171]Boulliau, *Ad Astronomos*, pp. 16–17.

[172]Boulliau, *Ad Astronomos*, p. 17. The variation in brightness of M31 was noted by a number of seventeenth-century observers; at present there is no adequate physical explanation, unless one assumes that a super nova caused the galaxy to appear unusually bright in 1664.

[173]Boulliau grew increasingly bitter and routinely challenged the Académie's findings. On 19 Jun 1676 Des Noyers wrote Hevelius: "Nostre bon amy M. Boulliau me prie de vous faire entendre que les astronomes de l'academie de Paris disent qu'ils ont observé l'equinoxe et pretend en avoir trouvé que le Calcul de M. Hecker s'esloigne du Ciel de 7.′ en longitude pour le mouvement du soleil ce qui va a 5. heure 48′ de temps. M. Boulliau doute que l'observation fait bien juste et bien exacte et vous prie pour en pouvoir mieux juger, que sy vous avez observeé cet equinoxe, de luy communiquer vostre observation. . . ." Epistolae Hevelium, AC 1 (12) No. 1717, BO.

[174]On 11 Jul 1661 Boulliau offered the following opinions to Huygens: "Messieurs de l'Academie d'Angleterre nous produiront quelque chose de beau auec temps, & je croy quils nous donneront de belles inuentions en matiere de Lunettes, Monsieur Heuel y trauaille aussi de son costé. Si l'on pouuoit persuader nos Messieurs de Paris, qui sont riches, curieux des belles choses, & desireux d'vne gloire jmmortelle, de faire quelque desponse pour acquerir des cognoissances par les experiences, on pourroit esperer quelque chose. Mais ils veulent sans peine (sinon que celle que l'on met a faire de beaux discours & philosopher en lair) sans application & sans despense s'acquerir du plus beau & du meilleur. Vous l'auez veu par experience, & je ne vous en diray pas dauantage par escrit." HO III: p. 293.

[175]On 14 Oct 1645 Boulliau wrote Dupuy: "Je voudrois auoir veu des Moustaches, dont vous me parles, autour de ces MSS, je leur aurois dict un petit mot de galanterie. Cette sotte canaille auec son dictionaire reformé pretend de juger de choses qu'elle n'entend pas, Vous verrez que cette Academie sera vn seminaire de Barbarie, & qu'autant qu'elle pourra elle estouffera la cognoissance des langues & des bonnes lettres, et ce dautant plus quil y a aujourdhuy peu de personnes qui veuillent prendre la peine d'estudier, La science & capacité d'vn bel esprit consistera a bien arrondir une periode & a faire vn rondeau, une meta[mor]phose d'yeux &c ou quelque autre badinerie qui se fera en trois semaines ou vn mois, & puis sera promenee par les ruelles des coquettes de Paris, qui donneront credit a vn homme selon leur caprice. Si le chef de cette celebre compagnie . . . eust bien faict, il l'auroit dissipee, & il auroit raison poussé d'vne juste indignation, apres auoir voulu faire la despense & enuoier expres vn homme pour luy rapporter des liures, qui peuuent seruir au public plus que tout ce qui est

friends were supported by the Académie, and his correspondents were always eager for information.[176]

In September 1667 Ismaël Boulliau was sixty-two years old. His health was still generally good[177] (records show that he made astronomical observations well past his eightieth birthday), and there is evidence that his mental powers never faltered; Boulliau's massive *Arithemeticam infinitorum* of 1682, for example, was published in his seventy-seventh year. Nonetheless, though he continued a vigorous international correspondence, Boulliau's creative years had passed. His longtime correspondence with Hevelius, Des Noyers, Portner, Leopold, and Oldenburg continued with fair regularity, and he added a number of new correspondents, among them Viviani,[178]

---

jamais sorti des mains de ces beaux exprits. Mais tout ce que je vous dis-la c'est soubs le sceau de confession, & j'en excepte les amis du cabinet. . . ." CD 18, fol. 21r.

[176]In his letter to Hevelius of 18 Feb 1667, FF. 13026, fol. 153, Boulliau wrote: "Since I have not yet been able to learn all the names of those enrolled upon the list of the new academy, I cannot at the present moment, honoured sir, comply with your request. Physicists, anatomists, chemists, mathematicians, botanists, and others unnamed so mix in one coop that varied and different progenies are without doubt to be expected from such diverse and mutually distant interests and kinds of talent. A fortnight ago they dissected the body of a woman found guilty of infanticide. From this dissection they derived, on their own admission, no, or at least very little profit. However, that the principal names of this college may not be unknown to you, I shall list those known to me. Chief of all is the renown Christiaan Huygens, who, called from Holland, enjoys an annual stipend of 6,000 Polish florins. Next are Roberval (who, regarding your quadrants and sextants of less importance, promises himself that he will define and accurately learn, in fact more accurately than you, the position of the stars, planets, and comets from cocks set on weather vanes, the rafters of belfries, and from distant signs and observations); Auzout; Frenicle, a wonderful man in mathematics, for if you knew him you would admit that all the forms, kinds, and properties of all numbers were known to him; and a man by the name of Buot, ignorant of Latin and Greek, endowed with a rough and unpolished disposition, but not unsuited to geometry. The first three of these receive only a fourth part of the allowance of Huygens. Du Hamel is given money generously inasmuch as he has devoted his attention to translating into Latin the works which have been composed in the Duchy of Brabant on the king's law. And this Buot receives only four-fifths of the stipend of the afore-mentioned first three. There has been enrolled another man by the name of Picard, who is engaged in astronomy and enjoys it, but no money has been paid him thus far, (though) they give him hopes of getting some soon." Quoted and trans. in Albert J. George, "The Genesis of the Académie des Sciences," *Annals of Science* 3 (1938): 394–395.

[177]For Boulliau's assessments see Boulliau to Portner, 28 Sep 1667, FF. 13027, fol. 212r, and Boulliau to J. G. Graevius, 9 Dec 1672, FF. 13026, fol. 55r. In later years Boulliau suffered from rheumatism, Boulliau to Graevius, 9 Jan 1683, MS Thott 1259, fol. 7r., Det Kongelige Bibliotek. On 16 Jul 1683 Des Noyers wrote Hevelius: "Il m'escrit qu'il [Boulliau] commence a se mieux porter du cruel Rhumatisme dont il a esté travaillé tout l'hiver," Epistolae Hevelium, AC 1 (15) No. 2256, BO.

[178]Vincenzio Viviani (1622–1703), mathematician and classicist, was Galileo's last student. Boulliau and Viviani exchanged about a dozen letters.

de Gruter,[179] Burattini,[180] and Graevius.[181] In terms of sheer bulk, however, the greater share of Boulliau's extant correspondence during these years was with Bernart,[182] Gratta,[183] and Pieter and Philip Pels.[184]

Abbaye Saint-Victor, 1689–1694

In spring 1689, after living at the Collège de Laon for twenty-three years, Boulliau established himself in new quarters in the Abbaye Saint-Victor.[185] He was now in his eighty-fourth year. Pierre Des Noyers (then eighty-two), Boulliau's correspondent of thirty-five years, observed in a letter of 13 May 1689:

> Jay aujourdhuy Monsieur deux de vos lettres qui sont des 15. et 22. d'avril dont je vous rend tres humbles graces, dans la premiere vous me ditte vostre transport a St. Victor ou jespere que vous serez mieux pour l'air et pour la compagnie qu'au College de laon, vostre seconde lettre que Je supose estre escrite de St. Victor est desia plus lisible que la p^{re}.[186]

[179]Isaac de Gruter (1610–1680), Dutch scholar and classicist from Rotterdam.

[180]Tito Livio Burattini (1617–1681), also known as Giovanni Michele, was an Italian scientist and engineer, best known for his long-view telescope lenses and telescope riggings. Burattini was a friend of Paolo del Buono, and Des Noyers kept Boulliau abreast of their activities in Poland.

[181]Born and educated in Germany, Johannes Georgius Graevius (1632–1703) became one of the outstanding classical scholars in late seventeenth-century Holland.

[182]Bernart's correspondence runs almost continuously from late 1657 to 1674, when he temporarily moved to Paris in May. He returned to The Hague in 1679 where he died on 11 Oct 1680 (see Boulliau's note, FF. 13048, fol. 358r). Bernart's correspondence was concerned almost exclusively with diplomatic, political, and commercial news. Bernart was well-known to Des Noyers and the Pels brothers.

[183]Francesco Gratta, an Italian, served as postmaster in Danzig. His letters were received almost weekly from 1660–1669 and contain information on some of Hevelius' activities. See FF. 13035–13036.

[184]Situated in The Hague, Pieter Pels corresponded regularly with Boulliau from 1661 to 1672; FF. 13032 and 13033 also contain letters of his brother, Philip, who lived in Danzig. The Pels' correspondence is more valuable than that of Bernart and Gratta. Both seemed to be in touch with intellectual activities in Holland and Poland, and their correspondence casts light on the activities of Hevelius and Des Noyers. Elsa Zilberbogen is currently completing a detailed study of their correspondence.

[185]The Abbaye Saint-Victor was located on the left bank east of Saint-Michel. It was much more pleasant than the Collège de Laon; four years after Boulliau's death the English physician Martin Lister observed: "I see the *Library of St. Victor*: This most Ancient Convent is the best seated of any in *Paris*; has very large Gardens, with shady Walks, well kept. The Library is a fair and large Gallery. . . ." Martin Lister, *A Journey to Paris in the Year 1698*, p. 129.

[186]Des Noyers to Boulliau, 13 May 1689, FF. 13023, fol. 28r.

If the contents of the Collection Boulliau are any indication, Des Noyers was Boulliau's only regular correspondent during their final years.[187] Together they continued to exchange the news of the day—as they had for decades—only occasionally troubling over the legibility of their writing.

Pierre Des Noyers died in 1693. By this time, of course, Hevelius had been dead six years, Leopold twenty, and Kepler, Galileo, and Gassendi several generations. Isaac Newton was just in his fifties but slowly losing interest in science. It was almost Christmas in Paris when, on 10 December 1694 OS, Flamsteed advised Newton in his postscript: "Bullialdus is lately dead in Paris."[188] He died at the Abbaye Saint-Victor on Thursday, 25 November 1694.[189] Three years later Charles Perrault, in an effort to avoid eulogy, quoted Boulliau's own words:

> Il n'y a rien au monde que j'apprehende tant que les loüanges, si ce que je fais est approuvé par les honnestes gens intelligens dans les matieres que j'ay traittées, cette approbation pure & simple sans éloges, vaut mieux que tous les panegyriques.[190]

I would like to conclude this biographical sketch with an introduction to the Collection Boulliau.

Correspondence—like biography itself—is a personal affair, and each manuscript collection is as unique as the personality it represents. The Collection Boulliau is no exception; it stands as an autobiography of sorts, a record of ideas and activities addressed (by intention at least) to an audience of one. But in cutting across the boundaries of science, philosophy, and practical affairs of state, the

[187]Boulliau's nephew, Urbain Boulliau, sent two letters, 22 Sep 1688 and 18 Jul 1690; FF. 13040, fols. 27–32.

[188]John Flamsteed to Newton, *Correspondence of Isaac Newton*, 4: 59.

[189]*Gazette de France*, 4 Dec 1694, p. 588: "Le sieur Ismaël Boulliauld connu dans toute l'Europe par plusieurs ouvrages, & par sa grande capacité dans les Mathématiques, particuliérement dans l'Astronomie, mourut le 25 du mois dernier, âgé de quatre vingt-sept [sic] ans, à l'Abbaye de S. Victor où il s'estoit retiré."

[190]Charles Perrault, *Les hommes illustres qui ont paru en France pendant ce siècle . . .*, 2 vols. in 1 (Paris, 1696–1700) pp. 73–74: p. 74.

"Correspondance et papiers politiques et astronomiques d'Ismaël Boulliau (1605–1694)" collects the thoughts of men long dead, individuals divided by generation, class, temperament, and ability. The modern reader might be wary of contradiction, of bold opinions and honeyed words; correspondence (like conversation itself) is tailored to a known audience. But inevitably correspondence reveals. Let the letters speak as they will.

Plan & Procedure

The Collection Boulliau, or Correspondance et papiers politiques et astronomiques d'Ismaël Boulliau (1605–1694), is conserved in the Salle des Manuscrits of the Bibliothèque Nationale in Paris. One of the largest manuscript sources of its kind, the Collection is bound in forty-one folio volumes containing over 22,000 pages of unedited correspondence; it is formally identified in the *Catalogue général des manuscrits* as Fonds Français 13019–13059.

The present Inventory and Name Index serve as a guide and companion to the Collection Boulliau.[191] Reflecting the organization and physical arrangement of the Collection itself, the Inventory is divided into forty-one sections indicated by Roman numerals, I-XLI. Within each section individual letters and manuscripts appear as separate entries designated by folio number. Immediately following the Inventory is the Name Index providing the precise location of documents.

Entry Format

Within each of the forty-one sections, individual letters and manuscripts are identified by a separate *entry*. Unless there are contrary indications, each entry represents a distinct letter, manuscript, or group of documents. In the case of correspondence, each letter is identified by correspondent, city, date, language, folio size, number of pages, and (where appropriate) source of publication. As an illustration, the following sample entry is taken from the Collection Boulliau, volume XVI:

98rv	From Oldenburg at London	22 Jul 1675 OS
	½p + ad (15×20) Latin	[1 Aug 1675]
	Publ, OC, Vol XI, pp 413–414, No 2707	

In the far left column of each entry is the *folio number*. This number

[191]Inventories are designed to put the scholar in contact with the document. In constructing an inventory it becomes possible to distinguish manuscripts by handwriting and other means, thus eliminating identification and paleographical problems associated with authorship, dates, and sequence of foliation—all of which are exacerbated on microfilm or photocopy. For readers with specific interests and limited endurance, a name index reduces thousands of pages of manuscripts to an afternoon of leisured thumbing.

indicates the folio or inclusive folios of the entry, in this case, folio 98rv. Standard abbreviations for recto (r) and verso (v) are employed throughout.

To the right of the folio number is the *correspondent*. Unless otherwise stated, all letters are assumed to be either *to* or *from* Boulliau. In the above example, therefore, the Oldenburg letter is addressed *to* Boulliau. If, on the other hand, the entry was to read: To Oldenburg at London, the letter would be *from* Boulliau. A letter neither *to* nor *from* Boulliau is clearly indicated with a two-line description, for example: From Oldenburg at London, followed by: To Hevelius at Danzig.

(In the interest of space, correspondents are normally identified by last name only. To avoid ambiguity, I have occasionally added initials or abbreviations to a correspondent's last name, for example, Philip Pels is indicated as: Ph Pels, thus distinguishing him from his brother, Pieter Pels. Should any question arise concerning the identity of a correspondent, the reader is directed to the Name Index.)

Following the correspondent is the correspondent's *city;* in the example above Oldenburg writes from London. In all letters to or from Boulliau, the correspondent's city is given. Generally speaking, Boulliau is assumed to be at Paris; on the basis of direct evidence, exceptions are clearly noted with a two-line description, for example: From Mersenne at Paris, followed by: To Boulliau at Venice. Editorial additions concerning a correspondent's location are placed in square brackets. As a rule, city names have been Anglicized except where seventeenth-century forms are preferred or foreign forms are required. For example, Lugdunum Batavorum is given as Leiden, Danzig is preferred to Gdańsk, and Wolbourg remains unchanged.[192]

The final element in the first line of most entries is the *date*. Dates are found in the far right column in the sequence: day, month, year; for example, 22 Jul 1675. Standard three-letter abbreviations indicate the month. For further particulars, see *Calendars and Dates* below.

The second line of the entry generally presents three types of in-

[192]An indispensable guide to city names in their various Latin forms is J. G. Th. Graesse (with F. Benedict), *Orbis latinus, oder Verzeichnis der wichtigsten lateinischen Orts- und Landernamen*, Berlin, 1909.

formation: document length (given in pages); approximate size of the folio(s), indicated in centimeters;[193] and language, other than French (all documents are in French unless otherwise noted). When several languages appear (e.g., French/Latin), the primary language is placed in the first position. Standard abbreviations indicate page (p), pages (pp), and address (ad).

A number of letters in the Collection Boulliau have been published. These letters are clearly identified in their appropriate place. In the Oldenburg example discussed above, we read: Publ, OC Vol XI, pp 413–414, No 2707. This indicates that the letter is published in the *Correspondence of Henry Oldenburg* (see *Abbreviations & Symbols*), volume 11, pages 413–414, letter number 2707. Letters published in unusually rare books have been omitted.

Editorial Procedures

Because the Collection Boulliau contains a variety of documents other than correspondence, the basic entry format outlined above has been carefully modified to circumstance. In addition to traditional categories of correspondence, the Collection contains copies of letters, extracts of letters, newsletters, news accounts, holograph treatises, mémoires, notes, illustrations, and printed documents. Procedures for identifying and distinguishing these documents are discussed below.

The most basic distinction is between signed and unsigned documents. As explained above, all correspondence (unless otherwise indicated) is assumed to be either *to* or *from* Boulliau; moreover, the reader is to assume that correspondence *from* Boulliau is signed. Letters *to* Boulliau are clearly indicated by the sender's name, as illustrated in the Oldenburg example above. Unsigned correspon-

[193]I depart from accepted practice in indicating document dimensions; for a general introduction see Morris L. Radoff, "A Guide to Practical Calendaring," *American Archivist* 11 (1948): pp. 123–140, 203–222. Since handwriting varies by individual, document size bears no fixed relation to length of text. Nonetheless, I have included these dimensions not only to make the traditional indication of length of document (pages) more meaningful, but because dimensions and text length were in good conformity during the seventeenth century, particularly for Boulliau. I also depart from accepted practice by indicating document dimensions in centimeters, rather than in inches; and in giving document length in fractions (e.g. 2 1/3pp) or number of lines (1p & 12 lines), rather than rounding to the higher whole number. Other modifications in format and procedure have been carefully considered in light of manuscript and reader.

dence, by contrast, may be indicated as: Anonymous; or, more often than not, the sender's name appears in brackets indicating an editorial addition, for example: From [Des Noyers] at Warsaw. Unsigned letters have been identified by handwriting, format, internal evidence (content, usage, style) or watermark. When a signature appears all but illegible, I have occasionally placed the name in brackets with a question mark. This practice was hesitatingly undertaken in the simple belief that a qualified indication was better than none at all.

Archivists traditionally divide correspondence into three categories: autograph drafts; autograph letters; and letters.[194] For reasons outlined below, the present Inventory does not uniformly distinguish these documents in each individual entry. With relatively few exceptions, however, the reader may assume that letters *to* Boulliau are autograph letters, while correspondence *from* Boulliau is ordinarily in the form of signed autograph drafts.[195] Strictly defined, all letters are not autograph letters; the Inventory does not identify these letters individually since they are few in number and readily recognized. In sum, signed letters written by an amanuensis—the predictable custom of Prince Leopold of Tuscany, Gaston d'Orléans, and other dignitaries—are not distinguished from signed autograph letters. Similarly, unsigned letters, for example those written for the ailing Jacques de Valois, are simply signaled by the sender's name in brackets. These exceptions aside, the reader may assume that letters to Boulliau are autograph originals.

Certain manuscripts defy simple classification.[196] A particularly

[194]Briefly: 1) autograph drafts, signed (ADS) and unsigned (ADuns), are usually written by the sender (often for his records) before the final fair copy; 2) autograph letters, signed (ALS) and unsigned (ALuns), are the fair copies (originals) written by the sender and dispatched; and 3) letters, signed (LS) and unsigned (Luns), are written by an amanuensis and presumably dispatched. For details see Radoff, "A Guide," pp. 136–140.

[195]Boulliau's drafts predictably contain marginal and interlinear additions as well as cancellations. While minor variations may occur between draft and original, the reader may safely assume that 1) textual changes are incidental and 2) Boulliau dispatched a fair copy to the addressee. Although most Boulliau originals are presumed lost, hundreds have been located; for details see *Sources*, Manuscript Sources of Ismaël Boulliau. I welcome information concerning other letters, drafts, and copies of letters to or from Boulliau.

[196]Extreme cases involve forged letters. For example, in the mid-nineteenth century Michel Chasles, an accomplished geometrician and Copley Medallist, unwittingly purchased some 27,000 forged manuscripts produced at the hand of one M. Vrain Lucas. (In 1867, on the strength of these forgeries, Chasles attempted to argue that Pascal had actually communicated the law of universal gravitation to Newton; after years of controversy, Chasles finally acknowledged he had been duped and swindled.) Among the Vrain Lucas forgeries were over

difficult problem is distinguishing between letters, scientific mémoires, and holograph newsletters. Prior to the regular publication of newspapers and scientific journals, correspondence carried a variety of scholarly information and news. This functional diversity often makes it impossible (on the basis of form, style, and content) to distinguish clearly a personal letter from a diplomatic newsletter. Scientific letters during this period are equally troublesome, often shading subtly into short treatises and formal mémoires. In the present Inventory, manuscripts written in letter form have been treated as letters. Newsletters, normally written in the hand of a regular correspondent, are distinguished by format and content; news accounts (usually written by an amanuensis) are easily identified, though they are often annotated, addressed, and dated by the correspondent, or bear his endorsement or postscript. Every effort has been made to distinguish and identify such documents.

One further difficulty. During the seventeenth century it was not uncommon for a sender to address his letters to an intermediary who would in turn forward the letter to a third party, that is, to a second recipient. Editors of published correspondence address this problem by adding an explanatory note to the text of the published letter. In the present Inventory, letters and documents clearly directed to (or received by) two or more recipients are identified with a self-explanatory note or a direct citation from the document.[197] It must be borne in mind, however, that a number of letters exchanged between Boulliau, Leopold, Des Noyers, Rivière, and others were often silently transmitted, copied, and shared with informal groups.

A traditional area of concern is the enclosure or the attachment. As a rule, enclosed letters, treatises, mémoires, and all titled documents (drawings, illustrations, astronomical data, copies and extracts of letters) are listed as separate entries. Other enclosures, often appearing as a physical extension of the letter or as a post-

---

100 forged letters of Boulliau to various contemporaries, including Louis XIV, Flamsteed, Huygens, Pascal, Viviani, and Newton. Of an estimated 365 forged letters addressed to Boulliau, there are nearly 200 from Louis XIV, and over 50 from Galileo—not to mention those of Pascal, Newton, Christina of Sweden, and others. There are no known forgeries in the Collection Boulliau. See Henri Bordier and Emile Mabilee, *Une fabrique de faux Autographs ou récit de l'affaire Vrain Lucas*, Paris, 1870, passim.

[197]A good example is Hevelius' letter to Mersenne (31 Aug 1648, FF. 13043, fols. 7r–9v), received in Paris after the latter's death on 1 Sep 1648. Boulliau's note is a direct citation, "Escrite au Pere Marin Mersenne/& apportee apres sa mort. arrivee/au commencement de Septemb./1648."

script, are ordinarily included in the same entry. Distinct copies of letters and extracts of letters are identified with a three-line description, for example: Copy of a letter, followed by: From G. B. Riccioli at Bologna; followed by: To Hevelius at Danzig. Wherever possible, titled documents and manuscripts are identified with a direct citation, for example: Extract of a letter in B's hand, followed by: "Fragmentum epistolae scriptae a viro Nobili/Tychone Brahe Dano ad virum clarissm./Dominum D Casparum Peucerum."

Other variations in the basic entry format should be self-explanatory. As mentioned above, the titles of mémoires, treatises, extracts, copies, and notes are given as direct citations wherever possible. When titles do not appear on the original, self-evident descriptions are supplied without quotation marks, for example: Newsletter; Astronomical data; Fragment. Editorial conventions employed in direct citations are discussed below.

Editorial Conventions

Since one function of the Inventory is to identify manuscripts with minimum effort, I have taken the liberty of making minor modifications in orthography and punctuation. As a rule, original spelling has been carefully preserved in all quoted material. The only consistent change has been in silently modernizing the orthography of the familiar "long form" *s* and in expanding most ligatures. Less frequently, when ambiguity threatened, I have regularized the seventeenth century practice of interchanging *v* and *u*, *i* and *j*; in these few instances, *i* and *u* are given as vowels, *v* and *j* as consonants. Every effort has been made to preserve the orthography of planetary symbols (also used for days of the week) which are reproduced as they appear in the originals. I have also attempted to retain raised letters (e.g., M^{r}) and symbols such as the ampersand (&) and its variants (&c.). French and Latin spelling during this period—even of proper names—was highly irregular.[198] Where titles and descriptions occur within quotation marks, I have attempted to retain the original spelling, only rarely modifying punctuation, capitalization, and accent marks. The modern reader will have no difficulty reading *evesque* for *évêque*, *escrit* for *écrit*; citations in other languages should be equally clear. Subject to these

[198] It is not uncommon to find three spellings of a proper name in the same letter.

conventions, idiosyncrasies of spelling, punctuation, and orthography have been preserved as far as possible, and where necessary editorial notes are appended.

As a rule I have allowed common abbreviations, contractions, and titles of position (V.E., S^{r}, P.) to stand as they appear in the original; where clarity was in question I have expanded these forms and indicated the change in brackets. Omitted words or lines appearing within quotation marks are of course indicated by ellipsis points, and distinct lines of quoted passages are signaled by the slash (/). When marginalia, annotations, or legible cancellations are quoted (or employed in an editorial addition), the source is identified in brackets or with an editorial note indicated by an asterisk (*). Drawings, etchings, diagrams, and other illustrations are identified by title or with a self-explanatory phrase; when illustrations are known to appear in seventeenth-century books or journals, the source is cited. For further information concerning editorial and printing conventions, see *Abbreviations & Symbols.*

Calendars and Dates

The most widely employed calendar in the period under consideration was the Gregorian or New Style (NS).[199] Introduced by Pope Gregory XIII in 1582, the New Style or *stilus novus* was officially adopted in Catholic countries (and areas of the Protestant Netherlands) by the late sixteenth century. In most cases, though excep-

[199]The Gregorian calendar quickly proved superior to its predecessor, the Julian or Old Style. Since the Julian calendar had erroneously assumed the solar year to be 365 days and 6 hours (it is actually some 11 minutes and 14 seconds short of that figure), by the late sixteenth century the cumulative error had shifted vernal equinox from 21 March to 11 March. Charged by the Council of Trent to correct the error, Pope Gregory XIII introduced the Gregorian calendar in his Bull of 24 Feb 1582. By the end of the year, the New Style spread from Italy to Spain, Portugal, and France, and shortly thereafter, to Catholic Switzerland, the Catholic Netherlands, and areas of the Protestant Netherlands—Rotterdam, Amsterdam, Leiden, Delft, Haarlem, and The Hague; other areas (Friesland, Gelderland, Gröningen, Overyssel, and Utrecht) did not adopt the New Style until about 1700. Roman Catholic states in Germany accepted the reform by 1583, followed by Poland (1586) and Hungary (1587); Denmark and Sweden adopted the New Style toward the end of the seventeenth century, England in 1752. By the early eighteenth century (1 Mar 1700–28 Feb 1800 NS) the Old Style had fallen eleven days behind the New Style. For an introduction to calendars and dates see John J. Bond, *Handy Book of Rules and Tables for Verifying Dates with the Christian Era; Giving an Account of the Chief Eras, and Systems used by Various Nations,* (London, 1889, reprint, New York, 1966).

tions are discussed below, those employing the New Style began the calendar year on 1 January.

The situation in Italy provides two exceptions. In Florence (and certain areas of France) the *calculus florentinus* was widely employed. In this system the dates are identical to the Gregorian except the new year is reckoned not from 1 January, but from the Feast of Annunciation (or Incarnation of the Word) on the following 25 March. As a consequence, letters dated between 1 January and 24 March in this style bear a date precisely one year behind the Gregorian or New Style. The *calculus florentinus* is generally signaled by the abbreviation "Ab Inc" (Ab Incarnatione) following the date. Once identified, this style is readily converted to the New Style. For example, a letter from Prince Leopold of Tuscany dated 15 Feb 1667 Ab Inc becomes 15 Feb 1668 NS.

In addition to the *calculus florentinus* some Italians (and others) employed the *calculus pisanus*. A style similar to the Gregorian and Florentine, the *calculus pisanus* began the new year not on 1 January or on the following 25 March, but on the preceding 25 March. This style is relatively rare, and I have not identified any instance of it in the Collection Boulliau.

The New Style was not officially recognized in England until 1752. During the seventeenth century most English correspondents therefore followed the Julian calendar or Old Style (OS) which lagged ten days behind the New Style. Conversion is straightforward. A letter from Seth Ward dated 27 Jul 1657 OS is easily converted, for example, to 6 Aug 1657 NS. To compound matters, however, the official year in England began not on 1 January but on the following 25 March (Annunciation). Letters dated from England between 1 January and 24 March consequently bore a date one year and ten days behind the New Style. To avoid confusion, English correspondents during this period often gave a double date, thus 6/16 Jan 1662/3 was readily converted to 16 Jan 1663 NS. In sum, letters dated in England during the seventeenth century—lacking contrary indications—are usually in the Old Style. But let the reader beware. When residing in France, for example, Henry Oldenburg understandably followed local custom and employed the New Style; in England, when corresponding with Englishmen, he reasonably followed the Old Style; but when corresponding with foreigners, Oldenburg frequently chose to reckon the new year from 1 January, not 25 March.

Finally, something must be said of the Roman calendar. A number of Boulliau's correspondents employed this style which divides the month into Kalends, Nones, and Ides. As an illustration, a letter of J. A. Portner is dated XIII Kal May MDCLXII. In an attempt to assist the modern reader, two dates appear in the far right column of the entry, in this case the Roman date has been converted to 19 Apr 1662 NS. The conversion table used in this Inventory is found in C. R. Cheney, *Handbook of Dates for Students of English History* (London, 1970), pp. 75–81.

The present Inventory follows the traditional scholarly practice of reducing dates to a uniform style. To avoid confusion, all dates have been converted to the New Style with the new year reckoned from 1 January. For purposes of accuracy, clarity, and convenience, the converted New Style date has been placed in brackets beneath the original date. In the absence of contrary evidence the reader is to assume that correspondents residing in Protestant countries, states, and cities employed the Old Style, while those in Catholic countries used the New Style. As detailed above, however, the Protestant Netherlands generally followed the New Style while some Italians chose the *calculus florentinus* or the *calculus pisanus*.

Every effort has been made to identify and convert variant calendar styles, to clarify ambiguous dates, and to supply missing information.[200] As a rule, when more than one date occurs in a letter or manuscript (as is often the case with postscripts) both dates are given. With few exceptions, undated letters are clearly indicated and—wherever possible—I have given the date of receipt when noted by Boulliau. When deletions occur without a second date, I have placed the original date (when legible) in brackets followed by an editorial note. In the case of overwritten dates, I have usually preferred the final to the original date, that is, generally speaking, the later to the earlier date, and I have also indicated any correction or alteration made by Boulliau. More often than not, Boulliau carefully corrected slips of the pen—particularly those occurring in the first month of a new calendar year; in several instances, however,

[200]In checking undated drafts against their originals, I have supplied the date in brackets and identified the source. In certain instances I have determined the date of a letter from internal evidence found in other letters exchanged between the two parties, or from internal evidence in other letters exchanged with a third party. Systematic use of contemporaneous letters was facilitated by a calendar of Boulliau's complete correspondence now in preparation.

it has been necessary to correct misdated alterations. These and similar editorial changes and additions are noted in brackets or are followed by an asterisk and an explanatory note. Saints' days, religious holidays, and seasonal dates (e.g., vernal equinox) are clearly noted and converted to the New Style.

Name Index

Immediately following the Inventory is the Name Index, an alphabetical listing of the correspondents found in the Collection Boulliau.[201] Under each name there are as many as three *subdivisions:*

A. Letters *from* Boulliau to the individual named;
B. Letters *to* Boulliau from the individual named;
C. Letters neither from nor to Boulliau, or other documents/citations.

Under each subdivision, Roman numerals indicate *volume numbers* (I-XLI). These volume numbers correspond to the forty-one volumes of the Collection Boulliau as described in the Inventory. Following each volume number, Arabic numerals indicate the *folio number* on which distinct letters or documents begin. Thus volume XXV, folio 18 is indicated as XXV: 18. As a practical illustration, the Name Index lists three subdivisions under the name **HEVELIUS.** As explained above, letters from Hevelius are found in the second subdivision. The Name Index shows that letters from Hevelius to Boulliau are located in volumes XIII, XXV, and XXVI. It further indicates that volume XXV, for example, contains eleven letters, the first beginning on folio 18. With this information from the Name Index, the Inventory may be consulted for specific information. The Name Index does not uniformly distinguish between letters, drafts, copies, and extracts of letters.

[201]Since most letters are from or to Boulliau, the Name Index does not duplicate this information under his name.

Abbreviations & Symbols

Ab Inc	Ab Incarnatione (Feast of Annunciation, 25 March)
ad	address
Anonymous	unsigned; author not cited
B	Ismaël Boulliau (1605–1694)
BAF	Tito Livio Burattini. *Intorno alla vita ed ai lavori di Tito Livio Burattini*. Edited by Antonio Favaro. Memorie del Reale Istituto Veneto de Scienze, Lettere ed Arti, vol. 25, no. 8. Venice, 1896.
BC	Jacques-Bénigne de Bossuet. *Correspondance de Bossuet*. Edited by Charles Urbain and Eugène Levesque. 15 vols. Paris, 1909–1925.
BHO	Henri Omont. "Lettres d'Émeric Bigot à Gilles Ménage et à Ismaël Bouillaud au cours de son voyage en Allemagne lors de l'élection de l'empereur Léopold Ier (1657–1658)." *Annuaire bulletin du Société de l'Histoire de la France* 23(1886):227–255.
BN	Bibliothèque Nationale, Paris FF. Fonds Français N.a.f. Nouvelles acquisitions françaises Lat. Fonds Latin N.a.L. Nouvelles acquisitions Latines
BO	Bibliothèque de l'Observatoire, Paris.
BOO	Tycho Brahe. *Tychonis Brahe Dani Opera Omnia*. Edited by J. L. E. Dreyer. 15 vols. Copenhagen, 1913–1929.
ca	circa
CB	*Collection Boulliau*. BN, Salle des Manuscrits, FF. 13019–13059.
CD	*Collection Dupuy*. BN, Salle des Manuscrits.
DC	René Descartes. *Correspondance*. Edited by Charles Adam and Gaston Milhaud. 8 vols. Paris, 1936–1963.
DHO	Henri Omont. "Trois lettres de Boulliau à Dupuy (1645–1651)." *Archives Historiques, Artistiques, et Littéraires* 2(Paris, 1890–1891):134–138.

DL	Pierre Des Noyers. *Lettres de Pierre Des Noyers, secrétaire de la reine de Pologne, Marie-Louise de Gonzague, princesse de Montaue et de Nevers, pour servir à l'histoire de Pologne et de Suède de 1655 à 1659* (Lettres de Pierre Des Noyers à Ismael Boulliaud). Librairie de B. Behr (E. Bock), Berlin 1859.
DO	René Descartes. *Oeuvres de Descartes.* Edited by Charles Adam and Paul T. Tannery. 13 vols. Paris, 1897–1913.
F	Adolphe Franck. "Correspondance de Boulliau." Unpublished manuscript. Bureau inventory (58^6) Salle des Manuscrits (BN), 96 fols., 1885.
FO	Pierre Fermat. *Oeuvres de Fermat.* Edited by Paul Tannery, Charles Henry, and Cornélis de Waard. 5 vols. Paris, 1891–1922.
fol(s)	folio(s)
GGO	Galileo Galilei. *Le Opere di Galileo Galilei, Edizione Nazionale.* Edited by Antonio Favaro. 20 vols. Florence, 1890–1939.
GOO	Pierre Gassendi. *Petro Gassendi . . . Opera Omnia . . . hactenus edita auctor ante obit recensuit . . . posthuma vero . . . in lucem nunc primum prodeunt* . . . Edited by H. L. Habert de Montmor and F. Henry. 6 vols. Lyon, 1658–1675.
HO	Christiaan Huygens. *Oeuvres Complètes de Christiaan Huygens*, publiées par la Société Hollandaise des Sciences. 22 vols. The Hague, 1888–1950.
HSM	Guillaume Libri. *Histoire des sciences mathématiques en Italie.* 4 vols. Paris, 1838.
Incl	Includes
LIF	*Lettre inedite d'uomini illustri.* Edited by Angelo Fabroni. 2 vols. Florence, 1773 and 1776.
LL	François Luillier. "Lettres de Luillier, père de Chapelle." In *Les Historiettes de Tallemant des Réaux*, edited by MM. de Monmerqué and Paulin Paris, vol. 4, pp. 489–516. Paris, 1850–1856.
LLI	Isaac de La Peyrère. "Quelques Lettres inédites d'Isaac de la Peyrère à Boulliau publiées avec une

notice, des notes et un appendice." *Plaquettes Gontaudaises*, No. 2 (1878) Paris.

MC	P. Marin Mersenne. *Correspondance du P. Marin Mersenne*. Edited by Paul Tannery, Cornélis de Waard, and Armand Beaulieu. 14 vols. Paris, 1932– .
ML	Cardinal Mazarin. *Lettres du Cardinal Mazarin pendant son ministère (1642–1661)*. Edited by Chéruel and G. D'Avenel. 9 vols. Paris, 1872–1896.
MS or MSS	Manuscript(s)
ND	No day of month
NS	New Style (Gregorian calendar)
NY	No year
OC	Henry Oldenburg. *The Correspondence of Henry Oldenburg*. Translated and edited by A. Rupert Hall and Marie Boas Hall. 9 vols. Madison, 1965–1973; since vol. 10, London, 1975– .
OS	Old Style (Julian calendar)
p or pp	page(s)
Publ	Published
r	recto
SEB	*Sylloges Epistolarum a viris illustribus scriptarum*. . . . Edited by Pieter Burman. 2d ed. 5 vols. Leiden, 1727.
URI	Isaac Uri. *Un cercle savant au XVIIe siècle: François Guyet (1575–1655): d'après des documents inédits*. Paris, 1886.
v	verso
w	with
*	Explained below
[Brackets]	Editorial addition
?	Perhaps; uncertain
+	plus
&	and
/	End of quoted line
. . .	Segment of quoted text omitted

Inventory Pagination

THE INVENTORY

"...all the pieces must be used,
their plain sides turned down...

COLLECTION BOULLIAU, VOLUME I, FONDS FRANÇAIS 13019

1rv	Two library notices 1p (16x22)	undated
2r-4v	From Des Noyers at Glogów 5pp + ad (15x21) Publ, DL, pp 21-25	4 Dec 1655
5rv	News account 2pp (15x21) [annotated by Des Noyers]	20 Oct 1655
6rv	Copy of a letter 13 lines (15x21) "L[ett]re de la Reyne Christine/au Roy de Suede." [Inspruck, 2 Nov 1655] Publ, DL, p 45	
7r-9v	From Des Noyers at Glogów 6pp incl ad (15x21) Publ, DL, pp 25-28	12 Dec 1655
10r-14v	From Des Noyers at Glogów 10pp (15x21) Publ, DL, pp 31-38	19 Dec 1655
15rv	Copy of a letter 1p (15x21) Italian "Copia della l[ette]ra del gran maresciale/Lubomirski alla ser[enissi]ma regina di Polonia." [12 Dec 1655, Lubownia] Publ, DL, pp 45-46	12 Dec 1655
15v-16r	Copy of a letter 1p (15x21) "Coppie de la lettre du Chan de Tartarie au/Seren.me Roy de Pologne." [27 Nov 1655, Halicz] Publ, DL, pp 41-42	
17r	"Coppie de la lettre du Chan de Tartarie au/Grand Marechal de Pologne." [27 Nov 1655, Halicz] ½p (20x31) Publ, DL, pp 42-43	

17rv	"Autre Coppie de lettre du meme au Grand General/de Pologne" [27 Nov 1655, Halicz] 1p (21x31) Publ, DL, pp 43-44	
18rv	Copy of a letter 1½pp (20x31) "Copie traduite de la lettre du Grand General de Pologne/Escritte au Roy le 17 decembre 1655 de Sokal" Publ, DL, pp 28-30	
19r-21v	From Des Noyers at Glogów 6pp (15x21) Publ, DL, pp 38-40	28 Dec 1655
22r-25v	From Des Noyers at Glogów 7pp + ad (15x21) See DL, pp 47-51	3 Jan 1656
26r-28v	From Des Noyers at Glogów 5pp (15x21) See DL, pp 51-54	9 Jan 1656
29r-30v	From Des Noyers at Glogów 4pp (15x21) Publ, DL, pp 54-57	17 Jan 1656
31r-32v	News account Glogów 2½pp (20x31) Publ, DL, pp 57-63	17 Jan 1656
33r-34v	From Des Noyers at Glogów 4pp (15x21) Publ, DL, pp 63-73	26 Jan 1656
35r-36v	News account Glogów 4pp (20x31) Publ, DL, see letter XXII, pp 63-73	26 Jan 1656
37r-38v	From Des Noyers at Glogów 4pp (15x21) Publ, DL, pp 73-76	2 Feb 1656

39rv	News account Glogów 1½pp (15x21) See DL, pp 76-77	3 Feb 1656
40r-44v	From Des Noyers at Glogów 6pp & 7 lines (15x21)* fol 44, Latin enclosure *fol 43, (14x30) See DL, pp 80-83	10 Feb 1656
45r-46v	News account Glogów 2½pp (15x21) Publ, DL, pp 77-79	10 Feb 1656
47r-48v	From Des Noyers at Glogów 4pp (15x21) Publ, DL pp 83-85	17 Feb 1656
49r-50v	From Des Noyers at Glogów 4pp (20x31) Publ, DL, pp 86-89	24 Feb 1656
51r-52v	News account Glogów 4pp (15x21) Publ, DL, pp 89-92	24 Feb 1656
53r-56v	From Des Noyers at Glogów 7pp (15x21) Publ, DL, pp 92-95	2 Mar 1656
57r-58v	News account Glogów 2½pp (15x21) Publ, DL, pp 96-98	2 Mar 1656
59rv	Copy of a verse, Anglicana Soboles 2pp (20x31) Latin	undated
60rv	From [Cunitz at Breslau?] To Anonymous 2pp (20x31)	17 [Jan?] 1656
61r-62v	From Des Noyers at Glogów 4pp (15x21) Publ, DL, pp 98-100	9 Mar 1656

63r-64v	News account Glogów 3pp & 2 lines (15x21) Publ, DL, pp 101-103	 9 Mar 1656
65r-66v	From Des Noyers at Glogów B at Amsterdam 3pp + ad (15x21) attached Publ, DL, pp 103-105	16 Mar 1656
67rv	News account Glogów 2pp (15x21) Publ, DL, pp 105-106	 16 Mar 1656
68r-70v	From Des Noyers at Glogów 5pp (15x21) Publ, DL, pp 113-119	23 Mar 1656
71r-74v	News account Glogów 6½pp (15x21)	 23 Mar 1656
75r-76r	Extract of a letter 3pp (20x31) "Extrait d'une Lettre de monsieur Czarnecky, un des g[e]n[er]aux/de l'armée du Roy, escritte a la Reine le 18e mars a/Grackouec, aupres de Premislie" Publ, DL, pp 107-109	
76v	"Extrait de quelques lettres de Leopol le 18 mars" Publ, DL, pp 110-111	
77r-78v	Copies of letters 4pp (20x31) Léopol Lantzut Publ, DL, pp 120-124	 23 Mar 1656 24 Mar 1656
79r-81v	From Des Noyers at Glogów 6pp (15x21) Publ, DL, pp 124-127	30 Mar 1656

82r-86v	From Des Noyers at Glogów 6pp (15x21) fol 85, note (10x15) fol 86, blank Publ, DL, pp 127-129	6 Apr 1656 7 Apr 1656
87r-90v	Copies of letters 3½pp (15x21) French/Latin "Copie d'une lettre traduitte de Mr. Czarnesky a La Reyne du 9 apuril/ 1656 de Warka." Publ, DL, pp 130-132 "Copie d'une autre lettre Escritte de/ Mr Woisky de Nowomiasto le 13 avril/ 1656" Publ, DL, pp 132-135 (3½pp) "Copie litterae scriptae a Magni Tartariae/Chami Vesiro ad Chmielniscium" Publ, DL, pp 135-136 (1p)	 undated
91r-93v	From Desnoyers at Glogów 6pp (15x21) Publ, DL, pp 136-140	13 Apr 1656
94r-97v	Copy of a letter 7½pp (15x21) "Traduction d'une lettre de Sandomirie" Publ, DL, pp 140-144	[13 Apr 1656]
98r-99v	From Des Noyers at Glogów 3pp (15x21) Publ, DL, pp 147-151	20 Apr 1656
100rv&103rv	News account 2pp (20x31) fol 103, 4 lines (6x20) Publ, DL, p 149	
101r-105v	From Des Noyers at Glogów 7pp (15x21) French/Latin Publ, DL, pp 153-156	27 Apr 1656
106r-108v	News account Glogów 5pp (15x21) Publ, DL, pp 156-159	 27 Apr 1656

109rv	Copy of a letter From Marie Cunitz at Breslau To Des Noyers at [Glogów?] 2pp (20x31) Latin Publ, DL, pp 144-147	25 Apr 1656
110r-112v	From Des Noyers at Glogów 3pp & drawing w/7 lines (15x21) Publ, DL, pp 159-161	4 May 1656
113r-114v	Copy of a letter 2pp + ad (15x21) German/French Copie d'une lettre de madame la princesse de Brieg à madame Dönhof, [Charanton?, 25 Apr 1656] Publ, DL, pp 151-152	
115rv	Copy of a letter From Mohamed at Baccesarai To Frederic-Guillaume, Elect. of Brandenburg 1p (20x31) Latin Publ, DL, pp 111-112	20 Mar 1656
116r-117v	From Des Noyers at Glogów 3pp (15x21) Publ, DL, pp 161-163	11 May 1656
118r-120v	From Des Noyers at Glogów 5pp (15x21) Publ, DL, pp 163-166	18 May 1656
121r-124v	News account Glogów 6½pp (15x21) Publ, DL, pp 166-171	18 May 1656
125r-127v	From Des Noyers at Glogów 5pp + ad (15x21) Publ, DL, pp 171-173	25 May 1656
128r-131v	From Des Noyers at Glogów 7pp (15x21) Publ, DL, pp 174-177	1 Jun 1656

132rv	From Des Noyers at Glogów 2pp (15x21) Publ, DL, pp 178-179	8 Jun 1656
133rv	Account of military affairs 1½pp (20x31) Latin	24 May 1656
134r-137v	News account Glogów 6pp & 4 lines (15x21) Publ, DL, pp 179-183	8 Jun 1656
138r-139v	From Des Noyers at Glogów 3pp (15x21) Publ, DL, pp 183-184	15 Jun 1656
140r-141v	News account Glogów 2pp & 8 lines (15x21) Publ, DL, pp 185-186	15 Jun 1656
142rv	From Des Noyers at Glogów 2pp (15x21) Publ, DL, pp 186-187	22 Jun 1656
143r-146v	News account Glogów 7pp (15x21) Publ, DL, pp 188-192	22 Jun 1656
147r-148v	From Des Noyers at Glogów 3pp + ad (15x21) Publ, DL, pp 192-194	27 Jun 1656
149r-150v	From Des Noyers at Czestochowa 3pp (15x21) Publ, DL, pp 194-196	3 Jul 1656
151r-154v	News account Czestochowa 7½pp (15x21) Publ, DL, pp 196-201	3 Jul 1656
155r-156v	From Des Noyers at Warsaw 4pp (15x21) Publ, DL, pp 201-203	20 Jul 1656

157r-159v	News account Warsaw 5pp (15x21) Publ, DL, pp 203-207	20 Jul 1656
160r-162v	From Des Noyers at Warsaw 5pp (15x21) Publ, DL, pp 207-210	27 Jul 1656
163r-164v	News account Warsaw 3pp (15x21) Publ, DL, pp 210-212	27 Jul 1656
165r-168v	From Des Noyers at Lantzut 8pp (20x31) Publ, DL, pp 212-219	11 Aug 1656
169r-170v	From Des Noyers at Lantzut 3pp + ad (15x21) Publ, DL, pp 226-227	18 Aug 1656
171rv	From Des Noyers at Lantzut 2pp (15x21) Publ, DL, p 228	27 Aug 1656
172r-175v	News account Lantzut 6pp (15x21) fol 175, blank Publ, DL, pp 229-232	27 Aug 1656
176r-177v	News account Lantzut 4pp (15x21) Publ, DL, pp 233-235	31 Aug 1656
178r-179v	From Des Noyers at Lantzut 3pp (15x21) Publ, DL, pp 236-237	1 Sep 1656
180r-181v	From Des Noyers at Lublin 4pp (15x21) Publ, DL, pp 237-239	5 Sep 1656
182r-185v	From Des Noyers at Lublin 7pp + ad (15x21) Publ, DL, pp 239-242	10 Sep 1656

Folios	Description	Date
186r-188v	From Des Noyers 5pp (15x21) Publ, DL, pp 243-245	undated [mid Sep 1656]
189r-190v	From Des Noyers at Wolbourg 3pp + ad (15x21) Publ, DL, pp 246-248	5 Oct 1656
191r-192v	From Des Noyers at Wolbourg 4pp (15x21) Publ, DL, pp 248-250	14 Oct 1656
193r-196v	News account Wolbourg 7½pp (15x21) Publ, DL, pp 250-256	14 Oct 1656
197r-198v	From Des Noyers at Wolbourg 4pp (15x21) Publ, DL, pp 256-257	21 Oct 1656
199rv	Latin List "Ex parte Suecorum," etc. 1p (15x21)	undated
200rv	News account Wolbourg 1p (15x21) Publ, DL, pp 245-246	27 Sep 1656
201r-202v	From Des Noyers at Wolbourg 3pp (15x21) Publ, DL, pp 258-259	27 Oct 1656
203r-205v	News account Wolbourg 5½pp (15x21) Cf. DL, pp 259-263	28 Oct 1656
206r-207v	From Des Noyers at Wolbourg 4pp (15x21) Publ, DL, pp 263-265	4 Nov 1656
208r-210v	News account Wolbourg 5pp & 7 lines (15x21) Publ, DL, pp 265-269	5 Nov 1656

211r-212v	From Des Noyers at Wolbourg 3pp (15x21) Publ, DL, pp 269-270	11 Nov 1656
213rv	News account Wolbourg 2pp (15x21) Publ, DL, pp 271-272	 12 Nov 1656
214r-215v	From Des Noyers at Wolbourg 2½pp (15x21) Publ, DL, pp 275-276	18 Nov 1656
216r-218v	News account Wolbourg 4pp & 3 lines (15x21) Publ, DL, pp 272-275	 18 Nov 1656
219r-220v	From Des Noyers at Wolbourg 4pp (15x21) Publ, DL, pp 279-281	25 Nov 1656
221r-222v	News account Wolbourg 3pp (15x21) Publ, DL, pp 277-278	 25 Nov 1656
223r-224v	From Des Noyers at Wolbourg 4pp (15x21) Publ, DL, pp 281-283	30 Nov 1656
225r-228v	From Des Noyers at Wolbourg 7pp (15x21) *dated by B Publ, DL, pp 283-288	[7 Dec 1656]*
229r-233v	News account and Copy of a letter 6pp & 2 lines (15x21) Italian fol 230r, Drawn map (20x22)	29 Mar 1656 25 Mar 1656
234r-235v	From Des Noyers at Kalisz 3pp + ad (15x21) Publ, DL, pp 290-292	20 Jan 1657
236r-237v	From Des Noyers at Dankow 3pp (15x21) Publ, DL, pp 294-295	4 Feb 1657

Folio	Description	Date
238rv	Drawing w/description in B's hand 1p	[ca 30 Nov 1656]
239r-240v	From Des Noyers at Dankow 3pp + ad (15x21) Publ, DL, pp 297-298	12 Feb 1657
241rv	News account Dankow 1½pp (20x31) Cf. DL, pp 295-297	4 Feb 1657
242r-243v	From Des Noyers at Kalisz 3pp + ad (15x21) Publ, DL, pp 292-293	27 Jan 1657
244r-245v	From Des Noyers at Dankow 3pp + ad (15x21) Publ, DL, pp 298-300	19 Feb 1657
246r-247v	From Des Noyers at Czestochowa 3pp + ad (15x21) Publ, DL, pp 300-301	26 Feb 1657
248r-250v	From Des Noyers at Czestochowa 5pp + ad (15x21) Publ, DL, pp 301-303	6 Mar 1657
251rv	News account Königsberg 18 lines (15x17) Latin	16 Feb [1657]
252r-253v	From Des Noyers at Czestochowa 4pp (15x21) Publ, DL, pp 304-306	12 Mar 1657
254r-255v	From Des Noyers at Czestochowa 3½pp + ad (15x21) Publ, DL, pp 306-308	19 Mar 1657
256r-257v	From Des Noyers at Warsaw 3pp + ad (15x21) Publ, DL, pp 391-392	24 Mar 1658
258rv	Latin verse 2pp (20x31) Latin "Charnetzkius Miraculosus/sive/Carolus Gustavus poenitens." etc.	Mense ante Calendas novi anni

259r-260v	From Des Noyers at Czestochowa 3pp (15x21) Publ, DL, pp 308-309	26 Mar 1657
261rv	Copy of a letter To Pope [Alexander VII] 1½pp (20x31) Latin	28 Apr 1656
262r-263v	From Des Noyers at Czestochowa 4pp (15x21) Publ, DL, pp 309-311	2 Apr 1657
264r-267v	From Des Noyers at Opole B at The Hague 7pp + ad (15x21) Publ, DL, pp 316-319	12 Apr 1657
268rv	From Des Noyers at Opole 2pp (15x21) Publ, DL, p 320	16 Apr 1657
269rv	Copy of a letter 1¼pp (20x31) Latin "Copia Hani Crimensis Litterarum ad Ser[enissi]mum Regem Poloniarum" [Mehmet Gherey] Publ, DL, pp 311-314	
270rv	Copy of a letter From Subhankazy Aga To King of Poland 1p (20x31) Latin Publ, DL, pp 314-315	
271r-273v	From Des Noyers at Krepitz 5pp (15x21) Publ, DL, pp 322-326	9 May 1657
274rv	From Morstin at Vienna To [Des Noyers?] 1p (15x21) Publ, DL, p 321	25 Apr 1657
275r-276v	From Des Noyers at Krepitz B at The Hague 3pp + ad (15x21) Publ, DL, pp 326-327	15 May 1657

277r-279v	From Des Noyers at Dankow 4pp & 4 lines (15x21) Publ, DL, pp 328-330	13 Jun 1657
280rv	Drawing and notes 1p (16x21) Latin	ca 20 Apr 1657
281r-282v	From Des Noyers at Czestochowa B at The Hague 3pp + ad (15x21) Publ, DL, pp 330-332	20 Jun 1657
283r-285v	From Des Noyers at Czestochowa B at The Hague 5pp + ad (15x21) Publ, DL, pp 332-334	27 Jun 1657
286r-287v	From Des Noyers at Czestochowa B at The Hague 2½pp + ad (15x21) Publ, DL, pp 335-336	4 Jul 1657
288r-289v	From Des Noyers at Czestochowa B at The Hague 3pp + ad (15x21) Publ, DL, pp 336-338	11 Jul 1657
290r-292v	From Des Noyers at Cracow 5pp + ad (15x21) Publ, DL, pp 338-341	19 Aug 1657
293r-294v	From Des Noyers at Cracow 4pp (15x21) Publ, DL, pp 341-343 Publ, BAF, p 80	26 Aug 1657
295r-296v	From Des Noyers at Cracow 3pp + ad (15x21) Publ, DL, pp 343-345	1 Sep 1657
297r-299v	From Des Noyers at Cracow 5pp + ad (15x21) Publ, DL, pp 345-346	8 Sep 1657
300rv	From Des Noyers at Cracow 1½pp (15x21) Publ, DL, pp 347	15 Sep 1657

301r-304v	From Des Noyers at Bydgoszcz 7pp + ad (15x21) Publ, DL, pp 348-352	1 Nov 1657
305r-308v	From Des Noyers at Poznań 7pp + ad (15x21) fol 308r, small drawing attached Publ, DL, pp 352-356	17 Nov 1657
309r-311v	From Des Noyers at Poznań 4½pp (15x21) Publ, DL, pp 356-359	26 Nov 1657
312r-313v	From Des Noyers at Poznań 3pp + ad (15x21) Publ, DL, pp 359-360	1 Dec 1657
314r-315v	From Des Noyers at Poznań 4pp (15x21) Publ, DL, pp 361-363	8 Dec 1657
316r-317v	From Des Noyers at Poznań 4pp (15x21) Publ, DL, pp 364-366	17 Dec 1657
318r-320v	From Des Noyers at Poznań 5pp + ad (15x21) Publ, DL, pp 366-368	23 Dec 1657

▼ ▼ ▼

COLLECTION BOULLIAU, VOLUME II, FONDS FRANÇAIS 13020

1r-2v	From Des Noyers at Poznań 3pp + ad (15x21) Publ, DL, pp 368-370	30 Dec 1657
3r-6v	From Des Noyers at Poznań 7¼pp (15x21) Publ, DL, pp 370-374	6 Jan 1658
7r-8v	From Des Noyers at Poznań 4pp (15x21) Publ, DL, pp 374-376	13 Jan 1658
9r-11v	From Des Noyers at Poznań 5pp (15x21) Publ, DL, pp 376-378	20 Jan 1658
12rv	Extract of a letter 1p (15x21) "Extrait d'une lettre de M^{r}. Neuman escritte/a son Altesse de Brandeburg l'11 Jan^{r}. 1658." Publ, DL, pp 379	
13r-14v	From Des Noyers at Poznań 3pp + ad (15x21) Publ, DL, pp 379-381	27 Jan 1658
15rv	From Des Noyers at Poznań 2pp (15x21) Publ, DL, pp 381-382	1 Feb 1658
16r-17v	From Des Noyers at Warsaw 3pp + ad (15x21) Publ, DL, pp 382-383	14 Feb 1658
18r-19v	From Des Noyers at Warsaw 3pp + ad (15x21) Publ, DL, pp 384-385	21 Feb 1658
20r-21v	From Des Noyers at Warsaw 4pp (15x21) Publ, DL, pp 385-387	[1 Mar 1658]

22r-23v	From Des Noyers at Warsaw 3pp + ad (15x21) Publ, DL, pp 387-388	10 Mar 1658
24r-25v	From Des Noyers at Warsaw 3½pp (15x21) Publ, DL, pp 389-390	17 Mar 1658
26r-28v	From Des Noyers at Poznań 6pp (15x21) Publ, DL, pp 393-396	8 Apr 1658
29rv	Latin verse 1½pp (16x33) Latin Publ, DL, pp 396-397	undated
30r-31v	From Des Noyers at Poznań 4pp (15x21) Publ, DL, pp 400-401	15 Apr 1658
32r-34v	From Des Noyers at Poznań 6pp (15x21) Publ, DL, pp 398-399	Le Jour de Pasque 1658 [21 Apr 1658]
35r-36v	From Des Noyers at Poznań 3pp (15x21) Publ, DL, pp 401-402	28 Apr 1658
37r-38v	From Des Noyers at Poznań 3pp (15x21) Publ, DL, pp 403-404	5 May 1658
39r-41v	From Des Noyers at [Boguszow] 4½pp (15x21) Publ, DL, pp 404-406	[12] May 1658
42r-44v	From Des Noyers at Boguszow 4pp & 7 lines (15x21) fol 43, 7 lines (8x10) Italian Publ, DL, pp 407-409	19 May 1658
45r-46v	From Des Noyers at Boguszow 3pp & 7 lines (15x21) fol 46, 7 line note (6x14) Publ, DL, pp 409-411	26 May 1658
48r-49v	From Des Noyers at Boguszow 4pp (15x21) Publ, DL, pp 411-413	2 Jun 1658

50r-51v	From Des Noyers at Sierakow 3pp (15x21) Publ, DL, pp 413-415	9 Jun 1658
52r-53v	From Des Noyers at Sierakow 3pp + ad (15x21) Publ, DL, pp 415-416	16 Jun 1658
54r-55v	From [Des Noyers] at Sierakow 4pp (15x21) See DL, pp 416-418	23 Jun 1658
56r-58v	From Des Noyers at Sierakow 5pp (15x21) Publ, DL, pp 418-419	7 Jul 1658
59rv	From [Hendt] to Anonymous 2pp (15x20)	30 Jun 1658
60r-64v	From Des Noyers at Warsaw 7pp (15x21) fol 64, blank Publ, DL, pp 421-424	21 Jul 1658
65rv	From Hendt at Danzig To Des Noyers at Warsaw 1p + ad (15x21) Publ, DL, p 420	13 Jul 1658
66r-67v	From Titi at [Pavia?] To M. [Bisaccioni?] at Venice 1p + ad (15x21) Italian	11 Jun 1658
68r-69v	From Des Noyers at Warsaw 3pp (15x21) Publ, DL, pp 424-426	28 Jul 1658
70r-72v	From Des Noyers at Warsaw 4pp (15x21) Publ, DL, pp 426-427	4 Aug 1658
73r-74v	From Des Noyers at Warsaw 4pp (15x21) Publ, DL, pp 428-430	11 Aug 1658
75r-76v	From Des Noyers at Warsaw 4pp (15x21) Publ, DL, pp 430-431	18 Aug 1658

77r-79v	From Des Noyers at Warsaw 4½pp (15x21) Publ, DL, pp 432-434	25 Aug 1658
80r-81v	From Des Noyers at Warsaw 4pp (15x21) Publ, DL, pp 434-436	1 Sep 1658
82r-83v	Latin verse, untitled 2½pp (20x33) Latin	undated
84rv&87rv	From Des Noyers at Niepolomice 1p & 6 lines + ad (15x21) Publ, DL, pp 436-437	5 Sep 1658
85r-86v	From Des Noyers at Niepolomice 3pp & 9 lines (15x21) Publ, DL, pp 437-438	8 Sep 1658
88r-91v	From Des Noyers at [Plock] 6½pp (15x21) Publ, DL, pp 439-442	[15 Sep 1658]
92r-96v	From Des Noyers at [Gotab] 9pp (15x21) Publ, DL, pp 442-445	[23 Sep 1658]
97r-100v	From Des Noyers at Torun 7pp (15x21) Publ, DL, pp 445-449	1 Oct 1658
101r-102v	News account 2½pp (17x22) Italian	undated
103rv	News account Milan 1p (20x31) Italian	 14 Aug 1658 [24 Aug 1658]
104r-106v	From Des Noyers at Torun 5pp (15x21) Publ, DL, pp 449-451	8 Oct 1658
107r-108v	News account 2pp (15x20) Italian fol 108, blank	8 Sep 1658

109rv	News account Venice 1p (20x26) Italian	14 Sep 1658
110r-111v	From Des Noyers at Torun 4pp (15x21) Publ, DL, pp 454-455	15 Oct 1658
112rv	From Hendt at Danzig To Des Noyers 2pp (15x21) Publ, DL, pp 452-453	12 Oct 1658
113rv	From [Hendt] To Des Noyers 1p (14x20) Publ, DL, p 453	[15 Oct 1658?]
114r-115v	From Des Noyers at Torun 4pp (15x21) Publ, DL, pp 457-459	22 Oct 1658
116rv	From [Hendt] at Danzig To [Des Noyers] 2pp (16x20) Publ, DL, pp 455-456	19 Oct 1658
117r-118v	From Des Noyers at Torun 2½pp (15x21) Publ, DL, pp 459-460	29 Oct 1658
119r-120v	From Des Noyers at Torun 3pp (15x21) See DL, pp 462-463	5 Nov 1658
121rv	From Hendt at Danzig To [Des Noyers] 1½pp (15x20) Publ, DL, p 461	2 Nov 1658
122r-124v	From Des Noyers at Torun 3pp + ad (15x21) fol 124, illustration, "Hic Finis, Carolus Gustavus Rex" Publ, DL, pp 463-464	12 Nov 1658
125r-128v	From Des Noyers at Torun 8pp (15x21) Publ, DL, pp 465-469	19 Nov 1658

129r-130v	News account Torun 4pp (20x31) Italian	17 Nov 1658
131r-132v	Two printed illustrations [Monstrous births, Danzig] 2pp (14x19) Latin	1656, 1658
133r-134v	From Des Noyers at Torun 3½pp (15x21) Publ, DL, pp 469-471	26 Nov 1658
135r-138v	From Des Noyers at Torun 7pp (15x21) Publ, DL, pp 471-473	3 Dec 1658
139r-142v	From [Des Noyers] at Torun 6pp & 7 lines (15x21) Publ, DL, pp 473-476	10 Dec 1658
143r-144v	From Des Noyers at Torun 4pp (15x21) Publ, DL, pp 476-477	17 Dec 1658
145r-147v	From Des Noyers at Torun 5pp (15x21) fol 147, Italian note (15x15) Publ, DL, pp 477-479	24 Dec 1658
148r-150v	From Des Noyers at Torun 4pp (15x21) Publ, DL, pp 479-480	31 Dec 1658
151r-152v	From Des Noyers at Torun 4pp (15x21) Publ, DL, pp 481-483	7 Jan 1659
153r-154v	From Des Noyers at Torun 4pp (15x21) Publ, DL, pp 483-484	14 Jan 1659
155r-156v	From Des Noyers at Torun 4pp (15x21) Publ, DL, pp 485-486	26 Jan 1659
157r-159v	From Des Noyers at Warsaw 5pp (15x21) Publ, DL, pp 486-487	3 Feb 1659

160r-161v	From Des Noyers at Warsaw 3pp (15x21) Publ, DL, p 488	10 Feb 1659
162r-167v	From Des Noyers at Warsaw 10pp (16x21) French/Latin Publ, LPD, pp 489-493 [fol 167, Latin enclosure (20x33)] Publ, DL, pp 493-494	20 Feb 1659
168r-169v	From Des Noyers at Warsaw 4pp (15x21) Publ, DL, pp 495-497	6 Mar 1659
170r-172v	From Des Noyers at Warsaw 5pp (15x21) Publ, DL, pp 494-495	27 Feb 1659
173r-174v	From Des Noyers at Warsaw 4pp (15x21) Publ, DL, pp 498-500	13 Mar 1659
175r-177v	From Des Noyers at Warsaw 4pp & 13 lines (15x21) fol 177, brief note (7x10) Publ, DL, pp 500-501	20 Mar 1659
178rv	From [Hendt] at Danzig To Des Noyers 1p (15x20) French/Latin Publ, DL, p 498	8 Mar 1659
179r-180v	From Des Noyers at Warsaw 4pp (15x21) See DL, pp 501-503	27 Mar 1659
181r-183v	From Des Noyers at Warsaw 6pp (15x21) Publ, DL, pp 503-504	3 Apr 1659
184r-188v	From Des Noyers at Warsaw 6pp (15x21) fol 186r, astrol. information (10x12) etc. Publ, DL, pp 504-506	9 Apr 1659
189r-191v	From Des Noyers at Warsaw 5pp (15x21) Publ, DL, pp 506-508	16 Apr 1659

192r-194v	From Des Noyers at Warsaw 5pp (15x21) See DL, pp 508-510	23 Apr 1659
195rv	Copy of a letter To Queen [Louise-Marie de Gonzague] 2pp (18x31) Latin	29 Mar 1659
196r-197v	From Des Noyers at Warsaw 3pp (15x21) Publ, DL, pp 510-511	30 Apr 1659
198r-199v	From Des Noyers at Warsaw 3pp & 9 lines (15x21) Publ, DL, pp 512-513	7 May 1659
200r-201v	From Des Noyers at Warsaw 4pp (15x21) Publ, DL, pp 513-514	14 May 1659
202rv	Anonymous note 1p (14x16) Latin/French	undated
203r-204v	From Des Noyers at Warsaw 4pp (15x21) Publ, DL, pp 514-516	21 May 1659
205r-210v	From Des Noyers at Warsaw 11pp (15x21) Publ, DL, pp 516-520	28 May 1659
211r-212v	From Des Noyers at Warsaw 3pp (15x21) See DL, pp 520-521	4 Jun 1659
213r-216v	From Des Noyers at Warsaw 7pp (15x21) See DL, pp 521-523	11 Jun 1659
217r-218v	From Des Noyers at Warsaw 4pp (15x21) Publ, DL, pp 523-525	21 Jun 1659
219r-222v	From Des Noyers at [Niepolomice] 7pp (15x21) Publ, DL, pp 525-527	25 Jun 1659
223rv	From Titi at Pavia To Anonymous 1p (10x10) Italian	26 May 1659

224r-225v	From Des Noyers at Warsaw 4pp (15x21) Publ, DL, pp 527-529	2 Jul 1659
226r-227v	From Des Noyers at Warsaw 3pp (15x21) Publ, DL, pp 529-530	12 Jul 1659
228r-229v	From [Des Noyers] at Warsaw 4pp (15x21) Publ, DL, pp 530-532	19 Jul 1659
230r-232v	From Des Noyers at Warsaw 5pp (15x21) Publ, DL, pp 532-533	26 Jul 1659
233r-234v	From Des Noyers at Warsaw 4pp (15x21) See DL, pp 533-534	2 Aug 1659
235r-237v	From Des Noyers at Warsaw 5pp (15x21) Publ, DL, pp 535-536	9 Aug 1659
238r-241v	From Des Noyers at Warsaw 8pp (15x21) Publ, DL, pp 536-538	16 Aug 1659
242r-244v	From Des Noyers at Warsaw 5pp (15x21) See DL, pp 538-539	23 Aug 1659
245r-246v	From Des Noyers at Warsaw 3pp (15x21) See DL, pp 539-540	30 Aug 1659
247r-248v	Document* "Nativité de roi Suede" 3pp (20x31) *translated from German	undated
249r-252v	From Des Noyers at Warsaw 7pp (15x21) Publ, DL, pp 542-544	6 Sep 1659
253r-254v	From Des Noyers at Warsaw 4pp (15x21) Publ, DL, pp 545-546	13 Sep 1659

255r-256v	From S. Niemirycz at Graudentz To the Queen of Poland 2pp (18x30) See DL, pp 540-542	31 Aug 1659
257r-259v	From Des Noyers at Warsaw 5pp (15x21) Publ, DL, pp 546-548	20 Sep 1659
260rv	Astrological nativities 1p (15x21)	undated
261r-264v	From Des Noyers at Warsaw 7pp (15x21) Publ, DL, pp 548-549	27 Sep 1659
265r-267v	From Des Noyers at Warsaw 5pp (15x21) fol 266, brief note (13x15) See DL, pp 549-550	4 Oct 1659
268r-270v	From Des Noyers at Warsaw 5pp (15x21) See DL, pp 550-551	11 Oct 1659
271r-274v	From Des Noyers at Warsaw 7pp (15x21) Publ, DL, pp 551-554	18 Oct 1659
275r-281v	From Des Noyers at Szubino 12pp (15x21) fol 280, brief note (7x21) Publ, DL, pp 554-555	8 Nov 1659
282r-283v	From Des Noyers at Szubino 3pp (15x21) Publ, DL, pp 555-556	15 Nov 1659
284r-289v	From Des Noyers at Tuchola 12pp (15x21) Publ, DL, pp 557-560	25 Nov 1659
290r-293v	From Des Noyers at Danzig 8pp (15x21) Publ, DL, pp 560-564	13 Dec 1659

294r-298v	From Des Noyers at Danzig 9pp (15x21) fol 297, brief note (10x17) See DL, pp 564-566	20 Dec 1659
299r-300v	From D. Roncalli at Rome To Anonymous 3pp (18x19) Italian	8 Nov 1659
301rv	From [?] To Des Noyers 1p (18x24)	undated
302r-303v	Copy of a letter w/list of casualties From Anonymous at Czorsztyn 2pp (20x31)	22 Nov 1659
304r-307v	From Des Noyers at Danzig 8pp (15x21) Publ, DL, pp 567-570	27 Dec 1659
308rv&311rv	From [illegible] 2½pp (16x22) Italian	[13?] Dec 1659
309r-310v	From Saint-Martin at Pressburg To [Des Noyers] 3pp (15x20)	24 Nov 1659
312r-313v	From Saint-Martin at Pressburg To [Des Noyers] 2½pp (20x31)	28 Nov 1659
314r-315v	Cipher Key in B's hand (315r) fol 314r, fragment (7x20) 2pp (16x20)	undated
316r-317v	From Des Noyers at Grokolino 3pp (15x21) Publ, DL, pp 289-290	8 Jan 1657
318rv	Printed page 1p (17x19) German	
319rv	From Des Noyers at Ratisbon 2pp (15x21) Publ, DL, pp 1-2	15 Oct 1655

320r-322v	From Des Noyers at Prague 3pp + ad (15x21) fol 322, blank Publ, DL, pp 2-4	18 Oct 1655
323r-324v	From Des Noyers at Glogów 3½pp + ad (15x21) Publ, DL, pp 4-6	27 Oct 1655
325r-326v	From Des Noyers at Glogów 3pp + ad (14x20) Publ, DL, pp 6-8	3 Nov 1655
327r-329v	From [Des Noyers] at Glogów 5pp + ad (14x20) Publ, DL, pp 8-11	8 Nov 1655
330r-331v	From Des Noyers at Glogów 3pp + ad (14x20) Publ, DL, pp 11-13	12 Nov 1655
332r-335v	From Des Noyers at Opole 7pp + ad (14x20) Publ, DL, pp 13-21	23 Nov 1655

▼ ▼ ▼

COLLECTION BOULLIAU, VOLUME III, FONDS FRANÇAIS 13021

1rv	From [Des Noyers] at Danzig 2pp (17x22)	6 Jan 1680
2r-3v	From [Des Noyers] at Warsaw 3pp + ad (17x22)	1 Mar 1680
4rv	From [Des Noyers] at Warsaw 2pp (13x17)	26 Jan 1680
5r-6v	From [Des Noyers] at Warsaw 2pp & 5 lines + ad + ad (13x17)	9 Feb 1680
8rv&11rv	From [Des Noyers] at Warsaw 2¼pp + ad (17x22)	8 Mar 1680
9r-10v	From [Des Noyers] at Warsaw 3pp + ad (17x22)	15 Mar 1680
12r-13v	From [Des Noyers] at Warsaw 3pp + ad (17x22)	22 Mar 1680
14r-15v	From [Des Noyers] at Warsaw 2½pp + ad (13x17)	29 Mar 1680
16rv&19rv	From [Des Noyers] at Warsaw 2pp & 4 lines + ad (17x22)	5 Apr 1680
17r-18v	From [Des Noyers] at Warsaw 2⅔pp + ad (17x22)	12 Apr 1680
20r-21v	From [Des Noyers] at Warsaw 3pp + ad (13x17)	19 Apr 1680
22r-23v	From [Des Noyers] at Warsaw 2pp & 11 lines + ad (17x21)	26 Apr 1680
24r-25v	From [Des Noyers] at Warsaw 2½pp + ad (17x22)	3 May 1680
26r-27v	From [Des Noyers] at Warsaw 2½pp + ad (13x17)	10 May 1680
28r-29v	From [Des Noyers] at Warsaw 3pp + ad (17x22)	17 May 1680

Folios	Description	Date
30r-31v	From [Des Noyers] at Warsaw 3pp + ad (13x17)	24 May 1680
32r-33v	From [Des Noyers] at Warsaw 3pp + ad (17x22)	31 May 1680
34rv	From [Des Noyers] at Warsaw 2pp (17x22)	7 Jun 1680
35r-36v	From [Des Noyers] at Warsaw 3½pp + ad (17x22)	14 Jun 1680
37rv	From [Des Noyers] at Warsaw 1p & 11 lines (17x22)	21 Jun 1680
38rv&41rv	From [Des Noyers] at Warsaw 2pp & 8 lines + ad (17x22)	28 Jun 1680
39r-40v	From [Des Noyers] at Warsaw 2½pp + ad (17x22)	5 Jul 1680
42r-43v	From [Des Noyers] at Warsaw 2pp & 9 lines + ad (17x22)	12 Jul 1680
44r-45v	From [Des Noyers] at Warsaw 2½pp + ad (17x22)	19 Jul 1680
46rv	From [Des Noyers] at Warsaw 1½pp (17x22)	26 Jul 1680
47rv&50rv	From [Des Noyers] at Warsaw 2⅓pp + ad (17x22)	2 Aug 1680
48r-49v	From [Des Noyers] at Warsaw 2½pp + ad (17x22)	9 Aug 1680
51r-52v	From [Des Noyers] at Warsaw 3pp + ad (17x22)	16 Aug 1680
53r-54v	From [Des Noyers] at Warsaw 3pp + ad (17x22)	23 Aug 1680
55r-56v	From [Des Noyers] at Warsaw 2½pp + ad (17x22)	30 Aug 1680
57r-58v	From [Des Noyers] at Warsaw 3pp + ad (17x22)	6 Sep 1680

59r-60v	From [Des Noyers] at Warsaw 3pp + ad (17x22)	13 Sep 1680
61r-62v	From [Des Noyers] at Warsaw 2¾pp + ad (17x22)	20 Sep 1680
63rv	From [Des Noyers] at Warsaw 2pp (17x22)	27 Sep 1680
64r-65v	From [Des Noyers] at Warsaw 3pp & 6 lines + ad (17x22)	4 Oct 1680
66r-67v	From [Des Noyers] at Warsaw 2½pp + ad (13x17)	28 Sep 1680
68r-69v	From [Des Noyers] at Warsaw 3pp + ad (17x22)	11 Oct 1680
70r-71v	From [Des Noyers] at Warsaw 3pp + ad (17x22)	18 Oct 1680
72r-73v	From [Des Noyers] at Warsaw 2¾pp + ad (17x22)	25 Oct 1680
74r-76v	From [Des Noyers] at Warsaw 2½pp + ad (17x22) fol 76, Latin note, 7 lines (8x15)	1 Nov 1680
77r-78v	From [Des Noyers] at Warsaw 2½pp + ad (17x22)	8 Nov 1680
79rv&82rv	From [Des Noyers] at Warsaw 2½pp + ad (17x22)	15 Nov 1680
80r-81v	From [Des Noyers] at Warsaw 3pp + ad (17x22)	22 Nov 1680
83r-84v	From [Des Noyers] at Warsaw 3½pp + ad (17x22)	29 Nov 1680
83rv bis	From [Des Noyers] at Warsaw 11 lines (8x17)	undated
85r-86v	From [Des Noyers] at Warsaw 3pp (17x22)	6 Dec 1680
87r-88v	From [Des Noyers] at Warsaw 3pp (17x22)	13 Dec 1680

89rv&92rv	From [Des Noyers] at Warsaw 2pp & 2 lines + ad (17x22)	20 Dec 1680
90r-91v	From [Des Noyers] at Warsaw 3pp + ad (17x22)	27 Dec 1680
93rv	From [Des Noyers] at Warsaw 2pp (17x22)	3 Jan 1681
94r-95v	From [Des Noyers] at Warsaw 2½pp + ad (17x22)	10 Jan 1681
96rv	From [Des Noyers] at Warsaw 2pp (17x22)	17 Jan 1681
97r-98v	From [Des Noyers] at Warsaw 3pp + ad (17x22) *[corrected from 24 Dec 1681]	[24 Jan 1681]*
99r-100v	From [Des Noyers] at Warsaw 3½pp + ad (17x22)	31 Jan 1681
101r-103v	From [Des Noyers] at Warsaw 5pp + ad (17x22)	7 Feb 1681
104r-105v	From [Des Noyers] at Warsaw 3½pp + ad (17x22)	14 Feb 1681
106r-107v	From [Des Noyers] at Warsaw 2pp & 13 lines + ad (17x22)	21 Feb 1681
108r-109v	From [Des Noyers] at Warsaw 2¾pp + ad (17x22)	28 Feb 1681
110r-111v	From [Des Noyers] at Warsaw 3¼pp + ad (17x22)	7 Mar 1681
112r-113v	From [Des Noyers] at Warsaw 2¾pp (17x22)	14 Mar 1681
114r-115v	From [Des Noyers] at Warsaw 3pp + ad (17x22)	21 Mar 1681
116r-117v	From [Des Noyers] at Warsaw 3pp + ad (17x22)	28 Mar 1681
118r-119v	From [Des Noyers] at Warsaw 3pp + ad (17x22)	4 Apr 1681

120r-121v	From [Des Noyers] at Warsaw 3pp + ad (17x22)	11 Apr 1681
122r-123v	From [Des Noyers] at Warsaw 3pp & 6 lines + ad (17x22)	18 Apr 1681
124r-125v	From [Des Noyers] at Warsaw 3pp & 8 lines + ad (17x22)	25 Apr 1681
126r-127v	From [Des Noyers] at Warsaw 3pp + ad (17x22)	2 May 1681
128r-129v	From [Des Noyers] at Warsaw 3pp & 11 lines + ad (17x22)	9 May 1681
130r-131v	From [Des Noyers] at Warsaw 2½pp + ad (17x22)	16 May 1681
132r-133v	From [Des Noyers] at Warsaw 3pp + ad (17x22)	23 May 1681
134r-135v	From [Des Noyers] at Warsaw 3½pp + ad (17x22)	30 May 1681
136r-137v	From [Des Noyers] at Warsaw 3½pp + ad (17x22)	6 Jun 1681
138r-139v	From [Des Noyers] at Warsaw 3pp & 8 lines + ad (17x22)	13 Jun 1681
140rv&143rv	From [Des Noyers] at Warsaw 2pp + ad (17x22)	20 Jun 1681
141r-142v	From [Des Noyers] at Warsaw 3pp & 7 lines + ad (17x22)	27 Jun 1681
144rv	From [Des Noyers] at Warsaw 2pp (17x22)	4 Jul 1681
145r-147v	From [Des Noyers] at Warsaw 3pp & 12 lines + ad (17x22) fol 147, Note (9x17) French/Latin	11 Jul 1681
148r-149v	From [Des Noyers] at Warsaw 3pp & 14 lines + ad (17x22)	18 Jul 1681
150r-151v	From [Des Noyers] at Warsaw 2½pp + ad (17x22)	25 Jul 1681

152r-153v	From [Des Noyers] at Warsaw 3pp + ad (17x22)	1 Aug 1681
154r-155v	From [Des Noyers] at Warsaw 2½pp + ad (17x22)	8 Aug 1681
156r-157v	From [Des Noyers] at Warsaw 3pp + ad (17x22)	15 Aug 1681
158r-159v	From [Des Noyers] at Warsaw 2¾pp + ad (17x22)	22 Aug 1681
160r-161v	From [Des Noyers] at Warsaw 2½pp + ad (17x22)	29 Aug 1681
162r-163v	From [Des Noyers] at Warsaw 2¾pp + ad (17x22)	5 Sep 1681
164r-165v	From [Des Noyers] at Warsaw 3pp & 8 lines + ad (17x22)	12 Sep 1681
166r-167v	From [Des Noyers] at Warsaw 3pp + ad (17x22)	19 Sep 1681
168r-169v	From [Des Noyers] at Warsaw 2½pp + ad (17x22)	26 Sep 1681
170r-171v	From [Des Noyers] at Warsaw 3pp + ad (17x22)	3 Oct 1681
172r-173v	From [Des Noyers] at Warsaw 3pp + ad (17x22)	10 Oct 1681
174r-175v	From [Des Noyers] at Warsaw 3pp & 9 lines + ad (17x22)	17 Oct 1681
176r-177v	From [Des Noyers] at Warsaw 3⅓pp + ad (17x22)	24 Oct 1681
178r-179v	From [Des Noyers] at Warsaw 3pp + ad (17x22)	31 Oct 1681
180rv	Translated extract, Anonymous 1½pp (17x22)	undated
181r-182v	From [Des Noyers] at Warsaw 2½pp + ad (17x22)	7 Nov 1681

183rv	From [Des Noyers] at Warsaw 2pp (17x22)	14 Nov 1681
184r-185v	From [Des Noyers] at Warsaw 3pp + ad (17x22)	21 Nov 1681
186rv	From [Des Noyers] at Warsaw 2pp (17x22)	28 Nov 1681
187r-188v	From [Des Noyers] at Warsaw 3pp + ad (17x22)	5 Dec 1681
189rv	From [Des Noyers] at Warsaw 2pp (17x22)	12 Dec 1681
190r-191v	From [Des Noyers] at Warsaw 2½pp + ad (17x22)	19 Dec 1681
192r-193v	From [Des Noyers] at Warsaw 3pp & 3 lines + ad (17x22)	26 Dec 1681
194r-195v	From [Des Noyers] at Warsaw 3pp + ad (17x22)	2 Jan 1682
196r-197v	From [Des Noyers] at Warsaw 3pp + ad (17x22)	9 Jan 1682
198r-199v	From [Des Noyers] at Warsaw 3pp + ad (17x22)	16 Jan 1682
200rv&204rv	From [Des Noyers] at Warsaw 2pp & 3 lines + ad (17x22)	23 Jan 1682
201r-202v	From [Des Noyers] at Warsaw 3pp + ad (17x22)	30 Jan 1682
203rv	From [Des Noyers] at Warsaw 2pp (13x17)	6 Feb 1682
205r-206v	From [Des Noyers] at Warsaw 2½pp + ad (17x22)	12 Feb 1682
207rv	From [Des Noyers] at Warsaw 2pp (13x17)	13 Feb 1682
208r-209v	From [Des Noyers] at Warsaw 2½pp + ad (17x22)	20 Feb 1682

210r-211v	From [Des Noyers] at Warsaw 2½pp + ad (17x22)	27 Feb 1682
212r-213v	From [Des Noyers] at Warsaw 2¾pp + ad (17x22)	6 Mar 1682
214r-215v	From [Des Noyers] at Warsaw 3pp & 3 lines (17x22)	13 Mar 1682
216r-217v	From [Des Noyers] at Warsaw 3pp (17x22)	20 Mar 1682
218rv	From [Des Noyers] at Warsaw 1½pp (17x22)	27 Mar 1682
219r-220v	From [Des Noyers] at Warsaw 2½pp + ad (17x22)	3 Apr 1682
221r-222v	From [Des Noyers] at Warsaw 3pp + ad (17x22)	10 Apr 1682
223r-224v	From [Des Noyers] at Warsaw 2½pp + ad (17x22)	17 Apr 1682
225r-226v	From [Des Noyers] at Warsaw 3pp + ad (17x22)	24 Apr 1682
227r-228v	From [Des Noyers] at Warsaw 2¾pp + ad (17x22)	2 May 1682
229r-230v	From [Des Noyers] at Warsaw 2¾pp + ad (17x22)	9 May 1682
231rv	From [Des Noyers] at Amsterdam 1p + ad (11x17)	8 Jun 1682
232rv	From [Des Noyers] at Amsterdam 2pp (10x13)	11 Sep 1682
233rv	From [Des Noyers] at Hamburg 2pp (10x15)	20 Sep 1682
234r-235v	From [Des Noyers] at Danzig 3pp + ad (17x22)	10 Oct 1682
236r-237v	From [Des Noyers] at Warsaw 3pp + ad (13x18)	23 Oct 1682

238r-239v	From [Des Noyers] at Warsaw 3pp + ad (17x22)	30 Oct 1682
240r-241v	From [Des Noyers] at Warsaw 2½pp + ad (17x22)	6 Nov 1682
242rv	From [Des Noyers] at Warsaw 2pp (17x22)	13 Nov 1682
243rv	From [Des Noyers] at Warsaw 1½pp (17x22)	20 Nov 1682
244r-245v	From [Des Noyers] at Warsaw 3pp + ad (17x22)	27 Nov 1682
246r-247v	From [Des Noyers] at Warsaw 3pp + ad (17x22)	4 Dec 1682
248r-249v	From [Des Noyers] at Warsaw 3pp + ad (17x22)	11 Dec 1682
250r-251v	From [Des Noyers] at Warsaw 3pp + ad (17x22)	25 Dec 1682
252r-253v	From [Des Noyers] at Warsaw 3pp + ad (17x22)	1 Jan [1683]
254r-255v	From [Des Noyers] at Warsaw 2¾pp + ad (17x22)	8 Jan 1683
256r-257v	From [Des Noyers] at Warsaw 2½pp + ad (17x22)	15 Jan 1683
258r-260v	From [Des Noyers] at Warsaw 2½pp + ad (17x22) fol 260, Latin enclosure, 1p (17x27)	22 [Jan 1683]
261r-262v	From [Des Noyers] at Warsaw 2pp & 3 lines + ad (17x22)	29 Jan 1683
263r-264v	From [Des Noyers] at Warsaw 3pp + ad (17x22)	[5 Feb 1683]
265rv&268rv	From [Des Noyers] at Warsaw 2pp & 7 lines + ad (17x22)	19 Feb 1683
266r-267v	From [Des Noyers] at Warsaw 3pp + ad (17x22)	26 Feb 1683

269r-270v	From [Des Noyers] at Warsaw 3pp + ad (17x22)	5 Mar 1683
271r-272v	From [Des Noyers] at Warsaw 2½pp + ad (17x22)	12 Mar 1683
273r-274v	From [Des Noyers] at Warsaw 2pp & 6 lines + ad (11x17)	19 Mar 1683
275r-276v	From [Des Noyers] at Warsaw 2½pp + ad (17x22)	26 Mar 1683
277r-278v	From [Des Noyers] at Warsaw 3pp (17x22)	2 Apr 1683
279rv	Printed document 2pp (13x30) Latin/French "Mercurius Panegyricus/Rerum/Contra Turcas gestarum, Autumno Mense,/Anno 1683"	dated 9 Dec 1683
280rv	From [Des Noyers] at Warsaw 2pp (17x22)	9 Apr 1683
281r-282v	From [Des Noyers] at Warsaw 3pp + ad (17x22)	16 Apr 1683
283r-284v	From [Des Noyers] at Warsaw 3pp + ad (17x22)	23 Apr 1683
285r-286v	From [Des Noyers] at Warsaw 2½pp + ad (17x22)	29 Apr 1683
287r-288v	From [Des Noyers] at Warsaw 3pp + ad (17x22)	7 May 1683
289r-290v	From [Des Noyers] at Warsaw 2½pp + ad (17x22)	14 May 1683
291rv&294rv	From [Des Noyers] at Warsaw 2pp & 7 lines + ad (17x22)	21 May 1683
292r-293v	From [Des Noyers] at Warsaw 3pp + ad (17x22)	28 May 1683
295rv&298rv	From [Des Noyers] at Warsaw 2¾pp + ad (17x22)	4 Jun 1683

296rv	Note & declaration $\frac{3}{4}$p (10x14)	[Jun 1683]
297rv	Note, Anonymous 1p (19x24)	28 May 1683
299r-300v	From [Des Noyers] at Warsaw 2$\frac{1}{2}$pp + ad (17x22)	11 Jun 1683
301r-302v	From [Des Noyers] at Warsaw 3pp & 14 lines + ad (17x22)	18 Jun 1683
303r-304v	From [Des Noyers] at Warsaw 3pp & 10 lines + ad (17x22)	25 Jun 1683
305r-306v	From [Des Noyers] at Warsaw 3pp & 8 lines + ad (17x22)	2 Jul 1683
307r-308v	From [Des Noyers] at Warsaw 3pp & 10 lines + ad (17x22)	9 Jul 1683
309r-310v	From [Des Noyers at Warsaw] 3pp & 12 lines + ad (17x22)	16 Jul 1683
311rv	From [Des Noyers] at Warsaw 1$\frac{1}{2}$pp (17x22)	23 Jul 1683
312r-313v	From [Des Noyers] at Warsaw 2$\frac{1}{2}$pp + ad (17x22)	30 Jul 1683
314r-315v	From [Des Noyers] at Warsaw 2$\frac{1}{2}$pp + ad (17x22)	6 Aug 1683
316r-317v	From [Des Noyers] at Warsaw 2$\frac{3}{4}$pp + ad (17x22)	13 Aug 1683
318r-319v	From [Des Noyers] at Warsaw 2$\frac{1}{2}$pp + ad (17x22)	20 Aug 1683
320r-321v	From [Des Noyers] at Warsaw 2$\frac{3}{4}$pp (17x22)	27 Aug 1683
322r-324v	From [Des Noyers] at Warsaw 3pp & 7 lines + ad (17x22) fol 323, 7 line note (6x10)	3 Sep 1683
325r-326v	From [Des Noyers] at Warsaw 2$\frac{3}{4}$pp + ad (17x22)	10 Sep 1683

327rv	From [Des Noyers] at Warsaw 2pp (17x22)	17 Sep 1683
328r-329v	From [Des Noyers] at Warsaw 3½pp + ad (17x22)	1 Oct 1683
330r-331v	From [Des Noyers] at Warsaw 2¾pp + ad (17x22)	8 Oct 1683
332r-333v	From [Des Noyers] at Warsaw 2pp & 9 lines + ad (17x22) fol 333r, copy of an account	15 Oct 1683
334r-335v	From [Des Noyers] at Warsaw 2pp & 4 lines + ad (17x22)	22 Oct 1683
336r-337v	From [Des Noyers] at Warsaw 2pp & 6 lines + ad (17x22)	29 Oct 1683
338r-339v	From [Des Noyers] at Warsaw 3pp (17x22)	5 Nov 1683
340rv	From [Des Noyers] at Warsaw 1¾pp (17x22)	12 Nov 1683
341r-342v	From [Des Noyers] at Warsaw 3pp + ad (17x22)	19 Nov 1683
343r-344v	From [Des Noyers] at Warsaw 2½pp + ad (17x22)	26 Nov 1683
345r-346v	From [Des Noyers] at Warsaw 3pp + ad (17x22)	10 Dec 1683
347r-348v	From [Des Noyers] at Warsaw 2pp & 7 lines + ad (17x22)	17 Dec 1683
349r-350v	From [Des Noyers] at Warsaw 2½pp + ad (17x22)	24 Dec 1683
351r-352v	From [Des Noyers] at Warsaw 2pp & 10 lines + ad (17x22)	31 Dec 1683

▼ ▼ ▼

COLLECTION BOULLIAU, VOLUME IV, FONDS FRANÇAIS 13022

1r-2v	From [Des Noyers] at Warsaw 3pp + ad (17x22)	7 Jan 1684
3r-4v	From [Des Noyers] at Warsaw 2½pp + ad (17x22)	14 Jan 1684
5r-6v	From [Des Noyers] at Warsaw 3pp + ad (17x22)	21 Jan 1684
7r-8v	From [Des Noyers] at Warsaw 3pp + ad (17x22)	28 Jan 1684
9r-10v	From [Des Noyers] at Warsaw 2pp & 6 lines + ad (17x22)	4 Feb 1684
11r-12v	From [Des Noyers] at Warsaw 3pp + ad (17x22)	11 Feb 1684
13r-14v	From [Des Noyers] at Warsaw 2½pp + ad (17x22)	18 Feb 1684
15rv&18rv	From [Des Noyers] at Warsaw 2pp + ad (17x22)	25 Feb 1684
16r-17v	From [Des Noyers] at Warsaw 3pp + ad (17x22)	3 Mar 1684
19rv	From [Des Noyers] at Warsaw 2pp (17x22)	10 Mar 1684
20r-21v	From [Des Noyers] at Warsaw 2½pp + ad (17x22)	17 Mar 1684
22r-23v	From [Des Noyers] at Warsaw 2pp & 10 lines + ad (11x17)	24 Mar 1684
24r-25v	From [Des Noyers] at Warsaw 3pp + ad (17x22)	31 Mar 1684
26r-27v	From [Des Noyers] at Warsaw 3pp + ad (17x22)	7 Apr 1684
28r-29v	From Anonymous 1p & 7 lines + ad (11x17)	[18 Jun 1684]

30rv	From [Des Noyers] at Warsaw $1\frac{1}{2}$pp (17x22) *[Corrected by B from 7 Apr 1684]	14* Apr 1684
31rv	From [Des Noyers] at Warsaw 2pp (17x22)	21 Apr 1684
32r-33v	From [Des Noyers] at Warsaw $2\frac{1}{2}$pp + ad (17x22)	28 Apr 1684
34r-35v	From [Des Noyers] at Warsaw $2\frac{1}{2}$pp + ad (17x22)	5 May 1684
36r-37v	From [Des Noyers] at Warsaw 3pp + ad (17x22)	12 May 1684
38r-39v	From [Des Noyers] at Warsaw 3pp + ad (17x22)	19 May 1684
40r-41v	From [Des Noyers] at Warsaw 2pp & 11 lines + ad (17x22)	26 May 1684
42r-43v	From [Des Noyers] at Warsaw $2\frac{1}{2}$pp + ad (17x22)	2 Jun 1684
44r-45v	From [Des Noyers] at Warsaw $2\frac{1}{2}$pp + ad (17x22)	9 Jun 1684
46r-47v	From [Des Noyers] at Warsaw $2\frac{3}{4}$pp + ad (17x22)	16 Jun 1684
48r-49v	From [Des Noyers] at Warsaw 3pp + ad (17x22)	23 Jun 1684
50r-51v	From [Des Noyers] at Warsaw 2pp & 11 lines + ad (17x22)	30 Jun 1684
52r-53v	From [Des Noyers] at Warsaw 3pp & 12 lines + ad (17x22)	7 Jul 1684
54r-55v	From [Des Noyers] at Warsaw 3pp + ad (17x22)	14 Jul 1684
56r-57v	From [Des Noyers] at Warsaw 3pp + ad (17x22)	21 Jul 1684
58r-59v	From [Des Noyers] at Warsaw $2\frac{3}{4}$pp + ad (17x22)	28 Jul 1684

60r-61v	From [Des Noyers] at Warsaw 2½pp + ad (17x22)	4 Aug 1684
62r-63v	From [Des Noyers] at Warsaw 2pp & 13 lines + ad (17x22)	11 Aug 1684
64r-65v	From [Des Noyers] at Warsaw 2pp & 14 lines + ad (17x22)	18 Aug 1684
66r-67v	From [Des Noyers] at Warsaw 3pp & 11 lines + ad (17x22)	25 Aug 1684
68r-69v	From [Des Noyers] at Warsaw 2½pp + ad (17x22)	1 Sep 1684
70rv&73rv	From [Des Noyers] at Warsaw 2pp + ad (17x22)	8 Sep 1684
71r-72v	From [Des Noyers] at Warsaw 2¾pp + ad (17x22)	15 Sep 1684
74r-75v	From [Des Noyers] at Warsaw 2½pp + ad (17x22)	22 Sep 1684
76rv&79rv	From [Des Noyers] at Warsaw 2pp + ad (17x22)	29 Sep 1684
77r-78v	From [Des Noyers] at Warsaw 2½pp + ad (17x22)	5 Oct 1684
80r-81v	From [Des Noyers] at Warsaw 3pp + ad (17x22)	12 Oct 1684
82r-83v	From [Des Noyers] at Warsaw 3pp & 12 lines + ad (17x22)	19 Oct 1684
84r-85v	From [Des Noyers] at Warsaw 4pp (17x22)	26 Oct 1684
86rv	Description of medals (6) of Poland 1p (17x22) French/Latin [not in Des Noyers' hand]	undated
87r-88v	From [Des Noyers] at Warsaw 2¾pp + ad (17x22)	2 Nov 1684
89r-90v	From [Des Noyers] at Warsaw 2½pp + ad (17x22)	9 Nov 1684

91r-92v	From [Des Noyers] at Warsaw 2½pp + ad (17x22)	16 Nov 1684
93rv	From [Des Noyers] at Warsaw 1p & 12 lines (17x22)	23 Nov 1684
94r-95v	From [Des Noyers] at Warsaw 2½pp + ad (17x22)	[30] Nov 1684
96r-97v	From [Des Noyers] at Warsaw 2½pp + ad (17x22) [see Vol III, fol 345, et seq.)	3 Dec 1683
98rv	From [Des Noyers] at Warsaw 1½pp (17x22)	7 Dec 1684
99rv	From [Des Noyers] at Warsaw 2pp (17x22)	14 Dec 1684
100rv	From [Des Noyers] at Warsaw 1½pp (17x22)	21 Dec 1684
101r-102v	From [Des Noyers] at Warsaw 2pp & 11 lines + ad (17x22)	28 Dec 1684
103r-104v	From [Des Noyers] at Warsaw 2¾pp + ad (17x22)	12 Jan 1685
105r-107v	From [Des Noyers] at Warsaw 2½pp & 5 lines + ad (17x22) fol 107, 5 line fragment (3x15)	19 Jan 1685
108rv&111rv	From [Des Noyers] at Warsaw 2pp & 3 lines + ad (17x22)	9 Feb 1685
109r-110v	From [Des Noyers] at Warsaw 2½pp + ad (17x22)	2 Feb 1685
112rv	From [Des Noyers] at Warsaw 1p (17x22) French/Latin	16 Feb 1685
113rv	Copy of a letter* From Hevelius at Danzig To Anonymous 2pp (11x17) Latin *Encl w/fol 112	19 Jan 1685

114r-115v	From [Des Noyers] at Warsaw 2½pp + ad (17x22)	23 Feb 1685
116rv&119rv	From [Des Noyers] at Warsaw 1½pp + ad (17x22)	26 Jan 1685
117r-118v	From [Des Noyers] at Warsaw 2¾pp + ad (17x22)	2 Mar 1685
120r-121v	From [Des Noyers] at Warsaw 3pp + ad (17x22)	9 Mar 1685
122rv	From [Des Noyers] at Warsaw 2pp (17x22)	16 Mar 1685
123r-124v	From [Des Noyers] at Warsaw 2pp & 13 lines + ad (17x22)	23 Mar 1685
125rv	From [Des Noyers] at Warsaw 1p & 8 lines (17x22)	30 Mar 1685
126rv	From J.E. Olhoff at Danzig 1p + ad (17x22) Latin	Mar 1685
127rv	From [Des Noyers] at Warsaw 2pp (17x22)	6 Apr 1685
128rv	From [Des Noyers] at Warsaw 2pp (17x22)	13 Apr 1685
129r-130v	From [Des Noyers] at Warsaw 2½pp + ad (17x22)	20 Apr 1685
131r-132v	From [Des Noyers] at Warsaw 3pp & 6 lines + ad (17x22)	27 Apr 1685
133rv&136rv	From [Des Noyers] at Warsaw 2pp & 5 lines + ad (17x22)	4 May 1685
134r-135v	From [Des Noyers] at Warsaw 2¾pp + ad (17x22)	11 May 1685
137rv	From [Des Noyers] at Warsaw 1¾pp (17x22)	18 May 1685
138r-139v	From [Des Noyers] at Warsaw 2½pp + ad (17x22)	25 May 1685

140r-143v	From [Des Noyers] at Warsaw 2pp & 13 lines + ad (17x22) fols 141&142, 4pp (10x17)[extracts from a book of G. Kuhlman discussed in letters of 9&23 Nov 1684]	1 Jun 1685
144rv&146rv	From [Des Noyers] at Warsaw 2½pp + ad (17x22)	8 Jun 1685
145rv	From Hevelius at Danzig To Des Noyers at Warsaw 1p + ad (16x20) Latin	18 May 1685
147r-148v	From [Des Noyers] at Warsaw 2½pp + ad (17x22)	15 Jun 1685
149rv	From [Des Noyers] at Warsaw 1¾pp (17x22)	22 Jun 1685
150r-151v	From [Des Noyers] at Warsaw 4pp (17x22)	29 Jun 1685
152rv	From [Des Noyers] at Warsaw 1⅓pp (17x22)	6 Jul 1685
153r-154v	From [Des Noyers] at Warsaw 3pp + ad (17x22)	13 Jul 1685
155r-156v	From [Des Noyers] at Warsaw 2½pp + ad (17x22)	20 Jul 1685
157rv	From [Des Noyers] at Warsaw 2pp (17x22)	27 Jul 1685
158r-160v	From [Des Noyers] at Warsaw 2¾pp & 4 lines + ad (17x22) fol 160v, note in B's hand	5 Jan 1685
161rv&163rv	From [Des Noyers] at Warsaw 3pp + ad (17x22)	22 Mar 1686
162rv	From [Des Noyers] at Warsaw 1½pp (11x17)	29 Mar 1686
164rv	From [Des Noyers] at Warsaw 1p & 13 lines (17x22)	5 Apr 1686

165rv	From [Des Noyers] at Warsaw 1p & 12 lines (17x22)	12 Apr 1686
166r-167v	From [Des Noyers] at Warsaw 3pp + ad (17x22)	19 Apr 1686
168rv	From [Des Noyers] at Warsaw 1½pp (13x17)	26 Apr 1686
169rv	From [Des Noyers] at Warsaw 1½pp (17x22)	3 May 1686
170r-171v	From [Des Noyers] at Warsaw 2½pp + ad (17x22)	10 May 1686
172r-173v	From [Des Noyers] at Warsaw 2½pp + ad (17x22)	17 May 1686
174r-175v	From [Des Noyers] at Warsaw 2pp & 8 lines + ad (17x22)	24 May 1686
176r-177v	From [Des Noyers] at Warsaw 2pp & 10 lines + ad (17x22)	31 May 1686
178r-179v	From [Des Noyers] at Warsaw 3pp + ad (17x22)	7 Jun 1686
180r-181v	From [Des Noyers] at Warsaw 3pp + ad (17x22) [contains extracts of letters of Queen Christine and le Chevalier de Terlon, Rome, 21 Feb 1686 and 18 May 1686]	14 Jun 1686
182r-183v	From [Des Noyers] at Warsaw 2½pp + ad (17x22)	21 Jun 1686
184r-185v	From [Des Noyers] at Warsaw 2½pp + ad (17x22)	28 Jun 1686
186r-187v	From [Des Noyers] at Warsaw 3pp + ad (17x22)	5 Jul 1686
188r-189v	From [Des Noyers] at Warsaw 2½pp + ad (17x22)	12 Jul 1686
190r-191v	From [Des Noyers] at Warsaw 3pp + ad (17x22)	19 Jul 1686

192r-193v	From [Des Noyers] at Warsaw 4pp (17x22)	26 Jul 1686
194r-195v	From [Des Noyers] at Warsaw 3pp + ad (17x22)	2 Aug 1686
196r-197v	From [Des Noyers] at Warsaw 2½pp + ad (17x22)	9 Aug 1686
198rv&203rv	From [Des Noyers] at Warsaw 2pp + ad (17x22)	16 Aug 1686
199rv&202rv	From [Des Noyers] at Warsaw 2pp + ad (17x22)	23 Aug 1686
200r-201v	From [Des Noyers] at Warsaw 2¾pp + ad (17x22)	30 Aug 1686
204r-205v	From [Des Noyers] at Warsaw 3pp + ad (17x22)	6 Sep 1686
206r-207v	From [Des Noyers] at Warsaw 3pp + ad (17x22)	13 Sep 1686
208r-209v	From [Des Noyers] at Warsaw 3pp + ad (17x22)	20 Sep 1686
210r-211v	From [Des Noyers] at Warsaw 3pp + ad (17x22)	27 Sep 1686
212r-213v	From [Des Noyers] at Warsaw 2½pp + ad (17x22)	4 Oct 1686
214r-215v	From [Des Noyers] at Warsaw 3pp + ad (17x22)	11 Oct 1686
216r-217v	From [Des Noyers] at Warsaw 3½pp + ad (17x22)	18 Oct 1686
218r-219v	From [Des Noyers] at Warsaw 3pp + ad (17x22)	[25 Oct 1686]
220rv	From Hevelius at Danzig To [Des Noyers at Warsaw] 1p (17x22) Latin	18 Oct 1686

221r-222v	From [Des Noyers] at Warsaw 2pp & 14 lines + ad (17x22)	1 Nov 1686
223rv	From [Des Noyers] at Warsaw 2pp (17x22)	8 Nov 1686
224r-225v	From [Des Noyers] at Warsaw 2½pp + ad (17x22) brief note pasted to fol 225	15 Nov 1686
226rv	From [Des Noyers] at Warsaw 1½pp (17x22)	22 Nov 1686
227rv	From [Des Noyers] at Warsaw 1p & 11 lines (17x22)	29 Nov 1686
228rv	From [Des Noyers] at Warsaw 1¾pp (17x22)	6 Dec 1686
229rv	From [Des Noyers] at Warsaw 1¾pp (17x22)	13 Dec 1686
230r-231v	From [Des Noyers] at Warsaw 2¾pp + ad (17x22)	20 Dec 1686
232r-233v	From [Des Noyers] at Warsaw 2½pp + ad (17x22)	27 Dec 1686
234r-235v	From [Des Noyers] at Warsaw 2½pp + ad (17x22)	3 Jan 1687
236r-237v	From [Des Noyers] at Warsaw 3pp + ad (17x22)	10 Jan 1687
236rv bis	Fragment, from [Des Noyers] 6 lines (3x17) [between fols 236&237]	undated
238r-239v	From [Des Noyers] at Warsaw 2¾pp + ad (17x22)	17 Jan 1687
240r-241v	From [Des Noyers] at Warsaw 2½pp + ad (17x22)	24 Jan 1687
242rv	From Adam Kochanski at Danzig To [Des Noyers at Warsaw] 1p (15x20) Italian	17 Jan 1687

243rv&247rv	From [Des Noyers] at Warsaw 2pp & 2 lines + ad (17x22)	31 Jan 1687
244rv	From [Des Noyers] at Warsaw 1p & 8 lines (17x22)	7 Feb 1687
245r-246v	Copy of a letter & announcement* 12 lines (17x22) & 6 lines + ad (6x13) Latin/Italian/French *[death of Hevelius]	31 Jan 1687
248rv	From [Des Noyers] at Warsaw 2pp (17x22)	14 Feb 1687
249rv&252rv	From [Des Noyers] at Warsaw 2pp & 9 lines + ad (17x22)	21 Feb 1687
250r-251v	From [Des Noyers] at Warsaw 2¾pp + ad (17x22)	28 Feb 1687
253rv	From [Des Noyers] at Warsaw 1½pp (17x22)	7 Mar 1687
254rv	From [Des Noyers] at Warsaw 1¾pp (17x22)	21 Mar 1687
255rv	From [Des Noyers] at Warsaw 1¾pp (17x22)	14 Mar 1687
256rv	From [Des Noyers] at Warsaw 1¾pp (17x22)	28 Mar 1687
257rv	From [Des Noyers] at Warsaw 1¾pp (17x22)	4 Apr 1687
258r-259v	From [Des Noyers] at Warsaw 2¾pp + ad (17x22)	11 Apr 1687
260r-261v	From Des Noyers at Warsaw 2½pp + ad (17x22)	18 Apr 1687
262rv	From [Des Noyers] at Warsaw 1¾pp (17x22)	25 Apr 1687
263r-264v	From [Des Noyers] at Warsaw 3pp + ad (17x22)	2 May 1687
265rv	From [Des Noyers] at Warsaw 1⅔pp (11x17)	9 May 1687

266r-267v	From [Des Noyers] at Warsaw 3pp + ad (17x22)	16 May 1687
268rv	From [Des Noyers] at Warsaw $1\frac{2}{3}$pp (17x22)	23 May 1687
269rv&272rv	From [Des Noyers] at Warsaw 2pp & 6 lines + ad (17x22)	30 May 1687
270r-271v	From [Des Noyers] at Warsaw 3pp + ad (17x22)	6 Jun 1687
273r-274v	From [Des Noyers] at Warsaw $2\frac{1}{2}$pp + ad (17x22)	13 Jun 1687
275r-276v	From [Des Noyers] at Warsaw 2pp & 8 lines + ad (17x22)	20 Jun 1687
277r-278v	From [Des Noyers] at Warsaw 3pp & 6 lines + ad (17x22)	27 Jun 1687
279rv	From [Des Noyers] at Warsaw $1\frac{1}{2}$pp (17x22)	4 Jul 1687
280rv	From [Des Noyers] at Warsaw $1\frac{1}{2}$pp (17x22)	11 Jul 1687
281r-282v	From [Des Noyers] at Warsaw $2\frac{1}{2}$pp + ad (17x22)	18 Jul 1687
283r-284v	From [Des Noyers] at Warsaw $2\frac{3}{4}$pp + ad (17x22)	25 Jul 1687
285rv	From [Des Noyers] at Warsaw 2pp (17x22)	1 Aug 1687
286r-287v	From [Des Noyers] at Warsaw $2\frac{1}{2}$pp + ad (17x22)	8 Aug 1687
288r-289v	From [Des Noyers] at Warsaw 3pp + ad (17x22)	15 Aug 1687
290r-291v	From [Des Noyers] at Warsaw $2\frac{1}{3}$pp + ad (17x22)	[22 Aug 1687]
292rv	From [Des Noyers] at Warsaw 2pp (17x22)	29 Aug 1687

293rv	From [Des Noyers] at Warsaw 1½pp (17x22)	12 Sep 1687
294r-295v	From [Des Noyers] at Warsaw 2pp & 6 lines + ad (17x22)	5 Sep 1687
296rv&299rv	From [Des Noyers] at Warsaw 1¾pp & 12 lines + ad (17x22)	26 Sep 1687
297r-298v	From [Des Noyers at Warsaw] 2¾pp + ad (17x22)	[19 Sep 1687]
300r-301v	From [Des Noyers] at Warsaw 2pp & 8 lines + ad (17x22)	24 Oct 1687
302r-303v	From [Des Noyers] at Warsaw 2½pp + ad (17x22)	[17 Sep 1687]
304rv	From [Des Noyers] at Warsaw 2pp (17x22)	10 Oct 1687
305rv	From Des Noyers at Warsaw 2pp (11x17)	5 Oct 1687
306r-307v	From [Des Noyers] at Warsaw 3pp + ad (17x22)	3 Oct 1687
308rv	From [Des Noyers] at Warsaw 1¾pp (17x22)	31 Oct 1687
309r-310v	From [Des Noyers] at Warsaw 3pp + ad (17x22)	7 Nov 1687
311r-312v	From [Des Noyers] at Warsaw 2½pp + ad (17x22)	14 Nov 1687
313rv	From [Des Noyers] at Warsaw 2pp (17x22)	21 Nov 1687
314rv	From [Des Noyers] at Warsaw 2pp (17x22)	28 Nov 1687
315r-316v	From [Des Noyers] at Warsaw 2⅓pp + ad (17x22)	5 Dec 1687
317rv	From [Des Noyers] at Warsaw 1p & 12 lines (17x22)	[12 Dec 1687]

318rv	From [Des Noyers] at Warsaw 2pp (12x17)	19 Dec 1687
319rv	From [Des Noyers] at Warsaw 2pp (17x22)	26 Dec 1687
320rv&322rv	From [Des Noyers] at Warsaw 2pp & 2 lines + ad (17x22)	2 Jan 1688
321rv	From [Des Noyers] at Warsaw 1¾pp (17x22)	9 Jan 1688
323rv	From [Des Noyers] at Warsaw 1½pp (17x22)	16 Jan 1688
324rv	From [Des Noyers] at Warsaw 1½pp (11x17)	23 Jan 1688
325rv&327rv	From [Des Noyers] at Warsaw 2pp & 4 lines + ad (17x22)	30 Jan 1688
326rv	From [Des Noyers] at Warsaw 1½pp (17x22)	6 Feb 1688
328rv	From [Des Noyers] at Warsaw 1½pp (17x22)	13 Feb 1688
329rv&331rv	From [Des Noyers] at Warsaw 2pp + ad (17x22)	20 Feb 1688
330rv&332rv	Latin verse, "Fata Germaniae" 1p & 6 lines + ad (17x22) Latin	undated
333rv	From [Des Noyers] at Warsaw 1¾pp (17x22)	27 Feb 1688
334r-335v	From [Des Noyers] at Warsaw 2¾pp + ad (17x22)	5 Mar 1688
336rv	From [Des Noyers] at Warsaw 2pp (17x22)	12 Mar 1688
337rv	From [Des Noyers] at Warsaw 1½pp (17x22)	19 Mar 1688
338r-339v	From Anonymous at The Hague To Des Noyers at Warsaw 4pp (16x20) [unsigned fragment]	23 Mar 1688

340rv	From [Des Noyers] at Warsaw 1½pp (17x22)	26 Mar 1688
341rv	From [Des Noyers] at Warsaw 2pp (17x22)	2 Apr 1688
342r-343v	From [Des Noyers] at Warsaw 1½pp (17x22)	9 Apr 1688
344rv	From [Des Noyers] at Warsaw 1¾pp (17x22)	16 Apr 1688
345rv	From [Des Noyers] at Warsaw 2pp (17x22)	23 Apr 1688
346r-347v	From [Des Noyers] at Warsaw 2pp & 3 lines + ad (17x22)	30 Apr 1688
348rv	From [Des Noyers] at Warsaw 1½pp (17x22)	7 May 1688
349rv	From [Des Noyers] at Warsaw 1p (17x22)	14 May 1688
350rv	From [Des Noyers] at Warsaw 1p & 3 lines (17x22)	21 May 1688
351r-352v	From [Des Noyers] at Warsaw 2pp & 8 lines + ad (11x17)	28 May 1688
353rv	From [Des Noyers] at Warsaw 1p & 4 lines (17x22)	4 Jun 1688
354r-355v	From [Des Noyers] at Warsaw 3pp + ad (17x22)	11 Jun 1688
356rv	From [Des Noyers] at Warsaw 1¾pp (17x22)	18 Jun 1688
357r-358v	From [Des Noyers] at Warsaw 1½pp + ad (17x22)	25 Jun 1688
359r-360v	From [Des Noyers] at Warsaw 2pp & 10 lines + ad (17x22)	2 Jul 1688
361r-362v	From [Des Noyers] at Warsaw 2pp & 8 lines + ad (17x22)	9 Jul 1688

363r-364v	From [Des Noyers] at Warsaw 2pp & 3 lines + ad (17x22)	16 Jul 1688
365rv&369rv	From [Des Noyers] at Warsaw 2pp & 2 lines + ad (17x22)	23 Jul 1688
366rv	From [Des Noyers] at Warsaw 1p & 11 lines (17x22)	30 Jul 1688
367r-368v	From [Des Noyers] at Warsaw 3pp + ad (17x22)	6 Aug 1688
370rv&373rv	From [Des Noyers] at Warsaw 2pp & ½p list + ad (17x22)	13 Aug 1688
371r-372v	From [Des Noyers] at Warsaw 3pp + ad (17x22)	20 Aug 1688
374rv	From [Des Noyers] at Warsaw 2pp (17x22)	29 Oct 1688
375rv	From [Des Noyers] at Warsaw 2pp (17x22)	5 Nov 1688
376r-377v	From [Des Noyers] at Warsaw 2¾pp + ad (17x22)	12 Nov 1688
378rv	From [Des Noyers] at Warsaw 1p (11x17)	19 Nov 1688
379rv&381rv	From [Des Noyers] at Warsaw 2pp & 9 lines + ad (17x22)	26 Nov 1688
380rv	From [Des Noyers] at Warsaw 1p + ad (17x22)	3 Dec 1688
382rv	From [Des Noyers] at Warsaw 2pp (17x22)	10 Dec 1688
383rv	From [Des Noyers] at Warsaw 1½pp (17x22)	17 Dec 1688
384rv	From [Des Noyers] at Warsaw 2pp (11x17)	24 Dec 1688
385rv	From [Des Noyers] at Warsaw 2pp (17x22)	31 Dec 1688

▼ ▼ ▼

COLLECTION BOULLIAU, VOLUME V, FONDS FRANÇAIS 13023

1rv	From [Des Noyers] at Warsaw 1¾pp (17x22)	7 Jan 1689
2r-3v	From [Des Noyers] at Warsaw 1½pp + ad (11x17)	14 Jan 1689
4r-5v	From [Des Noyers] at Warsaw 2½pp + ad (11x17)	28 Jan 1689
6r-7v	From [Des Noyers] at Warsaw 2pp & 9 lines + ad (11x17)	4 Feb 1689
8r-9v	From [Des Noyers] at Warsaw 2½pp + ad (11x17)	11 Feb 1689
10r-11v	From [Des Noyers] at Warsaw 2¾pp + ad (11x17)	18 Feb 1689
12r-13v	From [Des Noyers] at Warsaw 2pp & 8 lines + ad (11x17)	25 Feb 1689
14rv&17rv	From [Des Noyers] at Warsaw 1p & 7 lines + ad (11x17)	4 Mar 1689
15r-16v	From [Des Noyers] at Warsaw 2½pp + ad (11x17)	11 Mar 1689
18rv	From [Des Noyers] at Warsaw 1½pp (11x17)	18 Mar 1689
19rv	From [Des Noyers] at Warsaw 2pp (11x17)	25 Mar 1689
20r-21v	From [Des Noyers] at Warsaw 3pp + ad (11x17)	1 Apr 1689
22rv	From [Des Noyers] at Warsaw 2pp (11x17)	8 Apr 1689
23rv	From [Des Noyers] at Warsaw 2pp (11x17)	15 Apr 1689
24rv	From [Des Noyers] at Warsaw 1½pp (11x17)	22 Apr 1689

25r-26v	From [Des Noyers] at Warsaw 2pp + ad (11x17)	29 Apr 1689
27rv	From [Des Noyers] at Warsaw 1⅔pp (11x17)	6 May 1689
28rv	From [Des Noyers] at Warsaw 2pp (11x17)	13 May 1689
29r-30v	From [Des Noyers] at Warsaw 2⅔pp + ad (11x17)	20 May 1689
31rv	From [Des Noyers] at Warsaw 2pp (11x17)	27 May 1689
32rv	From [Des Noyers] at Warsaw 2pp (11x17)	3 Jun 1689
33r-34v	From [Des Noyers] at Warsaw 2½pp + ad (11x17)	10 Jun 1689
35rv	From [Des Noyers] at Warsaw 1½pp (11x17)	17 Jun 1689
36rv	From [Des Noyers] at Warsaw 1½pp (11x17)	24 Jun 1689
37r-38v	From [Des Noyers] at Warsaw 2½pp + ad (11x17)	1 Jul 1689
39rv&42rv	From [Des Noyers] at Warsaw 2pp (11x17)	8 Jul 1689
40r-41v	From [Des Noyers] at Warsaw 2½pp + ad (11x17)	15 Jul 1689
43rv	From [Des Noyers] at Warsaw 1¾pp (11x17)	22 Jul 1689
44rv	From [Des Noyers] at Warsaw 2pp (11x17)	19 Aug 1689
45rv	From [Des Noyers] at Danzig 1⅔pp (11x17)	6 Aug 1689
46r-47v	From [Des Noyers] at Warsaw 2½pp + ad (11x17)	26 Aug 1689

48r-49v	From [Des Noyers] at Warsaw 3pp + ad (12x18)	2 Sep 1689
50rv&53rv	From [Des Noyers] at Warsaw 1⅔pp (11x17)	9 Sep 1689
51r-52v	From [Des Noyers] at Warsaw 2⅔pp + ad (11x17)	16 Sep 1689
54r-55v	From [Des Noyers] at Warsaw 2½pp + ad (11x17)	23 Sep 1689
56r-57v	From [Des Noyers] at Warsaw 2pp & 8 lines + ad (11x17)	30 Sep 1689
58rv	From [Des Noyers] at Warsaw 1½pp (11x17)	7 Oct 1689
59rv	From [Des Noyers] at Warsaw 2pp (11x17)	14 Oct 1689
60rv	From [Des Noyers] at Warsaw 2pp (11x17)	21 Oct 1689
61rv	From [Des Noyers] at Warsaw 1p & 6 lines (11x17)	28 Oct 1689
62rv	From [Des Noyers] at Warsaw 1p & 10 lines (11x17)	4 Nov 1689
63rv	From [Des Noyers] at Warsaw 1p (11x17)	11 Nov 1689
64rv	From [Des Noyers] at Warsaw 1½pp (11x17)	18 Nov 1689
65rv	From [Des Noyers] at Warsaw 1p & 3 lines (11x17)	25 Nov 1689
66r-67v	From [Des Noyers] at Warsaw 2pp & 8 lines + ad (11x17)	2 Dec 1689
68rv	From [Des Noyers] at Warsaw 1p & 5 lines (11x17) [torn & mended]	16 Dec 1689
69rv	From [Des Noyers] at Warsaw 2pp (11x17)	23 Dec 1689

70rv	From [Des Noyers] at Warsaw 1½pp (11x17)	30 Dec 1689
71rv	From [Des Noyers] at Warsaw 2pp (11x17)	13 Jan 1690
72r-73v	From [Des Noyers] at Warsaw 2½pp + ad (11x17)	10 Feb 1690
74r-75v	From [Des Noyers] at Warsaw 3pp + ad (11x17)	17 Feb 1690
76rv	From [Des Noyers] at Warsaw 2pp (11x17)	24 Feb 1690
77rv	From [Des Noyers] at Warsaw 2pp (11x17)	3 Mar 1690
78rv	From [Des Noyers] at Warsaw 2pp (11x17)	10 Mar 1690
79r-80v	From [Des Noyers] at Warsaw 3pp + ad (11x17)	31 Mar 1690
81rv	From [Des Noyers] at Warsaw 2pp (11x17)	28 Apr 1690
82rv	From [Des Noyers] at Warsaw 1¾pp (11x17)	5 May 1690
83rv	From [Des Noyers] at Warsaw 2pp (11x17)	12 May 1690
84rv	From [Des Noyers] at Warsaw 1p & 7 lines (11x17)	26 May 1690
85rv&87rv	From [Des Noyers] at Warsaw 2pp & 8 lines (11x17)	2 Jun 1690
86rv	From Comte Maurice de Fieschi at Venice To [Des Noyers at Warsaw] 2pp (13x16) French/Italian Enclosed w/fols 85&87	6 May 1690
88rv&91rv	From [Des Noyers] at Warsaw 2pp & 6 lines + ad (11x17)	9 Jun 1690

89r-90v	From [Des Noyers] at Warsaw 2¼pp + ad (11x17)	16 Jun 1690
92rv	From [Des Noyers] at Warsaw 1½pp (11x17)	23 Jun 1690
93rv	From [Des Noyers] at Warsaw 2pp (11x17)	7 Jul 1690
94rv	From [Des Noyers] at Warsaw 2pp (11x17)	14 Jul 1690
95r-96v	From [Des Noyers] at Warsaw 2¼pp + ad (9x12)	21 Jul 1690
97rv	From [Des Noyers] at Warsaw 2pp (11x17)	28 Jul 1690
98rv	From [Des Noyers] at Warsaw 1½pp (11x17)	4 Aug 1690
99rv	From [Des Noyers] at Danzig 2pp (11x17)	15 Sep 1690
100r-101v	From [Des Noyers] at Danzig 2½pp + ad (11x17)	23 Sep 1690
102rv&105rv	From [Des Noyers] at Warsaw ½p + ad (11x17)	6 Oct 1690
103r-104v	From [Des Noyers] at Warsaw 3pp + ad (11x17)	13 Oct 1690
106r-107v	From [Des Noyers] at Warsaw 2½pp + ad (11x17)	27 Oct 1690
108r-109v	From [Des Noyers] at Warsaw 3pp + ad (11x17)	17 Nov 1690
110rv	From [Des Noyers] at Warsaw 1p & 3 lines (11x17)	1 Dec 1690
111rv	From [Des Noyers] at Warsaw 1p & 5 lines (11x17)	8 Dec 1690
112rv	From [Des Noyers] at Warsaw 2pp (11x17)	15 Dec 1690

113rv	From [Des Noyers] at Warsaw 1½pp (11x17)	12 Jan 1691
114r-115v	From Des Noyers at Warsaw 3pp + ad (11x17)	14 Sep 1691
116rv	From Des Noyers at Warsaw 2pp (11x17)	17 Oct 1692
117r-118v	From Vair at Grenoble 3pp + ad (11x17)	3 Jan 1655
119r-120v	From Vair at Grenoble 2½pp + ad (11x17)	10 Jan 1655
121rv	From Vair at Grenoble 2pp (11x17)	24 Jan 1655
122rv	From Vair at Grenoble 2pp (11x17)	31 Jan 1655
123r-124v	From Vair at Grenoble 2½pp + ad (11x17)	7 Feb 1655
125r-126v	From Vair at Grenoble 2½pp + ad (11x17)	14 Feb 1655
127r-128v	From Vair at Grenoble 3pp + ad (11x17)	21 Feb 1655
129rv&132rv	From Vair at Grenoble 2pp + ad (11x17)	28 Feb 1655
130r-131v	From Vair at Grenoble 2½pp + ad (11x17)	28 Mar 1655
133r-134v	From Vair at Grenoble 2½pp + ad (11x17)	7 Mar 1655
135r-136v	From Vair at Grenoble 2pp & 8 lines + ad (11x17)	14 Mar 1655
137r-138v	From Vair at Grenoble 2pp & 6 lines + ad (11x17)	4 Apr 1655
139r-140v	From Vair at Grenoble 3pp + ad (11x17)	18 Apr 1655

141r-142v	From Vair at Grenoble 2½pp + ad (11x17)	25 Apr 1655
143r-144v	From Vair at Grenoble 3¾pp (11x17)	9 May 1655
145r-146v	From Vair at Grenoble 2½pp + ad (11x17)	16 May 1655
147r-148v	From Vair at Grenoble 2½pp + ad (11x17)	23 May 1655
149r-150v	From Vair at Grenoble 2½pp + ad (11x17)	30 May 1655
151r-152v	From Vair at Grenoble 2½pp + ad (11x17)	6 Jun 1655
153r-154v	From Vair at Grenoble 2½pp + ad (11x17)	13 Jun 1655
155r-156v	From Vair at Grenoble 3pp + ad (11x17)	20 Jun 1655
157r-158v	From Vair at Grenoble 2½pp + ad (11x17)	27 Jun 1655
159r-160v	From Vair at Grenoble 2½pp + ad (11x17)	4 Jul 1655
161r-162v	From Vair at Grenoble 3pp + ad (11x17)	18 Jul 1655
163rv&166rv	From Vair at Grenoble 2pp + ad (11x17)	1 Aug 1655
164r-165v	From Vair at Grenoble 2¾pp + ad (11x17)	25 Jul 1655
167r-168v	From Vair at Grenoble 2½pp + ad (11x17)	15 Aug 1655
169r-170v	From Vair at Grenoble 2½pp + ad (11x17)	22 Aug 1655
171rv&174rv	From Vair at Grenoble 2pp + ad (11x17)	29 Aug 1655

172r-173v	From Vair at Grenoble 2pp & 9 lines + ad (11x17)	5 Sep 1655
175rv	From [Vair] at Grenoble 1p (11x17)	12 Sep 1655
176r-177v	From Vair at Grenoble 2pp & 8 lines + ad (11x17)	19 Sep 1655
178r-179v	From Vair at Grenoble 3pp + ad (11x17)	26 Sep 1655
180rv&183rv	From Vair at Grenoble 2pp + ad (11x17)	3 Oct 1655
181r-182v	From Vair at Grenoble 2½pp + ad (10x15)	10 Oct 1655
184r-185v	From Vair at Grenoble 2½pp + ad (11x17)	17 Oct 1655
186rv&189rv	From Vair at Grenoble 2pp + ad (11x17)	24 Oct 1655
187r-188v	From Vair at Grenoble 2pp & 8 lines + ad (11x17)	31 Oct 1655
190r-191v	From Vair at Grenoble 2pp & 10 lines + ad (11x17)	7 Nov 1655
192r-193v	From Vair at Grenoble 3pp + ad (10x14)	14 Nov 1655
194r-195v	From Vair at Grenoble 2pp & 9 lines + ad (11x17)	21 Nov 1655
196r-197v	From Vair at Grenoble 2pp & 10 lines + ad (11x17)	28 Nov 1655
198r-199v	From Vair at Grenoble 2¾pp + ad (11x17)	12 Dec 1655
200r-201v	From Vair at Grenoble 2½pp + ad (11x17)	19 Dec 1655
202r-203v	From Vair at Grenoble 2pp & 10 lines + ad (10x15)	26 Dec 1655

204r-205v	From Vair at Grenoble 2¾pp + ad (11x17)	9 Jan 1656
206r-207v	From Vair at Grenoble 3pp + ad (11x17)	16 Jan 1656
208r-209v	From Vair at Grenoble 2½pp + ad (11x17)	23 Jan 1656
210r-211v	From Vair at Grenoble 2½pp + ad (11x17)	30 Jan 1656
212r-213v	From Vair at Grenoble 2¾pp + ad (11x17)	20 Feb 1656
214r-215v	From Vair at Grenoble 3pp + ad (11x17)	27 Feb 1656
216r-217v	From Vair at Grenoble 2½pp + ad (11x17)	5 Mar 1656
218r-219v	From Vair at Grenoble 2pp & 7 lines + ad (11x17)	12 Mar 1656
220r-221v	From Vair at Grenoble 3pp + ad (11x17)	19 Mar 1656
222r-223v	From Vair at Grenoble 3pp + ad (11x17)	26 Mar 1656
224r-225v	From Vair at Grenoble 3pp + ad (11x17)	2 Apr 1656
226r-227v	From Vair at Grenoble 3pp + ad (11x17)	16 Apr 1656
228rv&231rv	From Vair at Apt 2pp + ad (11x17)	2 May 1656
229r-230v	From Vair at Apt 2½pp + ad (11x17)	9 May 1656
232r-233v	From Vair at Apt 2½pp + ad (11x17)	16 May 1656
234r-235v	From Vair "De La Tour [d'Aygues?]" 3pp + ad (11x17)	30 May 1656

236r-237v	From Vair "De La Tour [d'Aygues?]" 3pp + ad (11x17)	6 Jun 1656
238r-239v	From Vair at Grenoble 3pp + ad (11x17)	18 Jun 1656
240r-241v	From Vair at Grenoble 3pp + ad (11x17)	25 Jun 1656
242r-243v	From Vair at Grenoble 2½pp + ad (11x17)	2 Jul 1656
244r-245v	From Vair at Grenoble 2¾pp + ad (11x17)	9 Jul 1656
246r-247v	From Vair at Grenoble 3pp + ad (11x17)	16 Jul 1656
248r-249v	From Vair at Grenoble 4pp (11x17)	20 Aug 1656
250r-251v	From Vair at Grenoble 2½pp + ad (11x17)	3 Sep 1656
252r-253v	From Vair at Grenoble 2½pp + ad (11x17)	10 Sep 1656
254r-255v	From Vair at Grenoble 2pp & 4 lines + ad (11x17)	17 Sep 1656
256r-257v	From Vair at Grenoble 2pp & 5 lines + ad (11x17)	24 Sep 1656
258r-259v	From Vair at Grenoble 3pp + ad (11x17)	1 Oct 1656
260r-261v	From Vair at Grenoble 2pp & 9 lines + ad (11x17)	8 Oct 1656
262rv&265rv	From Vair at Grenoble 2pp + ad (11x17)	15 Oct 1656
263r-264v	From Vair at Grenoble 3pp + ad (11x17)	22 Oct 1656
266r-267v	From Vair at Grenoble 4pp (11x17)	29 Oct 1656

268r-269v	From Vair at Grenoble 2pp & 10 lines + ad (11x17)	5 Nov 1656
270r-271v	From Vair at Grenoble 2½pp + ad (11x17)	19 Nov 1656
272rv&275rv	From Vair at Grenoble 1p + ad (11x17)	26 Nov 1656
273r-274v	From Vair at Grenoble 2pp & 10 lines + ad + ad (11x17)	10 Dec 1656
276r-277v	From Vair at Grenoble 3pp + ad (11x17)	24 Dec 1656
278r-279v	From Vair at Grenoble 2pp & 8 lines + ad (11x17)	31 Dec 1656
280rv&283rv	From Vair at Grenoble 2pp + ad (11x17)	7 Jan 1657
281r-282v	From Vair at Grenoble 2pp & 8 lines + ad (11x17)	14 Jan 1657
284r-285v	From Vair at Grenoble 2¾pp + ad (11x17)	21 Jan 1657
286r-287v	From Vair at Grenoble 2½pp + ad (11x17)	28 Jan 1657
288r-289v	From Vair at Grenoble 2pp & 7 lines + ad (11x17)	4 Feb 1657
290rv&293rv	From Vair at Grenoble 2pp + ad (11x17)	11 Feb 1657
291r-292v	From Vair at Grenoble 2½pp + ad (11x17)	18 Feb 1657
294r-295v	From Vair at Grenoble 3pp + ad (11x17)	25 Feb 1657
296r-297v	From Vair at Grenoble 3pp + ad (11x17)	4 Mar 1657
298r-299v	From Vair at Grenoble 2¾pp + ad (11x17)	11 Mar 1657
300r-301v	From Vair at Grenoble 2½pp + ad (11x17)	18 Mar 1657

302r-303v	From Vair at Grenoble 3pp + ad (11x17)	30 Sep 1657
304r-305v	Catalogue of library (probably of J. de Valois) 3pp + ad (18x26)	

▼ ▼ ▼

COLLECTION BOULLIAU, VOLUME VI, FONDS FRANÇAIS 13024

Folio	Description	Date
1r-2v	From Bigot at Strasbourg 1p & 2 lines + ad (15x20)	5 Oct 1657
3r-4v	From Bigot at Strasbourg 2pp + ad (15x20) Publ, BHO, pp 232-233	11 Nov 1657
5rv	From Bigot at Frankfurt 1p + ad (15x20) Publ, BHO, pp 235-236	10 Feb 1658
6r-7v	From Bigot at Frankfurt 2¾pp + ad (17x22) Publ, BHO, pp 237-240	24 Mar 1658
8r-9v	From [Bigot] at Frankfurt 3pp + ad (15x20) Publ, BHO, pp 240-243	31 Mar 1658
10r-12v	From [Bigot] at Frankfurt 6pp (15x20) Publ, BHO, pp 243-248	7 Apr 1658
13r-14v	From Bigot at Frankfurt 3pp + ad (15x20)	14 Apr 1658
15r-16v	From Bigot at Frankfurt 3pp + ad (15x20)	Easter Day 1658 [21 Apr 1658]
17r-19v	From [Bigot] at Frankfurt 4pp & 8 lines + ad (15x20) Publ, BHO, pp 248-250	5 May 1658
20r-21v	From Bigot at Frankfurt 2pp + ad (15x20)	12 May 1658
22r-23v	From [Bigot] at Frankfurt 3pp + ad (15x20) Publ, BHO, pp 251-252	19 May 1658
24r-25v	From [Bigot] at Frankfurt 2⅔pp + ad (15x20)	26 May 1658
26r-27v	From [Bigot] at Frankfurt 2½pp + ad (15x20) French/Latin	2 Jun 1658

28r-29v	From [Bigot] at Frankfurt $1\frac{1}{2}$pp + ad (15x20) Publ, BHO, pp 252-253	[9 Jun] 1658
30r-31v	From Bigot at Frankfurt 2pp & 5 lines + ad (15x20)	16 Jun 1658
32r-33v	From Bigot at Frankfurt 2pp & 6 lines + ad (15x20)	23 Jun 1658
34r-35v	From [Bigot] at Frankfurt $3\frac{1}{2}$pp (15x20)	30 Jun 1658
36r-37v	From Bigot at Frankfurt $1\frac{1}{2}$pp + ad (15x20)	7 Jul 1658
38rv	From Bigot at Frankfurt 1p + ad (15x20)	14 Jul 1658
39r-40v	From Bigot at Frankfurt $1\frac{3}{4}$pp + ad (15x20) Publ, BHO, pp 253-254	21 Jul 1658
41rv	From Bigot at Frankfurt 1p + ad (15x20)	28 Jul 1658
42r-43v	From Bigot at Frankfurt 2pp & 2 lines + ad (15x20)	4 Aug 1658
44r-48v	From [Bigot] at Frankfurt $6\frac{1}{2}$pp (15x20) fol 48, blank Cf. BHO, pp 254-255	11 Aug 1658
49r-50v	From Bigot at The Hague 1p & 7 lines + ad (15x20)	22 Aug 1658
51r-52v	From Bigot at The Hague $1\frac{3}{4}$pp + ad (15x20)	29 Aug 1658
53r-54v	From Bigot at The Hague $1\frac{1}{2}$pp + ad (15x20)	5 Sep 1658
55rv	From Bigot at The Hague 9 lines + ad (15x20)	12 Sep 1658
56rv	From Bigot at The Hague 1p + ad (15x20)	19 Sep 1659

57r-58v	From Bigot at Frankfurt 1½pp + ad (15x20)	21 Sep 1657
59rv	From Bigot at The Hague 1p + ad (15x20)	26 Sep 1658
60rv	From Bigot at The Hague 1p + ad (15x20)	3 Oct 1658

▼ ▼ ▼

COLLECTION BOULLIAU, VOLUME VII, FONDS FRANÇAIS 13025

1r-2v	From [Gondi] at [Vise?] 2pp (13x20)	11 Jul 1673
3r-4v	List 1p (13x20) "Abbayes dont les Superiores/ne sont point amovible" fol 4, blank	undated
5r-6v	From Gondi at [Castello?] 3pp + ad (10x15)	5 Jun 1682
7rv&9rv	From Gondi at Florence 1p + ad (10x15)	5 Sep 1682
8rv	From [Gondi] at Nancy 2pp (15x20) [Fragment]	14 Aug 1673
10r-11v	From [Gondi] at Nancy 3½pp (15x20)	13 Sep 1673
12r-13v	From [Gondi] at Nancy 2pp (15x20)	23 Aug 1673
14rv	From [Gondi] at Nancy 2pp (15x20)	21 Aug 1673
15rv	From [Gondi] at Maestricht 2pp (15x20)	18 Jun 1673
16r-17v	From [Gondi] at Maestricht 3½pp (15x20)	12 Jun 1673
18rv	From [Gondi]* at [Dôle?] 2pp (17x22) [B writes "M. l'Abbé de Gondi Resid de Toscane"]	5 Jun 1674
19r-20v	From Gondi at [Castello?] 4pp (10x15)	25 Mar 1683
21r-22v	From [Gondi] at Florence 4pp (10x15)	14 Nov 1684

23r-24v&40rv	From Gondi at Florence 6pp (15x20)	2 Jul 1683
25r-26v	From Gondi at l'Ambrogiane 4pp (15x20)	19 Nov 1683
27r-28v	From Gondi at [?] 3pp + ad (10x15)	1 Aug 1682
29r-30v	From Gondi at Florence 4pp (10x15)	18 Jul 1682
31r-32v	From Gondi at Pisa 3pp + ad (10x15)	17 Jan 1682 Ab Inc [17 Jan 1683]
33r-34v	From Gondi at Florence 4pp (10x15)	8 Oct 1683
35r-36v	From Gondi at Florence 3pp + ad (10x15)	23 Jan 1684 Ab Inc [23 Jan 1685]
37r-38v	From Gondi at Fontainebleau 3pp (15x20)	18 Sep [1673?]
39rv	From Gondi at Florence 2pp (10x15) [Cf. fol 23, regarding fol 40]	[27?] Nov 1682
41r-42v	From Gondi at Florence 4pp (15x20)	25 Jul 1684
43rv	From Gondi at Fontainebleau 1p (15x20)	17 Sep [1678?]
44rv	From Gondi at Fontainebleau 1p (15x20)	[7?] Sep [1678?]
45rv	From Gondi at Florence 1p (15x20)	31 Dec 1686
46r-47v	From Gondi at Florence 4pp (15x20)	13 Sep 1686
48rv	From Gondi at Florence 1p (15x20)	5 Aug 1690
49r-51v	From Gondi at Florence 4¼pp + ad (15x20)	26 Sep 1690

52rv	From [De Boüillon] at Blois 2pp (17x22)	6 Dec 1652
53r-54v	From De Boüillon at Blois 3pp + ad (17x22)	3 Aug 1653
55rv	From [De Boüillon] at Blois 2pp (17x22)	2 Sep 1653
56rv	From [De Boüillon] at Blois 1p + ad (17x22)	18 Sep 1653
57r-58v	From [Loire?] at Blois 2½pp + ad (14x18)	2 [Aug?] 1653
59rv	From [De Boüillon] at Blois 2pp (17x22)	27 Apr 1653
60rv	From [De Boüillon] at Orléans 2pp (17x22)	13 Apr 1653
61rv	From [De Boüillon] at Blois 1½pp (17x22)	27 Mar 1653
62r-63v	From De Boüillon at Blois 3pp + ad (17x22)	15 Aug 1655
64r-65v	From De Boüillon at Blois 1p & 6 lines + ad (17x22)	29 Aug 1655
66rv	From [De Boüillon] at Blois 2pp (17x22) [fragment]	8 Sep 1655
67rv	From De Boüillon at Blois 2pp (17x22)	24 Feb 1656
68rv	From De Boüillon at Blois 2pp + ad (17x22)	9 Mar 1656
69rv	From [De Boüillon] at Blois 1p (17x22)	6 Apr 1656
70rv	From De Boüillon at Blois 2pp (17x22)	24 Dec 1656
71rv	From De Boüillon at Blois 1p & 6 lines (17x22)	10 Oct 1658

72r-73v	From De Boüillon at Blois 2½pp + ad (17x22)	8 May 1659
74rv	From De Boüillon at Blois 1½pp (17x22)	2 Nov 1659
75r-77v	From De Boüillon at [Blois] 2½pp (17x22) fol 77, blank	27 Nov 1659
78rv	From [De Boüillon] at Blois 1p (17x22)	11 May 1653
79r-80v	From De Boüillon at Orléans 3pp + ad (17x22)	15 Nov 1652
81r-82v	From [De Boüillon] at Blois 2½pp (17x22)	5 Apr 1653
83rv	From le Cardinal d'Estrées at Rome 1p (17x22)	11 Apr 1673
84rv	From Bonsy at St-Jean-de-Luz 1p + ad (17x22)	8 Sep 1655
85r-86v	From Bonsy at St-Jean-de-Luz 3pp (17x22)	7 Aug 1659
87rv	From Bonsy at St-Jean-de-Luz ½p (17x22)	4 Aug 1659
88rv	From Bonsy at St-Jean-de-Luz 2pp (17x22)	31 Jul 1659
89r-90v	From Bonsy at St-Jean-de-Luz 2pp & 4 lines (17x22)	14 Aug 1659
91rv&94rv	From Bonsy at St-Jean-de-Luz 2pp & 3 lines (17x22)	28 Aug [1659?]
92r-93v	From Bonsy at St-Jean-de-Luz 4pp (17x22)	25 Aug 1659
95rv	Newsletter* St-Jean-de-Luz 2pp (17x22) *Annotated	15 Sep [1659?]

96rv	From Bonsy at St-Jean-de-Luz 1p (17x22)	2 Sep 1659
97rv	From Bonsy at St-Jean-de-Luz 2pp (17x22)	18 Sep 1659
98rv	From Bonsy 2pp (17x22) [fragment]	undated
99rv	From Bonsy at Paris 1p (17x22)	19 Nov 1660
100r-101v	From Bonsy at Cadillac 2½pp (17x22)	16 Jul 1659
102rv&107rv	From Bonsy at St-Jean-de-Luz 2pp + ad (17x22)	13 Oct 1659
103r-106v	From Bonsy at Bayonne 6½pp (17x22)	29 Jul 1659
108rv	From Bonsy at St-Jean-de-Luz 1p (17x22)	13 Nov 1659
109rv	From Bonsy at St-Jean-de-Luz 1p (17x22)	10 [Nov?] 1659
110rv	From Bonsy at [St-Jean-de-Luz?] 1p (17x22)	6 Dec 1659
111rv	From Bonsy at Bayonne 1p (17x22)	4 May 1660
112rv	From Bonsy at Carcassonne 1p & 6 lines (17x22)	26 Apr 1660
113rv	From Bonsy at Avignon 2pp (17x22)	27 Mar 1660
114rv	From Bonsy at Montpellier 1p (17x22)	5 Apr 1660
115rv	From Bonsy at Toulouse 1p (17x22)	17 Feb 1660
116rv	From Bonsy at St-Jean-de-Luz 1p & 2 lines (17x22)	10 May 1660

117rv	From Bonsy at St-Jean-de-Luz 2pp (17x22)	23 Oct 1659
118r-119v	From Bonsy at Aix 1p & 5 lines (17x22) fol 119, blank	9 Mar 1660
120rv	From Bonsy at Avignon 1½pp (17x22)	20 Mar 1660
121rv	From Parenti at Amsterdam 1½pp (17x22)	25 Oct 1660
122rv	From Parenti at Amsterdam 2pp (17x22)	26 Feb 1661
123r-124v	From Parenti at Amsterdam [B at Warsaw] 2pp & 3 lines + ad (17x22)	2 Apr 1661
125rv	From Parenti at Amsterdam 1p (17x22)	15 Nov 1661
126rv	From Parenti at Amsterdam ½p (17x22)	20 Dec 1661
127rv	From Parenti at Amsterdam 1p (17x22)	28 Dec 1661
128rv	From Parenti at Amsterdam 2pp (12x17)	16 Jan 1662
129rv	From Parenti at Amsterdam 1p (17x22)	3 Feb 1662
130rv	From [Parenti] at Amsterdam 1p & 5 lines (17x22)	14 Feb 1662
131rv	From Parenti at Amsterdam 1p (17x22)	23 Jan 1662
132rv	From Parenti at Amsterdam ⅔p (17x22)	24 Jan 1662
133r-134v	From Parenti at Amsterdam [B at The Hague] 1½pp + ad (17x22)	26 Jan 1662

135rv	From Parenti at Amsterdam 1½pp (17x22)	8 Mar 1662
136rv	From Parenti at Amsterdam 1p & 6 lines (17x22)	14 Mar 1662
137rv	From Parenti at Amsterdam 1½pp (17x22)	20 Mar 1662
138rv	From Parenti at Amsterdam 1p (17x22)	27 Mar 1662
139rv	From Parenti at Amsterdam 1p (17x22) Italian/French	3 Apr 1662
140rv	From Parenti at Amsterdam [B at The Hague] 1p + ad (20x31)	17 Apr 1662
141rv	From Parenti at Amsterdam 1p (17x22)	24 Apr 1662
142rv	From Parenti at Amsterdam 1p (17x22)	26 Apr 1662
143rv	From Parenti at Amsterdam 1p (17x22)	29 Apr 1662
144r-145v	From Parenti at Amsterdam B at The Hague 1p + ad (17x22)	1 May 1662
146rv	From Parenti at Amsterdam 1p (17x22)	8 May 1662
147rv	From Parenti at Amsterdam 1p (17x22) French/Italian	13 Jun 1662
148r-149v	From Parenti at Amsterdam 3pp + ad (17x22)	3 Aug 1662
150r-151v	From Parenti at Amsterdam 3½pp (17x22)	28 Dec 1662
152r-153v	From Parenti at Amsterdam 3pp (17x22)	18 Jan 1663

154r-155v	From Parenti at Amsterdam 2pp & 6 lines + ad (17x22)	8 Feb 1663
156r-157v	From Parenti at Amsterdam 3pp + ad (17x22)	22 Feb 1663
158rv	From Parenti at Amsterdam 2pp (17x22)	15 Mar 1663
159r-161v	From [Parenti] at Amsterdam 3pp & 6 lines + ad (17x22)	5 Apr 1663
162rv&165rv	From [Parenti] at Amsterdam 2pp & 3 lines + ad (17x22)	19 Apr 1663
163r-164v	From Parenti at Amsterdam 3pp + ad (17x22)	10 May 1663
166r-167v	From Parenti at Amsterdam 2½pp + ad (17x22)	31 May 1663
168r-169v	From [Parenti] at Amsterdam 2½pp + ad (17x22)	15 Jun 1663
170r-171v	From Parenti at Amsterdam 2½pp + ad (17x22)	26 Jul 1663
172rv	From Parenti at Amsterdam 2pp (17x22)	20 Sep 1663
173rv	From Parenti at Amsterdam 2pp (17x22)	11 Oct 1663
174r-175v	News account 1p + ad (15x22) [fragment]	21 Sep 1663
176rv&181rv	From Parenti at Amsterdam 2pp & 5 lines + ad (17x22)	25 Oct 1663
177rv&180rv	From [Parenti] at Amsterdam 1p + ad (17x22)	9 Nov 1663
178r-179v	From Parenti at Amsterdam 2½pp + ad (17x22)	29 Nov 1663
182rv	From Parenti at Amsterdam 2pp (17x22)	13 Dec 1663

183rv	From Parenti at Amsterdam 1p (17x22)	27 Dec 1663
184r-185v	From Parenti at Amsterdam 2pp & 5 lines + ad (17x22)	10 Jan 1664
186rv	From Parenti at Amsterdam 1p & 6 lines (17x22)	24 Jan 1664
187r-188v	From Parenti at Amsterdam 2½pp + ad (17x22)	21 Feb 1664
189rv	From Parenti at Amsterdam 2pp (17x22)	6 Mar 1664
190r-191v	From Parenti at Amsterdam 2½pp + ad (17x22)	4 Apr 1664
192rv	From Parenti at Amsterdam 2pp (17x22)	17 Apr 1664
193rv	From Parenti at Amsterdam 1½pp (17x22)	1 May 1664
194r-195v	From Parenti at Amsterdam 2pp & 4 lines + ad (17x22)	29 May 1664
196rv	From Parenti at Amsterdam 1p (17x22)	12 Jun 1664
197rv	From Parenti at Amsterdam 1p & 5 lines (17x22)	26 Jun 1664
198rv	From Parenti at Amsterdam 1½pp (17x22)	10 Jul 1664
199rv	From Parenti at Amsterdam 1p (17x22)	17 Jul 1664
200rv&203rv	From Parenti at Amsterdam 2pp + ad (17x22)	31 Jul 1664
201r-202v	From Parenti at Amsterdam 2pp & 6 lines + ad (17x22)	14 Aug 1664
204rv&207rv	From Parenti at Amsterdam 1p + ad (17x22)	21 Aug 1664

205r-206v	From Parenti at Amsterdam 2pp & 7 lines + ad (17x22)	4 Sep 1664
208rv&213rv	From Parenti at Amsterdam $1\frac{1}{2}$pp + ad (17x22)	18 Sep 1664
209rv&212rv	From [Parenti] at The Hague 2pp & 4 lines + ad (17x22)	1 Oct 1664
210r-211v	From [Parenti] at Amsterdam-Hague 1p & 4 lines + ad (17x22)	15 Oct 1664
214rv&214rv bis	From Parenti at Amsterdam 1p & 6 lines + ad (17x22)	11 Dec 1664
215rv	From Parenti at Anvers 1p & 7 lines (17x22)	28 Nov 1664
216r-217v	From [Parenti] at Amsterdam 3pp + ad (17x22)	24 Dec 1664
218r-219v	From Parenti at Amsterdam 3pp + ad (17x22)	8 Jan 1665
220r-221v	From [Parenti] at Amsterdam 3pp + ad (17x22)	22 Jan 1665
222r-223v	From [Parenti] at Amsterdam 2pp & 6 lines + ad (17x22)	5 Feb 1665
224rv	From [Parenti] at Amsterdam 1p (17x22)	19 Feb 1665
225rv	From [Parenti] at Amsterdam 1p (17x22)	5 Mar 1665
226rv	From [Parenti] at Amsterdam 2pp (17x22)	19 Mar 1665
227r-228v	From [Parenti] at Amsterdam 2pp & 6 lines + ad (17x22)	9 Apr 1665
229rv	From [Parenti] at Amsterdam 1p (17x22)	23 Apr 1665
230rv	Newsletter Warsaw 1p (17x22) Italian	3 Apr 1665

231rv	From [Parenti] at Amsterdam 1p & 6 lines (17x22)	7 May 1665
232rv	From Parenti at Amsterdam 1p & 6 lines (17x22)	21 May 1665
233rv	From [Parenti] at Amsterdam 1½pp (17x22)	5 Jun 1665
234r-235v	From [Parenti] at Amsterdam 3½pp (17x22)	18 Jun 1665
236rv	From [Parenti] at Amsterdam 1½pp (17x22)	16 Jul 1665
237rv	From [Parenti] at Amsterdam 2pp (17x22)	30 Jul 1665
238rv	From [Parenti] at Amsterdam 2pp (17x22)	10 Sep 1665
239rv	From [Parenti] at Amsterdam 1½pp (17x22)	13 Aug 1665
240rv	From [Parenti] at Amsterdam 1p & 8 lines (17x22)	27 Aug 1665
241r-242v	From [Parenti] at Amsterdam 2pp & 9 lines + ad (17x22)	24 Sep 1665
243rv	Printed document 2pp (10x22) "Charge Generale de 11 Navires de Retour des/Indes Orientales, partis de Batavia le 24. Decembre/1664."	
244rv	Missing	
245r-246v	From [Parenti] at Amsterdam 2½pp + ad (17x22)	8 Oct 1665
247rv	From [Parenti] at Amsterdam 2pp (17x22)	22 Oct 1665
248r-249v	From [Parenti] at Amsterdam 2pp & 10 lines + ad (17x22)	5 Nov 1665

250r-251v	From [Parenti] at Amsterdam 2½pp + ad (17x22)	19 Nov 1665
252r-253v	Notes [news dated 10 May] 1p & 10 lines (7x22) (11x17)	
254r-255v	From [Parenti] at Amsterdam 3pp + ad (17x22)	3 Dec 1665
256r-257v	From [Parenti] at Amsterdam 3pp & 8 lines (17x22)	17 Dec 1665
258rv	From [Parenti] at Amsterdam 2pp (17x22)	14 Jan 1666
259rv&262rv	From [Parenti] at Amsterdam 2pp & 3 lines + ad (17x22)	28 Jan 1666
260rv	From [Parenti] at Amsterdam 2pp (17x22)	11 Feb 1666
261rv&263rv	From [Parenti] at Amsterdam 1p & 4 lines + ad (17x22)	25 Feb 1666
264r-265v	From [Parenti] at Amsterdam 2pp & 8 lines + ad (17x22)	11 Mar 1666
266rv	From [Parenti] at Amsterdam 2pp (17x22)	15 Jul 1666
267rv	From [Parenti] at Amsterdam 1p (17x22)	29 Jul 1666
268rv	From [Parenti] at Amsterdam 1p (17x22)	12 Aug 1666
269r-270v	From [Parenti] at Amsterdam 3pp + ad (17x22)	26 Aug 1666
271r-272v	Extract of a letter From Anonymous at The Hague 1½pp (17x22)	31 Jan 1667
273r-274v	News account/extract Léopol 2½pp (17x22) Extract of a letter of Mr Polanowski	 23 Aug [1672?]

275r-276v	News account Danzig 3½pp (17x22)	3 Sep 1672
277r-278v	News account Danzig 2pp & 4 lines (17x22)	17 Sep 1672
279r-280v	News account/extract 4pp (17x22)	1 Oct 1672 27 Aug 1672
281r-282v	News account 3½pp (17x22)	8 Oct 1672
283r-284v	News account Danzig 3pp (17x22)	29 [Oct?] 1672
285rv	Latin copy 2pp (17x22) Latin	undated
286r-287v	News account Danzig 4pp (17x22)	22 Oct 1672
288r-289v	News account Danzig 3½pp (17x22)	5 Nov 1672
290r-291v	News account Danzig 3½pp (17x22)	29 Nov 1672
292r-293v	News account Danzig 3½pp (17x22)	26 Nov 1672
294r-295v	News account Danzig 3pp (17x22)	2 Dec 1672
296r-297v	News account Danzig 4pp (17x22)	10 Dec 1672
298r-299v	News account Danzig 4pp (17x22)	17 Dec 1672

300r-301v	News account Danzig 3pp & 9 lines (17x22)	24 Dec 1672
302r-303v	News account Danzig 3½pp (17x22)	31 Dec 1672
304rv	News account Warsaw 1½pp (23x35) Italian	19 Aug 1672
305rv	News account Warsaw 2pp (23x35) Italian	12 Aug 1672
306r-307v	News account Danzig 4pp (17x22) [fragment]	14 Jan 1673
308r-309v	News account Danzig 2½pp (17x22)	21 Jan 1673
310r-311v	News account Danzig 2½pp (17x22)	7 Jan 1673
312r-313v	News account Lyon 3pp (17x22)	26 Oct 1663
314r-315v	News account Rome 4pp (17x22) [PS, Grenoble, 23 Oct 1663]	16 Oct 1663
316r-317v	News account Lyon 2pp & 9 lines (17x22)	1 Nov 1663
318r-319v	News account 3pp (17x22) Venice Lyon	 24 Nov 1663 15 Dec 1663

320r-321v	News account	
	4pp (17x22)	
	Rome	4 Dec 1663
	Genoa	12 Dec 1663
322r-323v	News account	
	4pp (17x22)	
	Rome	9 Oct 1663
	Genoa	15 Oct 1663
	Lyon	23 Oct 1663
324r-325v	News account	
	Lyon	20 Oct 1663
	4pp (17x22)	
326r-327v	News account	
	Lyon	29 Dec 1663
	2pp & 3 lines (17x22)	
328r-329v	From Viviani at Florence	15 Jan 1677 Ab Inc
	3¼pp (17x22) Italian	[15 Jan 1678]

▼ ▼ ▼

COLLECTION BOULLIAU, VOLUME VIII, FONDS FRANÇAIS 13026

1rv	To Peiresc at Aix 1p + ad (20x31)	13 Nov 1635
2r	To Hortensius at [Amsterdam] 1p (20x31) Latin	10 Jul 1636 OS [20 Jul 1636]
2v	To Blaeu at [Amsterdam] 1p (20x31) Latin	[10 Jul 1636]
3r-4v	Testament of Madame de Thou* 3pp (15x20) *Gasparde de La Chastre	5 Feb 1664
5rv	To Rivière at [Paris] B at [The Hague] ½p + ad (15x20)	undated
6rv&10rv	To [Rivière at Paris] B at The Hague 1¾pp + ad (17x22)	14 Jun 1657
7rv&11rv	To [Rivière at Paris] B at The Hague 2pp & 11 lines + ad (17x22)	17 May 1657
8r-9v	To Rivière at Paris B at The Hague 3pp + ad (17x22)	3 May 1657
12rv	To Anonymous B at Meslay 1½pp (13x19)	26 Oct 1649
13r-15v	To Rivière at Paris B at The Hague 4pp + ad (15x20)	20 Jun [1657]
16rv	To [Rivière at Paris] B at The Hague 2pp (17x22)	10 May 1657
17rv	To Rivière at Paris B at Le Havre de Grace 1p + ad (17x22)	9 Apr 1657

18rv	To [Rivière? at Paris] B at The Hague 1p & 9 lines (17x22)	7 Jun 1657
19rv	To Anonymous 1⅓pp (17x22)	30 Sep 1648
20rv	To Anonymous B at Meslay 1½pp (20x26)	20 Oct 1649
21rv	To Boecler at [Strasbourg] 1½pp (18x25) Latin/Greek	28 Jan 1654
22rv	To Boecler at [Strasbourg] 1p (13x17) Latin	13 Mar 1654
23rv	To Boecler at [Strasbourg] 2pp (18x25) Latin	17 Jul 1654
24r-25v	To Boecler at [Strasbourg] 2½pp (15x20) Latin	11 Sep 1654
26rv	To Boecler at [Strasbourg] 1p (17x22) Latin	Kal Jan 1655 [1 Jan 1655]
26rv bis	Torn fragment of fol 26 6 lines (5x15)	
27rv	To Boecler at [Strasbourg] 1½pp (17x22) Latin	23 Jul 1655
28rv	To Boecler at [Strasbourg] 2pp (13x17) Latin	25 Dec 1655
29rv	To Boecler at [Strasbourg] B at The Hague 1p & 6 lines (17x24) Latin	20 Jul 1657
30r-31v	To Boecler at [Strasbourg] 2½pp (15x20) Latin	19 Mar 1658
32r-33v	To Boecler at [Strasbourg] 2⅔pp (15x20) Latin	10 Jul 1658
34rv	To Boecler at [Strasbourg] ¾p (17x24) Latin	12 Apr 1659

35rv	To Boecler at [Strasbourg] 1p (17x22) Latin	15 Jul 1659
36rv	To Anonymous* 2pp (15x20) *[a Minister of England]	3 Oct 1674
37rv	To E Bernard at Oxford 1p (13x17) Latin	11 Nov 1683
38rv	Fragment 8 lines (7x8) Latin	
39rv	Greek transcription in B's hand 14 lines (17x21) Greek/Latin "Ex Eruditissimi Viri Isaaci Vossii MS. Collect. Mathem./Pappi Alexandrini Lib. 7..."	undated
40rv	To Hortensius at Amsterdam 2pp (20x31) Latin	13 Aug 1638
41rv	To Gradius at [Lyons?] 1p (17x21) Latin	12 Sep 1664
42rv	To Gradius at 1p (17x21) Latin	14 Jan 1667
43rv	From Gradius at Rome 1½pp (15x20) Latin	V Kal Oct 1666 [27 Sep 1666]
44rv	From Gradius at Rome 1½pp (15x20) Latin	28 Oct 1664
45r-46v	From Gradius at Lyons 2pp (15x20) Latin/Italian/Greek [fol 46, list]	VII Kal Sep 1664 [26 Aug 1664]
47rv	To Galileo at Florence 1p (17x24) Latin Publ, GGO, XVII, pp 207-208, No 3588	30 Oct 1637
48rv	To Forstner at [Montbeliard?] 1p (15x20) Latin	29 Sep 1656
49rv	To Forstner at [Montbeliard?] 1p & 6 lines (17x22) Latin	15 Jul 1659

Folio	Description	Date
50rv	To Forstner at [Montbeliard?] 2pp (15x20) Latin	26 Jan 1657
51r-52v	To Fermat at [Toulouse] 1p (15x20) Latin fol 52, blank Publ, FO, II, pp 336-337, No 81bis	Mar 1657
53r-54v	To Forstner at [Montbeliard?] 2pp & 8 lines (15x20) Latin	8 Jul 1658
55rv	To Graevius at [Rhenen?] 1p (15x22) Latin	9 Dec 1672
56r-57v	Printed Text* 2pp (17x22) Latin/Hebrew "Excerpta Germara" *[Chpt 9, pp 1033, 1034] fol 57, blank	
58r-63v	Manuscript dissertation Cadomi [Caen] Author: Etienne Morin 11½pp (11x17) Latin/Greek	6 Kal Mar 1673 [25 Feb 1673]
64rv	To Graevius at [Rhenen?] 1p & 7 lines (17x22) Latin Publ extract, BC, II, p 454	23 Mar 1674
65rv	To Graevius at Rhenen 1½pp (17x22) Latin	16 Aug 1679
66rv	To E Bernard at Oxford 1½pp (17x22) Latin	8 Jul 1679
67rv	To Graevius at [Rhenen] 1p (17x22) Latin	5 Jul 1679
68rv	To Graevius at [Rhenen] 1p (17x22) Latin	22 Oct 1680
69r-70v	To Gruter at Rotterdam 1p (17x22) Latin fol 70, blank	16 Jan 1665
71rv	To Gruter at [Rotterdam] 1p & 9 lines (17x22) Latin	3 Jan 1668

72rv	To Gruter at [Rotterdam] 1p & 5 lines (15x20) Latin	28 Sep 1671
73rv	To Ravius at Uppsala 2pp (17x22) Latin	28 Sep 1671
74rv	To Titel at Leipsig 1p (17x22) Latin	8 Aug 1676
75rv	To Langius at [Copenhagen] 1½pp (15x20) Latin	undated
76rv	To Müller at Hamburg B at Danzig 1p (15x20)	16 Apr 1661
77rv	To Pastorius at [Danzig] 1p & 12 lines (17x22) Latin	9 Mar 1663
78rv	To Schickard at Tübingen 1p (15x29) Latin	Dominica Quasimodo genti 1635 [15 Apr 1635]
79r-80v	To [J-A de Thou at The Hague] B at Danzig 4pp (15x20)	26 Mar 1661
81r	Copy of a letter in B's hand From Caspar Peucer To Martin Schilling 1p (17x22) Latin	28 Jul 1591
81v-82v	Copy of a letter in B's hand From "Erasmus" at Klausenburg To Caspar Peucer 2¼pp (18x24) Latin	9 Dec NY
82v-83v	Copy of a letter in B's hand From Caspar Peucer To "Erasmus" at Klausenburg 2pp (18x24) Latin	undated
83v-84v	Extracts of letters in B's hand 3pp (18x24) Latin "Ex epistola Melancthonis ad Petrum Vincentium gubernatorem/studii Lubecensis die x Junii (Julii)"	

"Ex ejusdem epistola ad Gasparum Aquilam pastori/Ecclesiae Lutheranae Salueldiensis, Jena scripta"
"Ex ejusdem epistola ad Christophorum Libium/pastorem Ecclesiae Lutheranae Tangermundae scripta/18. April. 1558"
"Ex alia ad eundem Mai 20. 1558."
"Ex epistola eiusdem Philippi ad eundem Libium Novembr./13. 1557"
"Ex epistola ad Abdiam Praetorium April. 3."
"Ex epistola ad Georgium Buccholzerum Martii 25 mundi 5521."
"Ad Matthaeum Collina Magistrum in Academia Pragensi/prid. Kal. Maii 1556" etc.
"Ex epistola Vitum Theodorum, Augusta Vindelic. scripta/anno 1530 per tempus comitiorum, post Dominicam Trinitatis"

84v-85r Copies of letters in B's hand
2pp (17x22) Latin
"Henricus Dei grati Angliae Rex et Franciae ...Philippo Melancthoni sacrae Theologiae professori..." [ex Regia nostra Wintoniae 1. Octob. 1535]
"Serenissimo Principi Dno Henrico VIII. Angliae et Franciae/Regi, Hiberniae Dno, Walliae et Cornubiae Principi/Domino Clementissimo" [mense Augusto 1535. R.M.T. addictissimus Ph. Melanchton]

85v-87v Extracts of letters in B's hand
4pp (17x22) Latin
"Clarissimo viro, eruditione, et virtute praestanti Thadaeo/Hageko verae Philosophiae cultori, amico suo carissimo" [Philippus Mel.]
"Andreae Kegelio amico suo" [26 Aprilis]
"ex epta ad Philippum Gluespies. Oct. 19. 1542"
"ex epistola Christoph. Meinburg ad fratrem Philippo praescrib. 1548"
"ex epistola Rectoris et Senatus Academiae Wittebergensis/Caspar Eberhardo in vallibus Joachimi"
"Ex epistola Phil. Mel. ad quendam scripta Gotae VIII Kal./April 1547"
"ex epta Phil. Melan. Danieli Greisero in Dresda die/Laurentii anno 1547"

	"ex epistola ejusdem ad Chilianum Goldstein, Augustae die Ascension/ 1530"	
88r-89v	Copy of a letter in B's hand From Godefroy To the Duke of Longueville 3pp (18x25)	undated
90r-91v	News of Poland in B's hand Danzig Danzig 2pp & 6 lines (15x20)	 22 Apr 1673 29 Apr 1673
92r-93v	Extract, Philolaic Tables 2pp & 5 lines (15x20)* Latin *fol 93, short note (4x13) cipher	29 Oct 1678
94rv	To J-A de Thou at [Paris?] 1p (20x28) Latin	III Kal Sep [1637] [30 Aug 1637]
95r-96v	Preface, <u>Theonis Smyrnaei</u>, in B's hand 2½pp (21x30&21x21) Latin	
97rv	Astronomical Tables in B's hand 1p (21x35) Latin "Anno 1660 Aprilis die 25 Hora 11.'16."45./ Sequens observatio Gedani facta est ab Illmo. Viro/Dno. Johanne Hevelio."	
98rv	Copy of a dissertation on theology at Wittenburg 1½pp (21x35) Latin/Greek	21 Mar 1560
99r-101v	To Hevelius at Danzig 4pp (21x30) Latin fol 101, blank	18 Apr 1650
102rv	To Hevelius at Danzig B at The Hague 1p (17x22) Latin	25 Oct 1660
103rv	To Hevelius at Danzig B at The Hague 1p & 6 lines (17x22) Latin	29 Dec 1660

103rv bis	To Hevelius at Danzig B at Amsterdam 14 lines (17x22) Latin	26 Jan 1661
104rv	To Hevelius at Danzig B at Warsaw 1⅔pp (17x22) Latin	9 May 1661
105rv	To Hevelius at Danzig B at Warsaw 1p & 11 lines (17x22) Latin	16 May 1661
106rv	To Hevelius at Danzig B at Warsaw 1p & 6 lines (17x22)	29 May 1661
107rv	To Hevelius at Danzig B at Warsaw 2pp (17x22) Latin	27 Jun 1661
108rv	To Hevelius at Danzig B at Warsaw 1p (17x22) Latin	11 Jul 1661
109rv	To Hevelius at Danzig B at The Hague 2pp (17x22) Latin	14 Nov 1661
110rv	To Hevelius at Danzig B at The Hague 2pp (17x22) Latin	12 Dec 1661
111rv	To Hevelius at Danzig B at The Hague 1½pp (17x22) Latin	26 Dec 1661
112rv	Copy of a letter in B's hand From Théodore Moret To Hevelius at Danzig 1½pp (17x22) Latin	6 Jan 1662
113r-114v	To Hevelius at Danzig B at The Hague 2pp & 15 lines (17x22) Latin	15 Mar 1662
115rv	To Hevelius at Danzig B at The Hague 1p (17x22) Latin	28 Mar 1662

116rv	To Hevelius at Danzig B at The Hague 1p (17x22) Latin	3 Apr 1662
117rv	To Hevelius at Danzig B at The Hague 2pp (17x22)	19 Jun 1662
118rv	To Hevelius at Danzig 1½pp (17x22) Latin	20 Jul 1662
119rv	To Hevelius at Danzig 1p & 12 lines (17x22) Latin	3 Aug 1662
120r-121v	To Hevelius at Danzig 2pp & 12 lines (17x22) Latin	6 Dec 1662
122r-123v	To Hevelius at Danzig 3½pp (17x22) Latin	9 Mar 1663
124rv	To Hevelius at Danzig 1p & 11 lines (17x22) Latin	27 Apr 1663
125rv	To Hevelius at Danzig 1p & 5 lines (17x22) Latin	19 Oct 1663
126rv	To Hevelius at Danzig 1½pp (17x22) Latin	6 Jul 1663
127r-128v	To Hevelius at Danzig 2½pp (17x22) Latin	10 Aug 1663
129rv	To [Hevelius] at Danzig 1⅔pp (17x22) Latin	22 Feb 1664
130rv	To Hevelius at Danzig 2pp (17x22) Latin	9 May 1664
131rv	To Hevelius at Danzig 1⅔pp (17x22) Latin	29 Aug 1664
132r-134v	To Hevelius at Danzig 5½pp (17x22) Latin *[final draft dated 18 Nov 1678]	undated*
135rv	To Hevelius at Danzig 1p (17x22) Latin	9 Jan 1665

136rv	To Hevelius at Danzig 1½pp (17x22) Latin	29 May 1665
137rv	To Hevelius at Danzig 1p (17x22) Latin	5 Jun 1665
138rv	To Hevelius at Danzig ¾p (17x22) Latin	19 Jun 1665
139r-140v	To Hevelius at Danzig 2⅓pp (17x22) Latin	17 Jul 1665
141r-142v	To Hevelius at Danzig 3pp (17x22) Latin	25 Sep 1665
143r-144v	To Hevelius at Danzig 2pp (17x22) Latin fol 144, illustration, 1p (15x50)	6 Nov 1665
145r-146v	To Hevelius at Danzig 2pp & 6 lines (17x22) Latin	15 Jan 1666
147rv	To Hevelius at Danzig 1p (17x22) Latin	5 Mar 1666
148rv	To Hevelius at Danzig 2pp (17x22) Latin	21 May 1666
149rv&152rv	To Hevelius at Danzig 2pp & 5 lines (17x22) Latin	16 Jul 1666
150r-151v	From Hevelius at Danzig 2pp & 3 lines (10x15) Latin	18 Dec 1666
153rv	To Hevelius at Danzig 2pp (17x22) Latin	18 Feb 1667
154r-155v	To Hevelius at Danzig 4pp (17x22) Latin	26 Nov 1666
156r-157v	To Hevelius at Danzig 3pp (17x22) Latin	15 Apr 1667
158r-159v	To Hevelius at Danzig 2½pp (17x22) Latin	2 Dec 1667
160rv	To Hevelius at Danzig 2pp (17x22) Latin	15 Jun 1668

161r-163v	To Hevelius at Danzig 3pp (10x15) Latin fols 161&162, pasted to fol 163 *w/PS, 24 Aug	17 Aug 1668*
164rv	To Hevelius at Danzig 1$\frac{3}{4}$pp (17x22) Latin	19 Oct 1668
165r-166v	To Hevelius at Danzig 3$\frac{1}{2}$pp (15x20) Latin	11 Jan 1669
167rv	To Hevelius at Danzig 2pp (20x29) Latin	12 Apr 1669
168r-169v	To Hevelius at Danzig 3$\frac{3}{4}$pp (15x20) Latin	26 Aug 1670
170r-174v	Astronomical observations & drawings Stellar diagram by B 1p (15x15) Latin 1670, 20 Aug - Saturn, drawing by Hevelius 1p (15x20) Star positions comp w/Tycho's observations; data in H's hand 3pp (10x15) Latin Drawing of Cygni, 25 Jul 1670; Novas of 1600, 1670 1p (15x20) Latin	undated
175r-176v	To Hevelius at Danzig 2$\frac{1}{2}$pp (15x20) Latin	31 Dec 1670
177r-178v	To Hevelius at Danzig 3pp (20x29) Latin	6 Oct 1671
179r-181v	Astronomical calculations in B's hand 2pp & 10 lines (21x29) Latin "Calculus ex Tabulis Philolaicis Loci ☾ veri ad diem 22 Aprilis H 10'47" 56. T.A. Dantisci 1671" "Loca ⊙ et ☾ computata ex Tabulis Philolaicis..." [22 Apr 1671]	
182rv	To Hevelius at Danzig 1$\frac{1}{2}$pp (17x22) Latin	31 Mar 1672
183r-184v	To Hevelius at Danzig 2pp & 4 lines (17x22) Latin	24 May 1672

185rv	To Hevelius at Danzig 1p (17x22) Latin	23 Jun 1672
186rv	[Copy of a letter?] in H's hand From Hevelius at Danzig To Anonymous 1½pp (10x15) Latin	23 Jun [1672]
187rv	To Hevelius at Danzig 1p (17x22) Latin	11 Aug 1672
188rv	To Hevelius at Danzig 2pp (17x22) Latin	9 Dec 1672
189r-190v	To Hevelius at Danzig 3pp & 9 lines (17x22) Latin	5 Jan 1674
191r-192v	To Hevelius at Danzig 3½pp (17x22) Latin	25 May 1674
193r-194v	To Hevelius at Danzig 2pp & 6 lines (17x22) Latin	18 Jan 1675
195r-196v	To Hevelius at Danzig 2⅔pp (17x22) Latin	19 Apr 1675
197r-198v	To Hevelius at Danzig 4pp (18x27) Latin	20 Dec 1675
199r-200v	To Hevelius at Danzig 3⅔pp (18x27) Latin	11 Jun 1677
201r-202v	To Hevelius at Danzig 3pp (17x22) Latin	7 Oct 1679
203rv	To Hevelius at Danzig 2pp (10x15) Latin	21 Jun 1680
204rv	To Hevelius at Danzig 1p (17x22) Latin	18 Apr 1681
205rv	To Hevelius at Danzig 1p & 9 lines (17x22) Latin	16 May 1681
206rv	List of books in B's hand 1p (13x17) French/Latin	[ca 1666]
207rv	To Hevelius at Danzig 1p (17x22) Latin	29 Aug 1684

208rv	To Hevelius at Danzig 1p & 4 lines (10x15) Latin	undated
209rv	To Madame [Anonymous] 1p (17x22)	24 Oct 1655
210rv	To [J-A de Thou at Meslay] 2pp (17x22)	28 Oct 1655
211r-213v	To J-A de Thou at Meslay 3pp & 4 lines + ad (17x22)	31 Oct 1655
214r-215v	To [J-A de Thou at Meslay] 3pp (17x22)	4 Nov 1655
216rv	To [J-A de Thou at Meslay] 2pp (17x22)	7 Nov 1655
217r-219v	To J-A de Thou at Meslay 3pp & 4 lines + ad (17x22)	11 Nov 1655
219rv bis	To [J-A de Thou at Meslay] 2pp (17x22)	14 Nov 1655
220r-221v	To J-A de Thou at Meslay 3pp + ad (17x22)	18 Nov 1655
222rv	To [J-A de Thou at Meslay] 1p (17x22)	20 Nov 1655
223rv	To [J-A de Thou at Meslay] 2pp (17x22)	21 Nov 1655
224r-225v	To J-A de Thou at Meslay 2½pp + ad (17x22)	25 Nov 1655
226r-227v	To J-A de Thou at Meslay 3pp + ad (17x22)	27 Nov 1655
228rv	To [J-A de Thou at Meslay] 1p (17x22)	28 Nov 1655
229r-230v	To [J-A de Thou at The Hague] 4pp (17x22)	21 Feb 1659
231r-232v	To [J-A de Thou at The Hague] 4pp (17x22)	7 Mar 1659

233r-234v	To J-A de Thou at The Hague B at Danzig 3pp + ad (17x22)	16 Apr 1661
235r-236v	To J-A de Thou at The Hague 2½pp + ad (17x22)	23 Apr 1661
237r-238v	To J-A de Thou at [The Hague] B at Warsaw 3pp (17x22)	4 Jul 1661
239r-240v	To Rivière at Paris B at The Hague 2½pp + ad (15x20)	28 Jun 1657
241rv	To [Rivière at Paris?] B at The Hague 1½pp (15x20)	19 Jul [1657]
242rv	To [Rivière at Paris?] B at The Hague 1p (17x22)	14 Jun 1657
243r-244v	To [Rivière at Paris?] B at The Hague 2pp & 5 lines (15x20)	12 Jul 1657
245rv	Book list ½p + ad (17x22)	undated
245rv bis 246rv ter	To J-A de Thou at The Hague B at Warsaw 3pp + ad (17x22)	20 Jun 1661
246r-247v	To J-A de Thou at The Hague B at Warsaw 1p & 11 lines (17x22)	11 Jul 1661
248r-249v	To J-A de Thou at The Hague B at Amsterdam 1p + ad (17x22)	13 Apr 1662
250r-251v	To Lubienietzki at [Hamburg] 2⅔pp (17x22) Latin	24 May 1663
252r-253v	To Lubienietzki at [Hamburg] 3pp (17x22) Latin	[28?] Jul 1663

254r-255v	To Lubienietzki at [Hamburg] 2½pp (15x20) Latin *[PS, 10 lines, 3 Apr 1663]	30 Mar 1663*
256rv	To Lubienietzki at [Hamburg] 1¾pp (17x22) Latin	28 Sep 1663
257rv	To Lubienietzki at [Hamburg] 2pp (17x22) Latin	14 Dec 1663
258rv	To Lubienietzki at [Hamburg] 1p & 6 lines (17x22) Latin	25 Jan 1664
259rv	To Lubienietzki at [Hamburg] 2pp (17x22) Latin	16 May 1664
260r-261v	To Lubienietzki at [Hamburg] 2½pp (17x22) Latin	4 Jul 1664
262r-263v	To Lubienietzki at [Hamburg] 1½pp (13x17) Latin fol 263, blank	8 Aug 1664
264rv	To Lubienietzki at [Hamburg] 1½pp (17x22) Latin	5 Sep 1664
265rv	From Lubienietzki at Hamburg 1⅔pp (13x17) Latin	16 Aug 1664
266rv	To Lubienietzki at [Hamburg] 1⅔pp (17x22) Latin	26 Sep 1664
267rv&270rv	To Lubienietzki at [Hamburg] 2pp & 4 lines (17x22) Latin	10 Oct 1664
268r-269v	From Lubienietzki at Hamburg 3pp & 6 lines (10x15) Latin	20 Sep 1664
271r-272v	To Lubienietzki at [Hamburg] 2½pp (17x22) Latin	[ca Jun 1664]
273r-274v	To Lubienietzki at [Hamburg] 2½pp (17x22) Latin	21 Nov 1664
275r-276v	To Lubienietzki at [Hamburg] 2pp & 2 lines (10x15) Latin	31 Oct 1664
277rv	To Lubienietzki at [Hamburg] 1p (17x22) Latin	5 Dec 1664

278rv	To Lubienietzki at [Hamburg] 1p (17x22) Latin	19 Dec 1664
279r-282v	To Lubienietzki at [Hamburg] 2pp & 5 lines (17x22) Latin fols 281 & 282, blank	6 Nov 1665
283rv	To Lubienietzki at [Hamburg] 1¾pp (17x22) Latin	23 Jan 1665
284rv	To Lubienietzki at [Hamburg] 1⅔pp (17x22) Latin	13 Feb 1665
285r-286v	To Lubienietzki at [Hamburg] 2⅔pp (17x22) Latin	6 Mar 1665
287r-288v	To Lubienietzki at [Hamburg] 2¼pp (17x22) Latin	3 Apr 1665
289r-290v	To Lubienietzki at [Hamburg] 2⅔pp (17x22) Latin	Kal May 1665 [1 May 1665]
291r-292v	To Lubienietzki at [Hamburg] 2½pp (17x22) Latin	22 May 1665
293rv	To Lubienietzki at [Hamburg] 1p (17x22) Latin	5 Jun 1665
294r-296v	To Lubienietzki at [Hamburg] 2pp & 13 lines (16x20) Latin fol 296, blank	26 Jun 1665
297rv	To Lubienietzki at [Hamburg] 2pp (17x22) Latin	31 Jul 1665
298rv	To Lubienietzki at [Hamburg] 1¾pp (17x22) Latin	14 Aug 1665
299rv	To Lubienietzki at [Hamburg] 1½pp (17x22) Latin	18 Sep 1665
300r-301v	To Lubienietzki at [Hamburg] 2pp & 4 lines (17x22) Latin	16 Oct 1665
302rv	To Lubienietzki at [Hamburg] 1p (17x22) Latin	11 Dec 1665

303rv	Note concerning observations of Hipparchus* 1½pp (17x22) Latin *enclosed w/fol 302	undated
304r-305v	To Lubienietzki at [Hamburg] 2pp & 4 lines (12x17) Latin	19 Feb 1666
306rv	To Lubienietzki at [Hamburg] 1p & 6 lines (17x22) Latin	16 Jul 1666
307rv	To Lubienietzki at [Hamburg] 2pp (17x22) Latin	13 Aug 1666
308r-310v	To Lubienietzki at [Hamburg] 2⅔pp (15x20) Latin fol 310, blank	31 Mar 1669
311rv	To Viviani at [Florence?] 1½pp (17x22) Latin	7 Jun 1683
312rv	To Viviani at [Florence?] ⅔p (15x20) Latin	Kal Oct 1683 [1 Oct 1683]
313r-318v	To Viviani at [Florence?] 4½pp (17x22) Latin fols 316, 317, & 318, blank Publ extract, MC, Vol II, p 390	16 Mar 1678
319r-320v	To [Mazarin at Paris] 3⅔pp (15x20)	[1659]
321rv	To Colbert at [Paris] 2pp (17x22)	[ca 1662?]
322rv	To Mazarin at [Paris] 1½pp (17x22)	17 Mar 1657
323rv	To [J-A de Thou at The Hague] B at Danzig 1p & 1 line (15x20)	26 Mar 1661
324r-326v	To Rivière at Paris B at The Hague 2⅔pp + ad (15x20) fol 326, blank	24 May 1657

327rv	To Mademoizelle* 1p (10x15) *[on behalf of J-A de Thou]	undated
328rv&330rv	To [Rivière] at Paris B at Pontoise 1p + ad (17x22)	2 Apr [1657]
329rv&331r-332v	To Rivière at Paris B at Rotterdam 2pp + ad (17x22) fol 332, blank	23 Apr 1657
333r-334v	Copy of a letter in B's hand From J.G. Gallet at Avignon To G.D. Cassini at [Paris] 4pp (17x26) Latin	21 Nov 1677
335rv	Fragment in B's hand 20 lines (15x20) Latin	undated
336rv	Fragment Extract from unidentified treatise. Lib. X, 375 1p (19x26) Italian	undated

▼ ▼ ▼

COLLECTION BOULLIAU, VOLUME IX, FONDS FRANÇAIS 13027

1r-2v	Dedication letter, *De Lineis Spiralibus*, in B's hand To [Henrico Borbonio] 1p & 3 lines* (15x20) Latin *fol 2, 4 line fragment (10x20)	Kal Dec 1654 [1 Dec 1654]
3r-4v	Ad Lectorem Praefatio, *De Lineis Spiralibus*, in B's hand 2½pp (15x20) Latin	
5rv	To Leopoldo at [Florence] ½p (12x16) French/Latin [minute fragment]	undated
6rv	To Leopoldo at Florence 1p (17x22) Latin Publ, BAF, pp 110-111; cf. fn 3, p 110	7 Jan 1667
7rv	To Leopoldo at Florence ¾p (17x22) Latin	14 Jan 1667
8r-9v	Observations on the *Exercitatione Geometrica* of Michelangelo Ricci by B, in his hand 3pp (17x22) Latin [enclosed w/fol 7] Publ, LIF, I, pp 145-150	undated
10r-11v	To Leopoldo at Florence 2⅔pp (17x22) Latin	23 Jul 1666
12rv	To Leopoldo at Florence 2pp (17x22) Latin	4 Sep 1665
13rv	To Leopoldo at Florence 1⅔pp (17x22) Latin	10 Jul 1665
14rv	To Leopoldo at Florence 1½pp (17x22) Latin	24 Apr 1665
15rv	To Leopoldo at Florence 1p (17x22) Latin	9 Mar 1665

16r-17v	To Leopoldo at Florence B at Danzig 2½pp (15x20) Latin Publ, LIF, II, pp 205-209	5 Apr 1661
18r-19v	To Leopoldo at Florence B at The Hague 2pp (15x20) fol 19, blank	15 Dec 1661
20rv	Extract of a letter in B's hand ⅔p (17x22)	29 Oct 1660
21rv	To Leopoldo at Florence ½p (17x22) Latin	1 Apr 1670
22rv	To Ferdinando at Florence 1p (15x20) Latin	3 Jan 1670
23rv	To Leopoldo at Florence ¾p (17x22) Latin	12 Feb 1665
24rv	To Leopoldo at Florence 1p & 11 lines (17x22) Latin	30 Jan 1665
25rv	To Leopoldo at Florence 1p & 2 lines (17x22) Latin	29 Aug 1664
26rv	To Leopoldo at Florence 1p & 10 lines (17x22) Latin	24 Oct 1664
27r-28v	To Leopoldo at Florence 2⅓pp (15x20) Latin	4 Jun 1664
29rv	To Leopoldo at Florence 1½pp (17x22) Latin Publ, LIF, II, pp 210-212	10 Aug 1663
30rv	To Leopoldo at Florence ⅔p (17x22) Latin	18 May 1663
31r-32v	To Leopoldo at Florence 1⅔pp (17x22) Latin fol 32, blank	30 Aug 1662
33r-34v	Copy of a letter in B's hand 1p (15x20) Latin 1p (8x14) Latin* *[fragment in pencil]	

	[B writes: "Po^{r} M. Daubray Lieutenant Civil respondant/aux Paranymphes des Bacheliers en Théologie/le 14e Janvier 1670"]	
35rv	To Leopoldo* at Florence 1½pp (20x28) Latin *[Cardinal de' Medici]	10 Apr 1669
36rv	From Leopoldo* at Florence B at Florence 1p (20x28) Italian *[Cardinal de' Medici]	26 Apr 1669
37rv	To Leopoldo at Florence 1p (17x22) Latin	Kal Jan 1668 [1 Jan 1668]
38rv	To Leopoldo* at Florence 1p (17x22) Latin *[Cardinal de' Medici]	14 Oct 1672
39rv	To Leopoldo* at Florence 1p & 10 lines (17x22) Latin *[Cardinal de' Medici]	12 Aug 1672
40rv	To Anonymous ¾p* (17x22) *data on verso	22 Oct 1676
41rv	News account in B's hand Danzig 1p & 3 lines (17x22)	 24 Jun 1673
42rv	News account in B's hand Warsaw 1⅔pp (17x22)	 22 May 1674
43rv	To [De Boüillon at Blois] 2pp (17x22)	10 Aug 1653
44r-45v	Printed text, "Epistola/ad Fautorem et Amicum/De/Duodecim Saturni occultationibus...1677...1678" 4pp (14x17) Latin/German [author: Gottfried Kirch]	
46rv	List of treaties (1623-1656) 1p (17x22)	

47rv	To Pinocci at The Hague 1¼pp (17x22)	19 Sep 1658
48r-49v	To Anonymous 2pp & 5 lines (11x17)	undated
50rv	Extracts in B's hand 2pp (17x22) Latin	
	From Elenora at Vienna To Petro Ernesto [à Mercy Abbati Asseyensi]	31 Dec 1641
	"Extraict d'une lettre de Mad.e la Duchesse de Mantoue escrite a M. Zucconi le 13 Juin 1642."	
	From Louis de Gonzague To Princess Marie at Vienna	9 Mar 1644
	"Du mesme dans laquelle il parle du mariage du Roy de Pologne Vladislas" at Vienna	22 Jun 1644
	"Extraict d'une lettre du mesme a la mesme" at Vienna	1 Nov 1644
51r-53v	Extracts in B's hand 5pp (17x22) French/Italian	
	"Lettre du Marquis de Mantoue [Louis de Gonzague] a la Princesse Marie" at Vienna	16 Aug 1645
	"Extraict des lettres du Sr Roncalli a Mad. la Princesse Marie" at Warsaw	27 Mar 1645
	"Extraict des lettres du Sr Roncalli a Mad. la Princesse Marie" at Warsaw	27 Apr 1645
	"Du Mesme a M. le Card. Mazzarini"	18 May 1645
	"Del Theophilo Forni" at Warsaw	14 Jun 1645
	"Del Roncalli a S.A. Serma" at Warsaw	21 Jun 1645
	"Del Roncalli a S.A. Serma" at Warsaw	21 Jun 1645
	"Del Roncalli a S.A. Serma" at Warsaw	28 Jun 1645
	"Del Roncalli a S.A. Serma" at Warsaw	4 Jul 1645
	"Del Roncalli a S.A. Serma" at Warsaw	12 Jul 1645
	"Del Roncalli a S.A. Serma" at Warsaw	14 Jul 1645
	"Del Roncalli a S.A. Serma" at Warsaw	19 Jul 1645
	"Del med.mo a un sig.re" at Warsaw	19 Jul 1645
54r-61v	Extracts in B's hand 14pp (17x22) French/Latin	
	"Lettera de Sigr Domenico Roncalli a Mad. la Prpẽssa Maria" at Danzig	1 Aug 1645
	"Lettera de Sigr Domenico Roncalli a Mad. la Prpẽssa Maria" at Danzig	1 Aug 1645
	"Extraict des lettres de M. De Bregi	

	Flecellas touchant la/négotiation quil a faicte en Pologne, pour le mariage de/la Ser.me Reine de Pologne avec le Roy Vladislas"	
	From De Brégy at The Hague To [Madame]	29 May 1645
	From De Brégy [De Flécelles De Brégy] at Hamburg To Madame	17 Jun 1645
	"Lettre du Comte Denhoff à M. De Bregy" at [Danzig]	[21 Jun 1645]
	From De Brégy [De Flécelles Brégy] at Danzig To Madame	21 Jun 1645
	From [De Brégy] at Warsaw To Madame	6 Jul 1645
	From [De Brégy] at Warsaw To Madame	13 Jul 1645
	From [De Brégy] at Warsaw To Madame	14 Jul 1645
	From [De Brégy] at Warsaw To Madame	19 Jul 1645
	From [De Brégy] at Warsaw To S. Em.ce	19 Jul 1645
	From [De Brégy] at Warsaw To Mme V.A.	31 Jul 1645
	"De Mr le Comte de Donhoff" [to Mme la Princesse Marie] "En une lettre de Madame la Princesse a Madame la Princesse Marie du 18 Septembre 1645..." "En un autre du 23. d'Aoust"	15 Sep 1645
62rv	Copy of a letter in B's hand 2pp (17x22) Latin/French "Copie prise sur l'original de la lettre/escrite par l'Emp.r a Madame la Grand/Duchesse de Toscane 9 [Avril] 1676// Sur la mort de l'Imp.ce" [Leopoldus] Copy of a letter in B's hand "Copie de la lettre de l'Imp.ce Douairiere Eleonora/escrite a Mad. la Gr. Duchesse de Toscane sur la/subject de la mort de l'Imp.ce Claudia Felica, aussi/prise sur l'original// .9 Avril 1676" [Eleonora, Vienna]	

63rv

Copy of a letter in B's hand
1⅓pp (17x22) Italian
"Parte presa nell' Eccellen.mo Senato alli 25 di Gennaro 1675. Venetia."
[Alessandro Businello Seg.o]

64r-67v

Copy of a letter
6⅓pp (17x22)
"Du•l•Juillet 1666. de la Rade de la Rochelle/escrit par Mr. Verjus Sec.re de la Ser.me de Portugal"

68r-71v

Extracts in B's hand
7⅓pp (17x22) Latin
"Ex Exegesi Historica rerum Suecicarum post obitum/Regis Johannis III"

72rv

Copy in B's hand
1p & 10 lines (17x22)
"Declaration du Roy d'Angleterre Charles II. ou/Proclamation contenant la proscription de tous Jesuites/& Prestres hors des Royaumes d'Angleterre, Irlande, et Escosse/publiee le Mars 1673."

73rv — Jun 1568

Copy of a letter in B's hand
From Johannes Sommerus
To Michaeli Petraeo Dantiscano
2pp (18x25) Latin/Greek
[B notes: "Huius Sommeri Poemata impressa sunt in vol. paruo in 12° cuius Titulus, Delitiae Poetarum Hungaricorum,...1619 apud Nicol. Hoffman."]

74rv

Copy of a letter in B's hand
From [Johannes Sommerus]
To Christophorum Hanisium
2pp (18x25) Latin/Greek
[Bistricio Transsylvanorum, die Andreae Apostoli 1568]

75rv

Copy of a letter in B's hand
From Francisus Hotomannus
To Melanchthoni
1p (18x25) Latin
"Melanchthoni parenti meo venerando/ Franciscus Hotomannus S.D." [Argentinae, IX Cal. Junias]

76r-77v	Copy of a letter in B's hand 3⅔pp (20x31) Latin "Johannes de Lipielipsky Archiepũs Gnesnensis Legatus natus Regni Poloniae ...[to] Claudio de Mesmes Comiti d'Avaux ...[Lubaviae, III Jan 1639]	
78r-79v	Copy of a letter From Abrahamus Johnsonius To Pomponio de Bolliture 2⅔pp + ad (20x31) Latin [Envelope reads:] "Nobilissimo Viro Pomponio de Bolliture/in Supremo Galliarum Senatu Praesidi/Regi Christianissimo a Secretioris. Consiliis/Et Eiusdem ad Serenissimum Magnae/Brittinnae..."	undated
80rv	Note in B's hand 1p & 4 lines (20x31) "Leonardo Donato Doze di Venetia" [concerns death of the Doge of Venice, Lenard Donato, at 6:00pm, 16 Jul]	
81r-82v	Copy of a letter in B's hand 2⅔pp (20x31) Latin "Caroli Ogerii epistola ad Nicolaum Borbonium/Mariaeburgi Prussiae VII. Kal. Octob. 1635."	
83rv	Copy of a letter in B's hand 1½pp (20x31) Latin/French "Copie d'une lettre escrite de Strasbourg à Mr Portner à Ratisbone par/un sien ami" [Dat. 23. Augusti St. Vet. anno 1655]	
84r-85v	To [Des Noyers at Warsaw] 1¾pp (17x22) Latin/French fol 85, blank	3 Jan 1681
86rv	Account in B's hand 1⅔pp (20x31) [concerns J-A de Thou and François-Auguste de Thou, the latter arrested 12 Sep 1642 at Lyon] fol 86v, astron calculations	Dec 1642
87rv	To Anonymous 1p (14x16) *[10 Sep at bottom]	2 Sep 1681*

88rv	Copy of a letter in B's hand 1¾pp (20x31) Latin "Thaddaei Memici Hayko ab Hagiek/Et pietate et eruditione Excellente D. Philippo Melanchthoni in/Academia Vitterbergensi Primario..." [19 Maii 1556]	
89r-90v	To Lantin at Dijon 2pp + ad (17x22) fol 90, astrological chart	9 Feb 1677
91rv	Copy of a letter in B's hand From Cardinal de' Medici [Leopoldo] at Florence To Anonymous 1p (20x31) Italian	21 Jun 1669
92rv	News account in B's hand Warsaw 1p (17x22) Italian	5 Sep 1665
93rv	To [Bernart?] 1½pp (17x22)	8 Apr 1672
94rv	To Anonymous B at The Hague 1p (17x22) Italian	29 Dec 1660
95rv	News account in B's hand Madrid 1¾pp (17x22) Italian	13 Jan 1677
96rv	Note in B's hand* ½p + ad (17x22) *[unrelated to ad on envelope]	4 Mar 1667
97rv	To [Sherburne?] 1p (17x22) Latin	31 Oct 1677
98r-99r	News account in B's hand Leyden 1p & 6 lines (16x21)	9 Apr 1640
99rv	Copy of a declaration in B's hand ¾p (16x21) Latin "Declaratio Domus inferioris, quae scilicet plebem regni/repraesentat, Parlamenti in Hibernia"	undated

100rv	To Anonymous 1p & 2 lines (16x21) [fragment]	11 Oct 1641
101r-102v	News account in B's hand Danzig 3¾pp (15x20)	 15 Oct 1672
103rv	News account in B's hand 1¼pp (17x22) Italian Naples Florence	 28 Jan 1672 3 Apr 1676
104r-105v	News account in B's hand Venice 2pp & 6 lines (17x22) Italian	 31 Dec 1659
106rv	Copy in B's hand 1⅓pp (17x22) "Breve Relation de 3. Imposteurs"	undated
107rv	Extract of a letter in B's hand ½p (20x26) "Extraict d'une lettre escrite de Bayonne a la fin de Janvier 1651/ Receve a Bordeaux le 2^{e}. Fevrier"	
108rv	News account in B's hand Turin 2pp (17x22) Italian	 18 Sep 1657
109rv	News account in B's hand Vienna 1¼pp (17x22) Italian/French	 4 Jul 1661
110rv	To [Blaeu] B at Danzig 1p (20x31)	12 Sep 1651
111rv	To Kirchmayer at Wittenburg ½p (17x22) Latin	3 Mar 1665
112rv	To Kirchmayer at Wittenburg B at The Hague 1⅔pp (17x22) Latin	26 Mar 1662
113rv	To J. Ravius at Brandenburg B at The Hague 2pp (17x22) Latin	26 Mar 1662

Folio	Description	Date
114rv	To N. Heinsius at [Leiden?] $1\frac{3}{4}$pp (20x31) Latin/Greek	V Non Mar 1648 [3 Mar 1648]
115r-116v	To N. Heinsius at [Florence] $2\frac{2}{3}$pp (17x22) Latin/Greek	III Kal Dec 1652 [29 Nov 1652]
117rv	To N. Heinsius at [Amsterdam] $1\frac{1}{2}$pp (17x22) Latin	23 Feb 1657
118rv	To N. Heinsius at [Amsterdam] $1\frac{1}{2}$pp (17x22) Latin	21 Dec 1657
119r-120v	To N. Heinsius at [Amsterdam] 3pp (17x22) Latin Cf. SEB, Epist. V, p 592	Kal Feb 1658 [1 Feb 1658]
121rv	To N. Heinsius at [Amsterdam] $1\frac{3}{4}$pp (17x22) Latin	12 Apr 1658
122r-123v	To N. Heinsius at [The Hague] $2\frac{3}{4}$pp (15x20) Latin	12 Jul 1658
124r-125v	To N. Heinsius at [The Hague] $2\frac{3}{4}$pp (17x22) Latin	5 Mar 1660
126r-127v	To N. Heinsius at [The Hague] $3\frac{3}{4}$pp (17x22) Latin Publ, HO, III, pp 507-511, No 762[a]	30 Jul 1660
128r-129v	To Fabricius at Danzig B at The Hague 3pp & 7 lines (17x22) Latin * w/PS 19 Mar	15 Mar 1662*
130r-131v	To Fabricius at Danzig B at The Hague $2\frac{1}{2}$pp (17x22) Latin	21 Nov 1661
132r-133v	To Fabricius at Danzig B at The Hague 2pp & 4 lines (15x20) Latin	8 May 1662
134rv	To Fabricius at Danzig $1\frac{1}{2}$pp (17x22) Latin	7 Dec 1662
135rv	To Hevelius at [Danzig] 6 lines* (17x22) Latin *[fragment]	undated

136rv	To Rautenstein at [Ratisbon] ½p (17x22) Latin	3 Apr 1665
137r-138v	To Rautenstein at [Ratisbon] 3pp (17x22) Latin	16 May 1664
139r-140v	To Rautenstein at [Ratisbon] 3pp (17x22) Latin	18 Jul 1664
141rv	To Rautenstein at [Ratisbon] 1p & 6 lines (17x22) Latin	2 Jan 1665
142rv	To Rautenstein at [Ratisbon] 1¾pp (17x22) Latin	28 Sep 1663
143rv	To Rautenstein at [Ratisbon] 1½pp (17x22) Latin	31 Oct 1664
144rv	To Rautenstein at [Ratisbon] 1½pp (17x22) Latin	14 Aug 1665
145r-146v	To Rautenstein at [Ratisbon] 2pp & 6 lines (17x22) Latin	25 Jan 1664
147r-148v	To Rautenstein at [Ratisbon] 2⅔pp (17x22) Latin	13 Jul 1663
149r-150v	To Rautenstein at [Frankfurt?] 4pp (15x20) Latin	30 Mar 1663
151rv	To Rautenstein at [Frankfurt] 2pp (17x22) Latin	19 Jan 1663
152r-153v	To Rautenstein at [Frankfurt] 3½pp (17x22) Latin	10 Nov 1662
154rv	To Rautenstein at [Neuburg?] 1½pp (17x22) Latin	28 Jul 1662
155r-156v	To Rautenstein at [Frankfurt?] B at The Hague 4pp (17x22) Latin	27 Mar 1662
157r-158v	To Rautenstein at [Frankfurt?] B at The Hague 2pp & 8 lines (15x20) Latin	8 May 1662
159r-160v	From [Rautenstein] at Frankfurt [on the Main] 3⅔pp (18x25) Latin	20 Apr 1662

161r-162v	To Rautenstein at [Frankfurt] B at The Hague 3pp (17x22) Latin	26 Feb 1662
163rv	To Rautenstein at [Frankfurt] B at The Hague 2pp (17x22) Latin	16 Jan 1662
164rv	To Rautenstein at [Frankfurt] B at The Hague 1½pp (17x22) Latin	26 Dec 1661
165r-166v	To Rautenstein at [Frankfurt] B at The Hague 2½pp (17x22) Latin	6 Feb 1662
167rv	To Rautenstein at [Frankfurt] B at The Hague 2pp (17x22) Latin	15 Dec 1661
168rv	To Rautenstein at [Frankfurt] B at The Hague 1½pp (17x22) Latin	28 Nov 1661
169rv	To Oldenburg at London 1p (17x22) Publ, OC, III, pp 409-411, No 638	16 May 1667
170rv	To The Royal Society of London at London 1p & 8 lines (17x22) Latin Cf. Royal Society MS. B1, No 123	16 May 1667
171rv	Note to B re: the acceptance of letter 170rv 1p + ad (11x11)	undated
172r-175v	Extracts from a letter of Burattini in B's hand 4½pp (17x22) Italian "Dichiaratione dello strumento" fol 175, blank [cf. CB, XXVI, fols 133-135; fol 248]	undated
176rv	To [Johanni] Curtz at Neuburg 1p (17x22) Latin	28 May 1664
177rv	To A Curtz at Neuburg 2pp (17x22) Latin	31 Oct 1664

178rv	Copy of a letter in B's hand From A Curtz at Neuburg 1p & 8 lines (17x22) Latin	28 Nov 1664
179rv	Copy of a letter in B's hand From A Curtz at [Neuburg?] B at [Paris] ½p (17x22) Latin	undated
180rv	To A Curtz at Neuburg 1p (17x22) Latin	7 Oct 1667
181r-182v	To A Curtz at Neuburg 1p & 10 lines (17x22) Latin fol 182, blank	19 Jul 1667
183rv	Inscription [copy in B's hand] 1p & 4 lines (17x22) Latin/French "Inscription sur un pied d'estail quarré de marbre rouge/trouvé dans le Chasteau de Thorigni..." [enclosed w/fol 184]	undated
184rv	To Graindorge at [Caen?] 1½pp (17x22) Latin	11 Dec 1667
185r-186v	To Graindorge at [Caen?] 1½pp (17x22) Latin fol 186, blank [duplicate of above letter]	11 Dec 1667
187r-188v	Inscription, copy in B's hand 2⅓pp (17x22) Latin/French	undated
189rv	To Wachschlager at Thorn B at Danzig 1p (17x22) Latin	30 Aug 1661
190rv	Copy of a letter in B's hand From Wachschlager at Thorn To Anonymous 1⅓pp (17x22) Latin	1 Jul 1661
191rv	To Siri B at The Hague 1¾pp (17x22)	25 Nov 1660
192r-193v	To Silhon 2pp & 5 lines (17x22)	11 Nov 1660

194r-195v	News account in B's hand 1p (17x22) fol 195, blank	20 Apr 1668
196rv	Draft or copy of a letter in B's hand 1p (17x22) [concerning death of the Grand Duke of Tuscany]	13 Jun 1670
197r-198v	Copy of a letter in B's hand From P Melanchthon To J. Kreslingo 1p (17x22) Latin fol 198, blank	undated
199r-202v	From Rinaldini at Florence 6½pp (20x27) Latin w/diagrams & illustrations	undated
203r-204v	From [Rinaldini at Florence] To [B at Paris] 3pp (17x20) Latin "Solutio Problematis/mihi propositi/ab/ Excellentissimo Domino Ismaele Bulialdo/ ..."	undated
205r-209v	To Rinaldini at [Florence] 6⅔pp (17x20) Latin fol 209, blank fol 206, "Problema a Domino Carolo Renaldino propositum" w/geom diagrams & calculations	30 Jul 1660
210rv	To Portner at Ratisbon 1¾pp (17x20) Latin	1 Mar 1667
211rv	To Portner at [Ratisbon] 2pp (17x20) Latin	31 May 1667
212rv	To Portner at [Ratisbon] 2pp (17x22) Latin	28 Sep 1667
213r-214v	To Portner at [Ratisbon] 3⅔pp (17x22) Latin	13 Jul 1659
215r-216v	To Portner at [Ratisbon] 2pp & 2 lines (17x22) Latin	28 Nov 1659

217r-218v	To Portner at [Ratisbon] 2pp & 8 lines (17x22) Latin *w/PS, 2 Jan 1660	19 Dec 1659*
219r-221v	To Portner at [Ratisbon] 3pp (17x22) Latin fol 220r, list fol 221, blank	30 Jan 1660
222r-223v	To Portner at [Ratisbon] 2⅔pp (17x22) Latin	23 Apr 1660
224r-225v	To Portner at [Vienna?] 2¾pp (17x22) Latin	25 May 1674
226rv	To Portner at [Ratisbon] 1½pp (15x20) Latin	10 Jul 1669
227rv	To Portner at [Ratisbon] B at The Hague 2pp (17x22) Latin	24 Feb 1662
228rv	To Portner at [Ratisbon] B at The Hague 2pp (17x22) Latin	6 Feb 1662
229r-230v	To Portner at [Ratisbon] B at The Hague 2⅔pp (17x22) Latin	[late 1665?]
231rv	To Portner at [Ratisbon] 2pp (17x22) Latin	19 Jan 1665
232r-234v	To Portner at [Ratisbon] 4pp (17x22) Latin fol 234, blank	5 Oct 1663
235rv	To Portner at [Ratisbon] 1p & 5 lines (17x22) Latin	30 Mar 1663
236rv	To Portner at [Ratisbon] B at The Hague 1½pp (15x20) Latin	6 May 1662
237rv	To Portner at [Ratisbon] 2pp (15x20) Latin	17 Apr 1669
238rv	To Portner at [Ratisbon] 1p (17x22) Latin	9 Dec 1665

239rv	To Portner at [Ratisbon] 2pp (17x22) Latin	Epiphaniae Domini anno 1666 [6 Jan 1666]
240rv	To Portner at [Ratisbon] 1½pp (17x22) Latin	12 Aug 1665
241rv	To Portner at [Ratisbon] 1½pp (17x22) Latin	14 Apr 1666
242rv	Note in B's hand 6 lines (17x22) [concerns Père Joseph Eméric, Capucin]	2 Aug 1680
243rv	To [Mazarin at Paris] 1½pp (17x22)	19 Feb 1657
244rv	To Montausier at [St-Germain] 1p (17x22)	20 Sep 1668
245rv	From Montausier at St-Germain 1p (17x22)	14 Oct 1668
246rv	To Mazarin at [Paris] 1p (15x20)	16 Oct 1656
247rv	To Mazarin at [Paris] 1½pp (17x22)	24 Jul 1656
248r-251v	From Mazarin at Paris 1p + ad (17x20) fols 250 & 251, blank See ML, VII, p 674	8 Sep 1656
252r-262v	"Conseil pour la Paix" by B, in his hand 20⅔pp (20x31)	undated

▼ ▼ ▼

COLLECTION BOULLIAU, VOLUME X, FONDS FRANÇAIS 13028

Volume X of the Collection Boulliau, Fonds Français 13028, contains an extensive selection of horoscopes drawn and calculated by Ismaël Boulliau. Because of their large number and intrinsic complexity, a complete catalogue of this volume would be redundant. A brief summary follows.

Volume X consists of 373 folios and several times as many horoscopes; there are as many as four nativities per page. In addition to the horoscopes themselves, a number of manuscripts bear biographical information, for example, dates of birth (exact hour and location) and, in certain instances, hour and place of death. Most of the horoscopes are for friends and acquaintances--savants and amateurs, diplomatic, political, and military figures. The crowns and nobility of Europe, however, are well represented. With few exceptions all nativities are the work of Boulliau himself.

While in no sense complete, the names of major figures appearing in Volume X have been included in the Name Index.

▼ ▼ ▼

COLLECTION BOULLIAU, VOLUME XI, FONDS FRANÇAIS 13029

Folios	Description	Date
1r-2v	From [Milet?] at Utrecht 3pp + ad (15x20)	5 Sep 1662
3r-5v	From [Milet?] at The Hague 3pp & 12 lines (10x14) + ad (15x20)	21 Sep 1662
6r-7v&10rv	From Lorenzo Magalotti at Florence 2½pp + ad (15x20) Italian	5 Dec 1670
8r-9v	Extract of a letter in B's hand 2pp & 6 lines (17x22) Italian "Extraict d'une lettre de M. Lorenzo Magalotti a M. Panciatichi/de Florence le 20 Fevrier 1670. ab Incarn. (1671 vulg.)"	
11r-12v	From Pastorius at Danzig 4pp (15x20) Latin	Proprid Kal Jan 1663 [30 Dec 1662]
13rv&15rv	From Pastorius at Danzig 3½pp (15x20)	III Non Oct 1663 [5 Oct 1663]
14rv	Printed catalogue 1p (15x20) Latin "Catalogus laborum literariorum et tractatuum, quospar-/tim perfectos, partim affectos, partim ante jam editos, partim cum augmen-/to denuò edendos habet JOACH: PASTORIUS/ab Hirtenberg"	
16rv	From Gruter at Rotterdam 1p + ad (17x22) Latin	8 Jul 1665
17r-19v	From Gruter at Rotterdam 4pp & 2 lines + ad (15x20) Latin	21 Nov 1667
20rv	From Gruter at Rotterdam 2pp (13x17) Latin	VI Kal Mar 1671 [24 Feb 1671]
21rv	From J. Müller at Hamburg 1½pp (15x20) Latin	[XI Kal] Jul 1661 OS [1 Jul 1661]
22rv	From J. Müller at Hamburg 1⅔pp (17x22) Latin	$\frac{2}{12}$ Sep 1665 [12 Sep 1665]

23rv	From J. G. Graevius at Rhenen $1\frac{2}{3}$pp (20x31) Latin	XI Kal Oct 1672 [OS] [1 Oct 1672]
24rv	From N. Grevius [Grew] at London 2pp (15x20) Latin	16 Jul 1678 [OS] [26 Jul 1678]
25rv	From J. G. Graevius at [Utrecht?] $1\frac{1}{2}$pp (15x20) Latin	IX Kal Nov 1679 OS [3 Nov 1679]
25rv bis	From J. G. Graevius at [Utrecht?] $1\frac{1}{3}$pp (15x20) Latin	XXV [?] 1679 NS
26r-27v	From [Agarrat?] at Paris B at The Hague 3pp + ad (17x22)	5 Nov 1660
28r-29v	From [Agarrat?] at Paris B at The Hague 3pp + ad (17x22)	12 Nov 1660
30rv	From [Agarrat?] at Paris B at [The Hague] $1\frac{2}{3}$pp (17x22)	19 Nov 1660
31r-32v	From [Agarrat?] at Paris B at The Hague $2\frac{1}{2}$pp + ad (17x22)	26 Nov 1660
33r-34v	From [Agarrat?] at Paris B at The Hague $1\frac{1}{2}$pp + ad (17x22)	3 Dec 1660
35rv	From [Agarrat?] at Paris B at [The Hague] 2pp (13x17)	6 Aug 1661
36r-37v	From [d'Aubonet?] at Danzig B at The Hague 1p & 4 lines + ad (13x17)	26 Nov 1661
38r-39v	From [d'Aubonet?] at [Danzig?] B at The Hague $\frac{1}{2}$p & 6 lines + ad (17x22)	undated

40rv	From Lumbres at Warsaw 1p (17x22)	15 Aug [1661?]
41rv	From Lumbres at [Warsaw] 1p (17x22)	[12?] Sep 1661
42r-43v	From Lumbres at [Warsaw] B at The Hague 2pp & 3 lines + ad (17x22)	15 Dec 1661
44rv	From Lumbres at Warsaw 1p (17x22)	13 Apr 1662
45rv	From Lumbres at Warsaw 1½pp (17x22)	[16?] Feb 1662
46rv	From Wermels Kerch at [Nordes?] 2pp (15x20)	14 Dec 1664 OS [24 Dec 1664]
47r-48v	From Wermels Kerch at [Nordes?] 1⅔pp + ad (15x20)	1 Feb 1665 OS [11 Feb 1665]
49r-50v	From Berruyer at Rome 4pp (13x17)	18 Aug 1665
51rv	From Berruyer at Rome 1p & 6 lines (13x17)	24 Nov 1665
52r-54v	From Berruyer at Rome 3½pp + ad (13x17)	10 Nov 1665
55rv	From Cefali at Paris 1p (17x22) Italian	1 Oct 1660
56rv	From Cefali at Paris To [Lubomirski] 1p (17x22) Italian	29 Sep 1660
57rv	From Cefali at Paris To [Lubomirski] 1p (17x22) Italian	29 Sep 1660
58r-59rv	From Cefali at Paris To Mr. Sorgo at Warsaw 1p + ad (17x22) Italian "Monsieur Sorgo Cap^{ne} de Infanterie et Sec^{re} de Monseigner le Prince Lubomierski Grand Mareschal et Gnal. de Pologne."	29 Sep 1660

59rv bis 59rv ter	From Fabricius at Danzig $3\frac{2}{3}$pp (17x22) Latin	28 Jan 1662
60r-61v	From Fabricius at Danzig B at The Hague 3pp + ad (17x22) Latin	8 Apr 1662
62rv	From Fabricius at Danzig 2pp (17x22) Latin	11 Nov 1662
63r-64v	From Boecler at Strasbourg 2pp & 6 lines (15x20) Latin	10 Feb 1654
65rv	From Boecler at Strasbourg 1p (15x20) Latin	[Prid Non] Mar 1654 [6 Mar 1654]
66rv	From Boecler at Strasbourg $1\frac{1}{2}$pp (15x20) Latin	$\frac{4\ \text{May}}{24\ \text{Apr}}$ 1654
67rv	From Boecler at Strasbourg $1\frac{1}{2}$pp (15x20) Latin	$\frac{21}{11}$ Aug 1654
68rv	From Boecler at [Strasbourg] 1p & 3 lines (15x20) Latin	12 Sep 1654
68rv bis	From Boecler at Strasbourg $1\frac{1}{2}$pp (15x20) Latin	$\frac{19}{9}$ Oct 1654
69rv	From Boecler at Strasbourg 1p (15x20) Latin	10 Feb 1655
70rv	From Boecler at Strasbourg 1p (15x20) Latin	$\frac{26}{16}$ Dec 1653
71rv	From Boecler at Strasbourg 1p (15x20) Latin	5 Nov 1655
72rv	From Boecler at Strasbourg $1\frac{1}{2}$pp (15x20) Latin	$\frac{18}{28}$ Sep 1657
73rv	From Boecler at Strasbourg $1\frac{2}{3}$pp (15x20) Latin	2 Apr 1658
74rv	From Boecler at Strasbourg $1\frac{1}{2}$pp (15x20) Latin	28 Apr 1659
75rv	From Boecler at Strasbourg $\frac{1}{2}$p (15x20) Latin	6 Oct 1662

76rv From Boecler at Strasbourg 24 Mar 1663
1p + ad (15x20) Latin

77rv From Janot at The Hague 16 Dec 1655
1½pp (17x22)

78rv From Janot at The Hague 6 Jan 1656
2pp (17x22)

79rv From Janot at Amsterdam 24 Feb 1656
1½pp (17x22)

80rv From Janot at The Hague 14 Dec 1656
1p (17x22)

81r-82v From Janot at The Hague 6 Apr 1656
3pp + ad (17x22)

83rv From Janot at Rotterdam 30 Mar 1656
2pp (17x22)

84rv From Janot at Amsterdam 16 Mar 1656
2pp (17x22)

85rv From Janot at Amsterdam 9 Mar 1656
1½pp (17x22)

86rv From Janot at The Hague 21 Dec 1656
1½pp (15x20)

87rv Newsletter 21 Dec 1656
¾p (15x20)

88rv From Janot at The Hague 22 Feb 1657
2pp (15x20)

89rv From Janot at The Hague 15 Mar 1657
1½pp (15x20)

90rv From Janot at The Hague 18 Jan 1657
1p (15x20)

91rv From Janot at The Hague 4 Jan 1657
1p (15x20)

92rv From Menier at [The Hague] 27 Sep 1657
1½pp (20x31)

93rv	From Menier at [The Hague] 2pp (20x31)	18 Oct 1657
94rv	From Menier at [The Hague] $1\frac{2}{3}$pp (20x31)	20 Sep 1657
95r-96v	From Menier at [The Hague] 4pp (20x31)	4 Oct 1657
97rv	From N. Heinsius at Florence 1p (20x31) Latin	Kal Jan 1653 [1 Jan 1653/54?]
98r-99v	Copy of a letter in B's hand From N. Heinsius at Amsterdam To Ménage $2\frac{1}{2}$pp (15x20) Latin	18 Oct 1657
100rv	From N. Heinsius at Amsterdam 2pp (17x22) Latin	14 Feb 1657
101rv	From N. Heinsius at The Hague $1\frac{1}{2}$pp (15x20) Latin	21 Feb 1658
102rv	From N. Heinsius at The Hague $1\frac{1}{2}$pp (15x20) Latin	30 Mar 1658
103rv	From N. Heinsius at The Hague 2pp (15x20) Latin	28 Nov 1658
103rv bis	From N. Heinsius at The Hague $1\frac{1}{3}$pp (15x20) Latin	27 Jun 1658
104rv	To N. Heinsius at [The Hague] $1\frac{2}{3}$pp (17x22) Latin	13 Dec 1658
105rv	From N. Heinsius at The Hague 2pp (17x22) Latin	16 Jan 1658
105rv bis	From N. Heinsius at The Hague $1\frac{2}{3}$pp (17x22) Latin	17 Apr 1659
106rv	To N. Heinsius at [The Hague] $1\frac{1}{2}$pp (17x22) Latin	25 Apr 1659
107r-108v	From N. Heinsius at The Hague $2\frac{1}{2}$pp + ad (15x20) Latin	22 Jul 1660

109rv	From N. Heinsius at The Hague 1½pp (15x20) Latin Publ, HO, III, pp 511-513, No 766[a]	V Sextil 1660 [5 Aug 1660]
110r-112v	From N. Heinsius at The Hague 2⅔pp + ad (17x22) Latin Publ, HO, III, pp 499-502, No 757[a]	26 Jun 1660
113rv	From A. Curt at [Neuburg] 1p (15x20) Latin	undated
114rv&117rv	From A. Curt at Neuburg 2pp (15x20) Latin fol 117, blank	22 Jul 1667
115r-116v	From A. Curt at Neuburg 2½pp + ad (20x31) Latin	28 Nov 1664
118rv	From A. Curt at Neuburg 1p (20x31) Latin	10 Apr 1667
119r-120v	From Forstner at Montbéliard 3pp (17x22) Latin	[1656]
121r-122v	From Forstner at Montbéliard 3pp (17x22) Latin	26 Oct 1656 [OS] [4 Nov 1656]
123rv	From Forstner at Montbéliard 1p & 5 lines (17x22) Latin	30 Dec 1656 [OS] [8 Jan 1657]
124r-125v	From Forstner at Montbéliard 2pp & 5 lines + ad (17x22) Latin	IV Id Mar 1657 OS [22 Mar 1657]
125rv bis	From Forstner at Montbéliard 1p (17x22) Latin	XV Kal Jul 1658 [OS] [27 Jun 1658]
126r-127v	From Forstner at Montbéliard 2½pp + ad (15x20) Latin	19 Sep 1658 [OS] [29 Sep 1658]
128rv	From Forstner at Montbéliard 1p & 4 lines (18x23) Latin	XIV Kal Jun 1659 [OS] [29 May 1659]
129rv	From Pinocci at The Hague 1p (15x20)	12 Sep 1658
130rv	From Pinocci at The Hague 1p (15x20)	10 Oct 1658

131rv	Newsletter in Pinocci's hand Camp near Torun 1p (15x20)	 25 Sep 1658
132rv	Newsletter in Pinocci's hand Camp near Torun 1p (17x22)	 16 Oct 1658
133rv	From Pinocci at The Hague 1p (15x20)	24 Oct 1658
134rv	From Pinocci at The Hague 1p (15x20)	31 Oct 1658
135r-137v	Newsletter in Pinocci's hand The Hague 1p + ad (15x20) Italian fol 137, blank	 1 Nov 1658
138rv	From Pinocci at The Hague 1p (15x20)	19 Dec 1658
139r-140v	From Pinocci at The Hague 2pp & 6 lines (10x15)	5 Dec 1658
141rv	Newsletters Rieydani Torun 1p (15x20)	 25 Oct 1658 29 Oct 1658
142rv	From Pinocci at The Hague 1p (15x20)	26 Sep 1658
143r-145v	From Pinocci at The Hague 3pp + ad (15x20)	16 Jan 1659
146r-147v	From Pinocci at Westminster 3pp (15x20)	7 Jul 1659
148rv	From [Pinocci] at London 1½pp (15x20)	19 May 1659
149r-150v	From Pinocci at Westminster 3pp (15x20)	7 Apr 1659
151rv	From [Pinocci] at Westminster 1½pp (15x20)	5 Jun 1659

152r-153v	From Pinocci at Westminster 3pp (15x20)	23 Jun 1659
154rv	From Pinocci at London 1p (17x22)	15 Mar 1659
155r-157v	From Pinocci at The Hague 2½pp (15x20) fol 157, blank	2 Jan 1659
158rv	From Pinocci at The Hague 2pp (15x20)	26 Dec 1658
159rv	Newsletter [Pinocci?] Camp near Torun 1p (18x25)	 17 Dec 1658
160r-161v	From [Pinocci] at Amsterdam 1p (15x20) fol 161, blank	9 Sep 1659
162r-164v	From [Rautenstein] at Frankfurt B at The Hague 3½pp + ad (15x20) Latin	28 Jan 1662
165r-166v	From [Pinocci] at Westminster 3pp (15x20)	11 Aug 1659
167rv	From Pinocci at Cracow 2pp (15x20)	22 May 1660
168rv	From Pinocci at Cracow 1p (15x20)	2 Jan 1660
169rv	From Pinocci at Cracow 1p & 8 lines (15x20)	14 Mar 1660
170rv	From Pinocci at The Hague 2pp (15x20)	14 Nov NY
171rv	From Pinocci at Bologna 1p (15x20)	24 Feb 1665
172r-174v	From de Valois at Grenoble 2½pp (15x20) fol 174, blank	14 Dec 1642
175r-176v	From de Valois at Grenoble 3pp + ad (15x20)	1 Mar 1643

177r-178v	From [de Valois] at Grenoble 3pp + ad (10x15)	12 Jun 1650
179r-180v	From [de Valois] at Grenoble 3pp + ad (10x15)	30 Jun 1652
181r-182v	From Vair at Grenoble 2½pp + ad (10x15)	4 Nov 1657
183r-184v	From [de Valois] at Grenoble 3¼pp (13x17)	31 Mar 1652
185r-186v	From [de Valois] at Grenoble 3pp + ad (15x20)	14 Apr 1652
187r-188v	From [de Valois] at Grenoble 3pp + ad (15x20)	7 Apr 1652
189rv	From Huygens at The Hague 2pp (20x31) Publ, HO, II, pp 108-110, No 443	26 Dec 1657
190rv	From Huygens at The Hague 1½pp (15x20) Publ, HO, II, pp 183-184, No 490	13 Jun 1658
191rv	From Huygens at [The Hague] ½p (15x20) Publ, HO, II, p 402, No 615	8 May 1659
192r-193v	From Huygens at [The Hague] 2pp (15x20) fol 193, blank Publ, HO, II, pp 269-270, No 545	[Oct 1658]
194r-195v	From Huygens at The Hague 3pp (15x20) Publ, HO, II, pp 453-454, No 648	7 Aug 1659
196rv	From Huygens at [The Hague] 1p + ad (15x15) Publ, HO, III, pp 99-100, No 763	[Jul 1660]
197r-198v	From Huygens at The Hague 4pp (15x20) Publ, HO, II, pp 441-442, No 640	24 Jul 1659

199rv&202rv	From Huygens at The Hague 1p (15x20) fol 202, blank Publ, HO, II, p 357, No 591	23 Feb 1659
200r-201v	From Huygens at The Hague 3¾pp (20x31) incl drawings Publ, HO, II, pp 361-364, No 593	5 Mar 1659
203rv	From Huygens at The Hague 1½pp (15x20) Publ, HO, II, p 372, No 598	13 Mar 1659
204rv	From Huygens at The Hague 1½pp (15x20) Publ, HO, II, pp 313-314, No 565	16 Jan 1659
205r-206v	From Huygens at The Hague 2pp & 3 lines + ad (15x20) Publ, HO, II, pp 276-277, No 548	21 Nov 1658
206Ar-206Bv	From Huygens at The Hague 2pp + ad (15x20) Publ, HO, II, pp 509-510, No 683	20 Nov 1659
206Cr-206Dv	Tract, "Pour l'Assemblee de Physique" 2½pp (20x31)	undated
207r-208v	From Huygens at The Hague 3pp + ad (20x31) Publ, HO, II, pp 523-525, No 692	11 Dec 1659
209r-210v	From Huygens at The Hague 2½ pp + ad (15x20) Publ, HO, III, pp 65-67, No 743	22 Apr 1660
211rv&214rv	From Huygens at The Hague 2pp + ad (15x20)* *fol 214 (10x20) Publ, HO, III, pp 25-26, No 721	26 Feb 1660
212r-213v	From Huygens at The Hague 3pp + ad (20x31) Publ, HO, III, pp 32-34, No 724	4 Mar 1660
215r-216v	From Huygens at The Hague 4pp (15x20) Publ, HO, III, pp 12-14, No 711	22 Jan 1660

217r-218v	From Huygens at The Hague 2½pp + ad (15x20) Publ, HO, III, pp 16-17, No 714	29 Jan 1660
219r-221v	From Huygens at The Hague 4½pp (18x25)* *fol 220, diagram (13x17) Publ, HO, III, pp 278-280, No 865; p 281, No 866	13 Jun 1661
222r-223v	From Huygens at The Hague 2pp + ad (15x20) Publ, HO, IV, pp 208-209, No 1050	24 Aug 1662

▼ ▼ ▼

COLLECTION BOULLIAU, VOLUME XII, FONDS FRANÇAIS 13030

1rv	Note in B's hand See fols 100-101 & 120-123 ¼p (16x24)	
2r-3v	From de Valois at Lyon 3pp + ad (17x22)	9 Sep 1642
4rv	From de Valois at Lyon 1¼pp + ad (17x22)	12 Sep 1642
5r-6v	From de Valois at Grenoble 3pp + ad (15x20)	26 Oct 1642
7r-8v	From de Valois at Grenoble 3pp + ad (15x20)	1 Feb 1643
9r-10v	From de Valois at Grenoble 3¼pp + ad (15x20) Publ, MC, XII, pp 240-241	19 Jul 1643
11r-12v	From de Valois at Grenoble 3pp + ad (15x20)	18 Oct 1643
13r-14v	From de Valois at Grenoble 3pp + ad (15x20) Publ, MC, XII, p 347	25 Oct 1643
15r-16v	From [de Valois] at Grenoble 3pp + ad (15x20)	15 Nov 1643
17r-18v	From [de Valois] at Grenoble 2pp + ad (15x20)	6 Dec 1643
19r-20v	From [de Valois] at Grenoble 2pp + ad (15x20) Publ, MC, XIII, pp 1-2, No 1241	3 Jan 1644
21r-22v	From [de Valois] at Grenoble 3pp + ad (15x20)	6 Jan 1644
23r-24v	From [de Valois] at Grenoble 3pp + ad (15x20)	24 Jan 1644
25rv	Account 1p (18x27)	undated

26r-27v	From [de Valois] at Grenoble 3pp + ad (15x20)	13 Mar 1644
28r-29v	From [de Valois] at Grenoble 3pp + ad (15x20)	22 May 1644
30r-31v	From [de Valois] at Grenoble 3pp + ad (15x20)	11 Dec 1644
32r-33v	From [de Valois] at Grenoble 3pp + ad (15x20)	15 Jan 1645
34r-35v	From [de Valois] at Grenoble 3pp + ad (15x20)	5 Feb 1645
36r-37v	From [de Valois] at Grenoble 3pp + ad (15x20)	12 Feb 1645
38r-39v	From [de Valois] at Grenoble 3pp + ad (15x20)	19 Feb 1645
40r-41v	From [de Valois] at Grenoble 3¼pp + ad (10x15)	17 Mar 1648
42r-43v	From [de Valois] at Grenoble 3pp + ad (10x15)	18 Jul 1649
44r-45v	From [de Valois] at Grenoble 3pp + ad (10x15)	8 Aug 1649
46r-47v	From [de Valois] at Grenoble 3pp + ad (10x15)	23 Apr 1651
48r-49v	From [de Valois] at Grenoble 3pp + ad (10x15)	30 Apr 1651
50r-51v	From [de Valois] at Grenoble 2½pp (15x20)	19 Jan 1652
52r-53v	From [de Valois] at Grenoble 3pp + ad (10x15)	11 Feb 1652
54r-55v	From [de Valois] at Grenoble 4pp (15x20)	24 Mar 1652
56r-57v	From [de Valois] at Grenoble 3pp + ad (15x20)	21 Apr 1652

58r-59v	From [de Valois] at Grenoble 3pp + ad (10x15)	24 Apr 1652
60r-61v	From [de Valois] at Grenoble 3pp + ad (15x20)	28 Apr 1652
62r-63v	From [de Valois] at Grenoble 3pp + ad (15x20)	5 May 1652
64r-66v	From [de Valois] at Grenoble 4⅔pp + ad (15x20)	11 Aug 1652
67r-68v	From [de Valois] at Grenoble 2½pp + ad (15x20)	18 Aug [1652]
69r-70v	From [de Valois] at Grenoble 3pp + ad (10x15)	24 Nov 1652
71r-72v	From [de Valois] at Grenoble 4pp (10x15)	22 Dec 1652
73r-74v	From [de Valois] at Grenoble 3pp + ad (10x15)	29 Dec 1652
75r-76v	From [de Valois] at Grenoble 3pp + ad (10x15)	25 May 1653
77rv	Missing	
78r-79v	From [de Valois] at Grenoble 3¾pp + ad (10x15)	15 Jun 1653
80r-81v	From [de Valois] at Grenoble 3pp + ad (10x15)	22 Jun 1653
82r-84v	From [de Valois] at Grenoble 3pp + ad (10x15) fol 84, astrological square, Mr. de Créqui	6 Jul 1653
85r-86v	From [de Valois] at Grenoble 3pp + ad (10x15)	13 Jul 1653
87r-88v	From [de Valois] at Grenoble 3pp + ad (10x15)	27 Jul 1653
89r-90v	From [de Valois] at Grenoble 3pp + ad (10x15)	3 Aug 1653

91r-93v	From [de Valois] at Grenoble 5pp + ad (10x15)	10 Aug 1653
94r-95v	From [de Valois] at Grenoble 3pp & 8 lines + ad (10x15)	17 Aug 1653
96r-97v	From [de Valois] at Grenoble 3pp + ad (10x15)	24 Aug 1653
98r-99v	From [de Valois] at Grenoble 3pp + ad (10x15)	31 Aug 1653
100r-101v	From [de Valois] at Grenoble 4pp (10x15)	14 Sep 1653
102r-103v	From [de Valois] at Grenoble 3pp & 6 lines + ad (10x15)	21 Sep 1653
104r-105v	From [de Valois] at Grenoble 3pp + ad (10x15)	28 Sep 1653
106r-107v	From [de Valois] at Grenoble 3pp & 4 lines + ad (10x15)	5 Oct 1653
108r-109v	From [de Valois] at Grenoble 3pp & 8 lines + ad (10x15)	12 Oct 1653
110r-111v	From [de Valois] at Grenoble 3pp + ad (10x15)	19 Oct 1653
112r-113v	From [de Valois] at Grenoble 3pp + ad (10x15)	26 Oct 1653
114r-115v	From [de Valois] at Grenoble 3pp + ad (10x15)	2 Nov 1653
116r-117v	From [de Valois] at Grenoble 3pp + ad (10x15)	9 Nov 1653
118r-119v	From [de Valois] at Grenoble 3pp + ad (10x15)	16 Nov 1653
120r-123v	From [de Valois] at Grenoble 6pp & 9 lines + ad (10x15)	23 Nov 1653
124r-125v	From [de Valois] at Grenoble 3pp + ad (10x15)	30 Nov 1653
126r-127v	From [de Valois] at Grenoble 3pp + ad (10x15)	7 Dec 1653

128r-130v	From [de Valois] at Grenoble 4½pp + ad (10x15)	14 Dec 1653
131r-133v	From [de Valois] at Grenoble 5pp & 5 lines + ad (13x20)	21 Dec 1653
134r-135v	From [de Valois] at Grenoble 3pp + ad (13x20)	28 Dec 1653
136r-137v	From [de Valois] at Grenoble 3pp + ad (13x20)	4 Jan 1654
138r-139v	Calculations of new moons, 1654 2½pp (13x20) Latin	undated
140r-141v	From [de Valois] at Grenoble 3pp + ad (13x20)	11 Jan 1654
142r-143v	From [de Valois] at Grenoble 4pp (13x20)	18 Jan 1654
144r-145v	From [de Valois] at Grenoble 3pp + ad (13x20)	25 Jan 1654
146r-147v	From [de Valois] at Grenoble 3pp + ad (13x20)	8 [Feb] 1654
148r-149v	From [de Valois] at Grenoble 3pp + ad (13x20)	15 Feb 1654
150r-151v	From [de Valois] at Grenoble 2¾pp + ad (13x20)	1 Mar 1654
152r-153v	From [de Valois] at Grenoble 3pp + ad (10x15)	8 Mar 1654
154r-155v	From [de Valois] at Grenoble 3pp & 6 lines + ad (10x15)	15 Mar 1654
156r-157v	From [de Valois] at Grenoble 3pp + ad (10x15)	22 Mar 1654
158r-159v	From [de Valois] at Grenoble 3pp + ad (10x15)	29 Mar 1654
160r-161v	From [de Valois] at Grenoble 3pp + ad (10x15)	Easter day 1654 [5 Apr 1654]
162r-163v	From [de Valois] at Grenoble 2¾pp + ad (10x15)	12 Apr 1654

164r-165v	From [de Valois] at Grenoble 3pp + ad (10x15)	19 Apr 1654
166r-167v	From [de Valois] at Grenoble 3pp + ad (10x15)	26 Apr 1654
168r-169v	From [de Valois] at Grenoble 4pp (10x15)	10 May 1654
170r-171v	From [de Valois] at Grenoble 3pp + ad (10x15)	17 May 1654
172r-173v	From [de Valois] at Grenoble 4pp (10x15)	20 May 1654
174r-176v	From [de Valois] at Grenoble 5pp + ad (10x15)	7 Jun 1654
177r-178v	From [de Valois] at Grenoble 3pp + ad (10x15)	14 Jun 1654
179r-180v	From [de Valois] at Grenoble 3pp + ad (10x15)	21 Jun 1654
181r-182v	From [de Valois] at Grenoble 3pp + ad (10x15)	28 Jun 1654
183r-184v	From [de Valois] at Grenoble 3pp + ad (10x15)	5 Jul 1654
185r-186v	From [de Valois] at Grenoble 3pp & 5 lines + ad (10x15)	12 Jul 1654
187r-188v	From [de Valois] at Grenoble 3pp + ad (10x15)	26 Jul 1654
189r-190v	From Vair at Grenoble 3pp + ad (13x17)	11 Oct 1654
191r-192v	From [de Valois] at Grenoble 3pp + ad (10x15)	2 Aug 1654
193r-194v	From [de Valois] at Grenoble 3pp & 6 lines + ad (10x15)	9 Aug 1654
195r-196v	From [de Valois] at Grenoble 3pp + ad (10x15)	16 Aug 1654
197r-198v	From [de Valois] at Grenoble 3pp + ad (10x15)	6 Sep 1654

199r-200v	From [de Valois] at Grenoble 3pp + ad (10x15)	16 Sep 1654
201r-202v	From [de Valois] at Grenoble 3pp & 5 lines + ad (10x15)	20 Sep 1654
203r-204v	From [de Valois] at Grenoble 3pp + ad (10x15)	27 Sep 1654
205r-206v	From Vair at Grenoble 2pp + ad (15x20)	9 Dec 1654
207r-208v	From Vair at Grenoble 3pp + ad (13x17)	13 Dec 1654
209rv&212rv	From [de Valois] at Grenoble 2½pp + ad (15x20)	19 Mar 1645
210r-211v	From [de Valois] at Grenoble 3pp + ad (10x15)	5 Jun 1650

▼ ▼ ▼

COLLECTION BOULLIAU, VOLUME XIII, FONDS FRANÇAIS 13031

1rv	Library note 4 lines (15x20)	undated
2rv	News account Warsaw $1\frac{1}{2}$pp (15x20)	3 Jun 1662
3rv	News account Warsaw 2pp (15x20)	27 May 1662
4rv	From Lubienietzki at Hamburg $1\frac{2}{3}$pp (15x20) Latin	5 May 1663
5r-7v	From Lubienietzki at Hamburg $2\frac{1}{2}$pp + ad (15x20) Latin fol 7, blank	28 Feb 1663
8rv	From Lubienietzki at Hamburg 2pp (15x20) Latin	21 Apr 1663
9r-10v	From Lubienietzki at Hamburg To Denis de la Rivière at Paris 2pp + ad (15x20) Latin fol 10, blank	27 Oct 1663
11r-12v	From Lubienietzki at [Iserum?] $3\frac{2}{3}$pp (15x20) Latin	$\frac{18}{28}$ Jul 1663
13rv&18rv	From Lubienietzki at [Hamburg?] 2pp + ad (15x20) Latin	$\frac{12}{22}$ Jun 1663
14rv&17rv	From Lubienietzki at Hamburg $2\frac{1}{2}$pp + ad (15x20) Latin	29 Dec 1663
15r-16v	News account Danzig, Hamburg 4pp (20x31) Latin	$\frac{15}{25}$ Dec 1663
19rv	From Lubienietzki at Hamburg $1\frac{1}{2}$pp (15x20) Latin	Kal Nov 1663 [1 Nov 1663]
20r-21v	From Lubienietzki at Hamburg 3pp (15x20) Latin	27 Oct 1663

22rv&25rv	From Lubienietzki at Hamburg 1p (10x15) Latin fol 25, blank	Kal Nov 1664 [1 Nov 1664]
23r-24v	News account Hamburg, etc 4pp (15x20)	 29 Nov 1664
26rv	News accounts 2pp (15x20) Latin Vienna Ratisbon Neuburg Berlin	 24 Nov 1664 27 Nov 1664 29 Nov 1664 3 Dec 1664
27rv	From Lubienietzki at Hamburg 1½pp (17x22) Latin	29 Nov 1664
28rv	News account Danzig 2pp (17x22)	 29 Nov 1664
29rv&32rv	From Lubienietzki at Hamburg 2pp & 3 lines (10x15) Latin	8 Nov 1664
30r-31v	News accounts 4pp (17x22) Latin Vienna Ratisbon Danzig	 6,8 Dec 1664 11 Dec 1664 13 Dec 1664
33rv&36rv	From Lubienietzki at Hamburg 2pp + ad (17x22) Latin	25 Oct 1664
34r-35v	News accounts 4pp (17x22) Hamburg Ratisbon Danzig	 15 Nov 1664 6 Nov 1664 8 Nov 1664
37rv&40rv	From Lubienietzki at Hamburg 1p & 7 lines (10x15) Latin fol 40, blank	15 Nov 1664
38rv	Newsletter Danzig, etc 2pp (17x22)	 1 Nov 1664
39rv	News account Hamburg 2pp (17x22)	 8 Nov 1664

41r-42v	News account	
	Hamburg	22 Nov 1664
	3pp & 5 lines + PS (17x22)	
43rv&46rv	From Lubienietzki at Hamburg	4 Oct 1664
	2pp + ad (10x15)* Latin	
	*fol 46 (9x14)	
44r-45v	News accounts	
	4pp + ad (17x22) Latin/French	
	Vienna	18 Oct 1664
	Nonnberga	25 Oct 1664
	Ratisbon	28 Oct 1664
	Danzig	25 Oct 1664
47r-50v	News accounts	
	8pp (17x22) Latin	
	Vienna	12 Oct 1664
	Ratisbon	16 Oct 1664
	Berlin	22 Oct 1664
	Danzig, etc.	18 Oct 1664
51rv&53rv	From Lubienietzki at Hamburg	29 Mar 1664
	1½pp (10x15) Latin	
	fol 53, blank	
52rv	News accounts	
	2pp (17x22) Latin	
	Cracow	22 Mar 1664
	Hamburg	18 Apr 1664
54r-55v	News account	
	Danzig	15 Mar 1664
	4pp (17x22) Latin	
56rv	News accounts	
	2pp (20x21) Latin/French	
	Groty [in Croatia]	14 Mar 1664
	Ratisbon	20 Mar 1664
	Danzig	22 Mar 1664
	Hamburg	29 Mar 1664
57rv&60rv	From Lubienietzki at Hamburg*	26 Apr 1664
	⅔p + ad (10x15) Latin	
	*Note, w/news accounts	
58r-59v	News accounts	
	4pp + ad (17x22) Latin/French	
	Danzig	26 Apr 1664

	Ratisbon	24 Apr 1664
	Hamburg	2 May 1664
61r-62v	From Lubienietzki at Hamburg	2 May 1664
	4pp (17x22) Latin	
63r-68v	News accounts	
	11pp (17x22) Latin/French	
	Ratisbon	3 Jul 1664
	Berlin	8 Jul 1664
	Danzig	5 Jul 1664
	Copy	7 Jul 1664
	Vienna	13 Jul 1664
	Ratisbon	17 Jul 1664
	Danzig	19 Jul 1664
	"In castris ad Serini castrum"	5 Jul 1664
	Vienna	5 Jul 1664
	Nürnberg	12 Jul 1664
	Ratisbon	7,10 Jul 1664
	Danzig	12 Jul 1664
69r-72v	From Lubienietzki at Hamburg	16 Feb 1664
	4pp (17x22) Latin	
	fol 70, enclosed account (Danzig, 9 Feb 1664)	
	fol 71, concerns eclipse of the sun of 1664 [28 Jan]	
73r-78v	News accounts	
	11pp (17x22) French/Latin	
	Vienna	26 Jul 1664
	Ratisbon	31 Jul 1664
	Danzig	2 Aug 1664
	Vienna	3 Aug 1664
	Ratisbon	7 Aug 1664
	Danzig	9 Aug 1664
	Prague	31 Jul 1664
	Vienna	24 Aug 1664
	Nürnberg	30 Aug 1664
	Ratisbon	28 Aug 1664
	Frontiers of Prussia	26 Aug 1664
	Danzig	30 Aug 1664
79rv&81rv	From Lubienietzki at Hamburg	6 Sep 1664
	1p + ad (10x15) Latin	
80rv	Extract of a letter	22 Aug 1664
	From A. Curtz at [Neuburg?]	
	To Rautenstein at [Ratisbon?]	
	1p (17x22) Latin	

82r-89v	News accounts	
	16pp (17x22) French/Latin	
	Vienna	9 Aug 1664
	Ratisbon	14 Aug 1664
	Danzig	16 Aug 1664
	Prague	9 Aug 1664
	Vienna	18 Aug 1664
	Nürnberg	23 Aug 1664
	Ratisbon	21 Aug 1664
	Danzig	23 Aug 1664
	Vienna	31 Aug 1664
	Berlin	10 Sep 1664
	Ratisbon	4 Sep 1664
	Danzig	6 Sep 1664
	Vienna	7 Sep 1664
	Nürnberg	13 Sep 1664
	Magdeburg	15 Sep 1664
	Ratisbon	11 Sep 1664
	Danzig	13 Sep 1664
90rv&93rv	From Lubienietzki at Hamburg	13 Sep 1664
	1p + ad (17x22) Latin	
91r-92v	News accounts	
	4pp (17x22) Latin/French	
	Prague	15 Sep 1664
	Ratisbon	18 Sep 1664
	Danzig	20 Sep 1664
	Stockholm	$\frac{3}{13}$ Sep 1664
94r-97v	From Lubienietzki at Hamburg*	8 Mar 1664
	6pp & 5 lines (13x18) Latin	
	Ratisbon [in L's hand]	25 Feb 1664
	Ratisbon	28 Feb 1664
	Ratisbon	
	*w/enclosed copies and newsletters	
98r-101v	News accounts	
	8pp (15x18) Latin	
	Vienna	13 Dec 1664
	Ratisbon	18 Dec 1664
	Danzig	20 Dec 1664
	Stockholm	$\frac{3}{13}$ Dec 1664
	Copenhagen	$\frac{9}{19}$ Dec 1664
	Hamburg	27 Dec 1664
	Vienna*	10 Mar 1664

	Klausenbourg (7 dates)*	Mar 1664
	Vienna*	11 Mar 1664
	Ratisbon*	13 Mar 1664
	Hamburg*	22 Mar 1664
	*(14x16)	
102rv	From Lubienietzki at Hamburg 1½pp (17x22) Latin	20 Dec 1664
103rv	Copies of letters*/astron. observ 2pp (17x22) Latin *[Joannes Müllerg and Joh. Alb. Huswedelig.]	17-19 Dec 1664
104rv&107rv	From Lubienietzki at Hamburg 1p (14x16) Latin fol 107, blank	27 Dec 1664
105r-106v& 108r-115v	News accounts 19pp (15x20) Latin/French	
	Vienna	20 Dec 1664
	Ratisbon	25 Dec 1664
	Ratisbon	
	Danzig	25 Dec 1664
	Danzig	3 Jan 1665
	Vienna	28 Dec 1664
	Nürnberg	3 Jan 1665
	Ratisbon	1 Jan 1665
	Danzig	27 Dec 1664
	Danzig	27 Dec 1664
	Ratisbon	16 Jun 1664
	Ratisbon	19 Jun 1664
	Danzig	21 Jun 1664
	Danzig	17 Jun 1664
	Vienna	22 Jun 1664
	Linz	22 Jun 1664
	Ratisbon	23 Jun 1664
	Ratisbon	26 Jun 1664
	Danzig	28 Jun 1664
116rv&119rv	From Lubienietzki at Hamburg ½p + ad (17x22) Latin	26 Jul 1664
117rv	News account Vienna 2pp (17x22) Latin	 13 Jun 1664
118rv	From Lubienietzki at Hamburg 1p + ad (17x22) Latin	28 Jun 1664

120r-127v	News accounts	
	14pp (17x22) Latin/French	
	Danzig	14 May 1664
	Danzig	24 May 1664
	Vienna	18 May 1664
	Ratisbon	15 May 1664
	Danzig	17 May 1664
	Ratisbon	21 May 1664
	Ratisbon	26 May 1664
	Ratisbon	29 May 1664
	Danzig	31 May 1664
	Hamburg	7 Jun 1664
	Ratisbon	2 Jun 1664
	Ratisbon	5 Jun 1664
	Danzig	7 Jun 1664
	Hamburg	$\frac{4}{14}$ Jun 1664
	fol 125 (13x16)	
128rv&132rv	From Lubienietzki at Hamburg	21 Jun 1664
	1½pp + ad (17x22) Latin	
129rv	From Lubienietzki at Hamburg	$\frac{\text{25 Jun}}{\text{5 Jul}}$ 1664
	1½pp (13x16) Latin	
130r-131v	News accounts	
	4pp (17x22)	
	Ratisbon	9 Jun 1664
	Ratisbon	12 Jun 1664
	Danzig	14 Jun 1664
133rv	From Lubienietzki at Hamburg	17 May 1664
	1½pp (17x22) Latin	
134r-135v&137rv	From Lubienietzki at Hamburg	24 May 1664
	4pp + ad (17x22) Latin	
136rv	From Lubienietzki at Hamburg	31 May 1664
	1½pp (13x16) Latin	
138r-139v	From Lubienietzki at Hamburg	7 Jun 1664
	2¾pp (17x22) Latin	
140r-142v	News accounts	
	6pp (17x22) Latin	
	Danzig	6 May 1664
	Greece	30 Apr 1664
	Ratisbon	5 May 1664

	Ratisbon	8 May 1664
	Danzig	10 May 1664
	fol 142, Latin verse	
143rv	From Lubienietzki at Hamburg	14 Jun 1664
	1p (10x15) Latin	
144rv&147rv	From Lubienietzki at Hamburg	12 Jul 1664
	2pp & 4 lines + ad (10x15) Latin	
145r-146v&148rv	News accounts	
	6pp (17x22) Latin	
	Vienna	28 Mar 1664
	Ratisbon	3 Apr 1664
	Danzig	5 Apr 1664
	Ratisbon	7,10 Apr 1664
	Königsberg	6 Apr 1664
149rv	From Lubienietzki at Hamburg	19 Apr 1664
	½p (17x22) Latin	
150r-151v	News accounts	
	4pp (17x22) Latin	
	Danzig	16 Apr 1664
	Danzig	19 Apr 1664
	Ratisbon	18 Apr 1664
152rv&155rv	From Lubienietzki at Hamburg	28 Nov 1665
	1p + ad (10x15) Latin	
153rv	Extracts of letters	13 May 1665
	From Hevelius at Danzig	
	1⅔pp (17x22) Latin	
154rv	From Lubienietzki at Hamburg	23 May 1665
	½p (17x22) Latin	
156r-157v	From Lubienietzki at Hamburg	16 May 1665
	2½pp (17x22) Latin	
158r-159v	Extract of a letter/newsletter	
	4pp (17x22) Latin	
	"Excerptum ex litteris ad Ampliss. Joan. Hevelium/d. 14 Julii 1665"	
	Ratisbon	9 Jul 1665
160rv	From Lubienietzki at [Hamburg]	undated
	1½pp (17x22) Latin	

161r-167v	News accounts 14pp (17x22) Latin	
	Ratisbon	13 Aug 1665
	Holmia	12 Aug 1665
	Vienna	15 Aug 1665
	Ratisbon	17 Aug 1665
	Ratisbon	19 Aug 1665
	Hamburg	19/29 Aug 1665
	Warsaw	31 Jul 1665
	Warsaw	29 Oct 1665
	Danzig	7 Nov 1665
168rv	From Lubienietzki at Hamburg 1p & 5 lines (17x22) Latin	24 Oct 1665
169r-177v	News accounts 16pp (17x22) Latin/French	
	Vienna	8 Mar 1665
	Danzig	14 Mar 1665
	Vienna	19 Jul 1665
	Danzig	25 Jul 1665
	Ratisbon	23 Jul 1665
	Vienna	26 Jul 1665
	Warsaw	21 Jul 1665
	[Grodna?]	23 Jul 1665
	Königsberg	29 Jul 1665
	Danzig	1 Aug 1665
	Ratisbon	30 Jul 1665
	Ratisbon	7 May 1665
	Danzig	9 May 1665
	Hamburg	16 May 1665
	fol 171, blank	
178rv	Latin verse J. E. Rautenstein at Düsseldorf 2pp (17x22) Latin [concerning Lubienietzki's work on comets]	1665
179r-180v	From Lubienietzki at Hamburg 3⅔pp (17x22) Latin fol 180, extract	21 Nov 1665
181r-182v	From Rautenstein at Ratisbon 3pp + ad (17x22) Latin	21 Sep 1665
183rv	From Lubienietzki at Hamburg 1½pp (13x16) Latin	26 Sep 1665

184r-188v	News accounts/extracts	
	8pp (17x22) Latin/French	
	Warsaw	29 Sep 1665
	Extract of a letter	6 Oct 1665
	Danzig	10 Oct 1665
	Vienna	14 May 1665
	Vienna	16 May 1665
	Frontiers of Poland	20 May 1665
	Warsaw	16 May 1665
	fol 186, blank	
189r-190v	From Lubienietzki at Hamburg	14 Mar 1665
	2⅓pp + ad (17x22) Latin	
191r-192v	News accounts	
	4pp (17x22) Latin	
	Raab	16 Jul 1664
	Vienna	20 Jul 1664
	Danzig	26 Jul 1664
193r-194v	From Lubienietzki at Hamburg	6 Jun 1665
	3½pp (17x22) Latin	
195r-202v	News accounts	
	16pp (17x22) Latin/French	
	Vienna	23 May 1665
	Warsaw	19 May 1665
	Danzig	30 May 1665
	Extracts of letters (2)	
	Vienna	30 May 1665
	Ratisbon	4 Jun 1665
	Danzig	6 Jun 1665
	Cracow	25 Dec 1665
	Warsaw	24 Dec 1665
	Warsaw	11 Sep 1665
	Danzig	19 Sep 1665
203r-204v	Copy of a letter	
	From Lubienietzki at Hamburg	24 Mar 1665 OS
	To Heinsius	[3 Apr 1665]
	3¾pp (20x31) Latin/Greek	
205r-206v	From Lubienietzki at Hamburg	Kal Aug 1665
	3pp (10x15) Latin	[1 Aug 1665]
207r-209v	From Lubienietzki at Hamburg	9 May 1665
	2pp + ad* Latin/French	
	*fol 207&209 (15x20)	
	*fol 208, List of books (11x11)	

210rv&213rv	From Lubienietzki at Hamburg 1p & 5 lines + ad (17x22) Latin	2 May 1665
211r-212v	Newsletter [in Lubienietzki's hand] 4pp (17x22) Warsaw Warsaw Danzig	 24 Sep 1665 25 Sep 1665 3 Oct 1665
214rv	From Lubienietzki at Hamburg 1½pp (13x16) Latin	10 Oct 1665
215rv	From Lubienietzki at Hamburg 2pp (17x22) Latin	8 May 1666
216r-217v	Extract of a letter "Excerptum ex litteris d. 18 Julii 1665" [concerns Auzout] ⅔p + ad (17x22) Latin fol 217, blank	18 Jul 1665
218rv&221rv	From Lubienietzki at Hamburg 1½pp + ad (17x22) Latin	18 Apr 1665
219r-220v	News accounts/extracts 5pp (17x22) Warsaw Copy of a declaration Danzig	 13 Nov 1665 21 Nov 1665
222r-223v	News accounts 4pp (17x22) Latin/French Vienna Ratisbon Ratisbon Danzig	 19 Apr 1665 20 Apr 1665 23 Apr 1665 25 Apr 1665
224rv&227rv	From Lubienietzki at Hamburg 1½pp (13x16) Latin fol 227, blank	12 Sep 1665
225r-226v	News accounts 4pp (17x22) Latin/French Danzig Vienna Ratisbon Neuburg Neuburg	 18 Apr 1665 13 Apr 1665 13 Apr 1665 3 Apr 1665 10 Apr 1665

228rv	From Lubienietzki at Hamburg	11 Apr 1665
	1p (17x22) Latin	
229rv&231rv	From Lubienietzki at [Hamburg]	25 Mar / 4 Apr 1665
	$1\frac{1}{3}$pp + ad (17x22) Latin/Greek	
230rv	News accounts	
	1p & 2 lines (17x22) Latin	
	Ratisbon	9 Apr 1665
	Ratisbon	12 Mar 1665
	Ratisbon	16 Mar 1665
	Ratisbon	19 Mar 1665
	Ratisbon	23 Mar 1665
	Ratisbon	26 Mar 1665
232r-235v	News accounts	
	8pp (17x22) Latin/French	
	Vienna	29 Mar 1665
	Ratisbon	30 Mar 1665
	Ratisbon	2 Apr 1665
	Berlin	6 Apr 1665
	Danzig	4 Apr 1665
	Vienna	15 Mar 1665
	Ratisbon	16,19 Mar 1665
	Berlin	25 Mar 1665
	Danzig	21 Mar 1665
	Stockholm	18 Mar 1665
236rv&239rv	From Lubienietzki at Hamburg	28 Mar 1665
	$1\frac{1}{2}$pp + ad (17x22) Latin	
237r-238v	News accounts	
	4pp (17x22) Latin/French	
	Warsaw	3 Jul 1665
	Danzig	11 Jul 1665
	Danzig	11 Jul 1665
240r-241v	Extract, Anonymous	undated
	$3\frac{2}{3}$pp (17x22)	
	"Quaesitio de Cometis circa Ludovici Pii excessum"	
242r-245v	News accounts	
	7pp (17x22) Latin/French	
	Vienna	22 Mar 1665
	Ratisbon	23 Mar 1665
	Ratisbon	26 Mar 1665
	Warsaw	ND 1665
	Vienna	28 Jun 1665

	Ratisbon	2 Jul 1665
	Danzig	4 Jul 1665
	Ratisbon	2 Jul 1665
246r-247v	Copy of a document, untitled	10 Jun 1665
	Hamburg-Ratisbon	
	4pp (20x31) Latin	
248r-253v	News accounts	
	12pp (17x22) Latin/French	
	Vienna	13 Jun 1665
	Danzig	20 Jun 1665
	Vienna	5 Jan 1665
	Frontiers of Poland	13 Jan 1665
	Warsaw	4 Jan 1665
	Stockholm	31 Dec 1664
	Vienna	18 Jan 1665
	Warsaw	16 Jan 1665
254rv	From Lubienietzki at Hamburg	10 Jan 1665
	1p (17x22) Latin	
255rv	From [Johannes Müller]* at Hamburg	28 Dec 1664 OS
	To Anonymous	[18 Dec 1664]
	1p & 6 lines (17x22) Latin	
	*[signed "J.M."]	
256rv	From Lubienietzki at Hamburg	13 Jan 1665
	½p (17x22) Latin	
257rv	From Lubienietzki at Hamburg	24 Jan 1665
	¾p (17x22) Latin	
258rv	Extracts concerning comets [in L's hand]	
	1¼pp (17x22) Latin	
	fol 258r, J. A. Huswedel; Johannes Müller	
	fol 258v, Henry Syvers (13/23 Jan 1665)	
259r-262v	News accounts	
	8pp (17x22) Latin/French	
	Vienna	11 Jan 1665
	Danzig	17 Jan 1665
	Ratisbon	12 Jan 1665
	Hamburg-Ratisbon	24 Jan 1665
	Hamburg	24 Jan 1665
	Vienna	1 Feb 1665
	Ratisbon	29 Jan 1665
	Ratisbon	2 Feb 1665
	Ratisbon	5 Feb 1665
	Danzig	6 Feb 1665

263r-264v	From Lubienietzki at Hamburg	14 Feb 1665
	2¾pp (17x22) Latin	
265rv	From Lubienietzki at Hamburg	21 Feb 1665
	1⅓pp (17x22) Latin	
266rv	Newsletter	
	Neuburg	30 Jan 1665
	¾p (17x22)	
267r-268v	News accounts	
	3pp (17x22) Latin/French	
	Vienna	9 May 1665
	Ratisbon	14 May 1665
	Danzig	16 May 1665
269rv	Newsletter	
	[Krenae?]	8 Feb [1663?]
	2pp (17x22)	
270rv	News accounts	
	2pp (17x22) Latin/French	
	Ratisbon	9 Feb 1665
	Danzig	14 Feb 1665
	Copenhagen	31 Jan/10 Feb 1665
271rv&274rv	From Lubienietzki at Hamburg	28 Feb 1665
	1p + ad (10x15) Latin	
272r-273v	News accounts	
	4pp. (17x22) Latin/French	
	Vienna	14 Feb 1665
	Warsaw	13 Feb 1665
	Königsberg	17 Feb 1665
275r-276v	News accounts	
	4pp (17x22) Latin/French	
	Stockholm	25 Feb 1665
	Vienna	26 Feb 1665
	Ratisbon	2 Mar 1665
	Danzig	7 Mar 1665
277r-278v	From Lubienietzki at [Hamburg]	13 Jun 1665
	3¾pp (17x22) Latin	
279rv	News account	
	Warsaw	20 Jun 1665
	1¾pp (17x22)	

280rv	From Lubienietzki at Hamburg 2pp (17x22) Latin	11 Jul 1665
281rv	Newsletter Ratisbon 1p + ad (17x22) Latin	 3 Dec 1665
282rv	From Lubienietzki at Hamburg ½p (13x16) Latin	12 Dec 1665
283rv	From Lubienietzki at Hamburg ½p + ad (13x16) Latin	25 Jul 1665
284r-285v	Newsletter 4pp (17x22) Latin	23 May 1665
286rv	Copy of a letter From Colonel [Niemirycz?] at Frankfurt To Anonymous 1½pp (17x22)	13 Oct 1665
287rv	From Lubienietzki at Hamburg 1½pp (17x22) Latin	8 Aug 1665
288r-294v	News accounts 14pp (17x22) French/Latin Warsaw Danzig Warsaw Warsaw Ratisbon Danzig fol 290, blank	 27 Nov 1665 5 Dec 1665 10 Dec 1665 18 Dec 1665 24 Dec 1665 26 Dec 1665
295rv&298rv	From Lubienietzki at Hamburg 1⅔pp + ad (13x16) Latin	26 Dec 1665
296rv	Extract of a letter From Hevelius at [Danzig] To Anonymous 2pp (17x22) Latin	28 Nov 1665
297rv	News accounts 2pp (17x22) Warsaw Fraustad Berlin	 20 Nov 1665 25 Nov 1665 2 Dec 1665

299rv	From Lubienietzki at Hamburg	5 Dec 1665
	1p + ad (17x22) Latin	
300r-303v	News accounts	
	8pp (17x22) French/Latin	
	Thorn	11 Nov 1665
	Danzig	14 Nov 1665
	The Frontiers	6 Nov 1665
	Berlin	19 Nov 1665
	Vienna	7 Jun 1665
	Ratisbon	11 Jun 1665
	Danzig	13 Jun 1665
304r-308v	News accounts/copy of a letter	
	9pp (17x22) French/Latin	
	From Anonymous at Poland	28 Aug 1665
	Warsaw	29 Aug 1665
	Copenhagen	3 Dec 1664
	Copenhagen	6 Dec 1664
	Vienna	4 Apr 1665
	Ratisbon	11 Apr 1665
	Ratisbon	6 Apr 1665
	Ratisbon	9 Apr 1665
	Danzig	11 Apr 1665
309r-310v	From Lubienietzki at Hamburg	18 Jul 1665
	3½pp (17x22) Latin	
311r-312v	News accounts	
	4pp (17x22)	
	Danzig	18 Jul 1665
	Ratisbon	16 Jul 1665
	Hamburg	25 Jul 1665
313rv	From Lubienietzki at Hamburg	$\frac{20}{30}$ May 1665
	1⅔pp (17x22) Latin/Greek	
314r-316v	From Lubienietzki at Hamburg	10 Apr 1666
	2⅔pp + ad (17x22) Latin	
	fol 316, blank	
317r-318v	News accounts/extracts	
	4pp (17x22)	
	Danzig	12 Sep 1665
	Frankfurt	4/14 Sep 1665
	Extract of a letter from Poland	7 Sep 1665
319rv&322rv	From Lubienietzki at Hamburg	29 Aug 1665
	1⅓pp + ad (13x16) Latin	

320r-321v	News accounts/extracts	
	4pp (17x22)	
	Danzig	26 Sep 1665
	Extract of a letter from Colonel	
	[Niemirycz?]	21 Sep 1665
	Danzig	26 Sep 1665
323r-324v	News accounts	
	4pp (17x22) French/Latin	
	Danzig	29 Aug 1665
	Frankfurt	30 Aug 1665
	Ratisbon	24 Aug 1665
	Ratisbon	27 Aug 1665
	Hamburg	5 Sep 1665
325rv	From Lubienietzki at Hamburg	19 Sep 1665
	1p & 7 lines (17x22) Latin	
326r-330v	News accounts	
	9pp (17x22) Latin/French	
	Ratisbon	7 Sep 1665
	Ratisbon	10 Sep 1665
	Stockholm	22 Apr 1665
	Warsaw	9 Oct 1665
	Hamburg	24 Oct 1665
	Warsaw	16 Oct 1665
	Danzig	24 Oct 1665
	The Frontiers	26 Oct 1665
331rv	From Lubienietzki at Hamburg	31 Oct 1665
	½p (17x22) Latin	
332r-333v	News accounts	
	4pp (17x22)	
	Warsaw	23 Feb 1665
	Königsberg	24 Feb 1665
334r-335v	News accounts	
	4pp (15x20) Latin	
	Vienna	22 Feb 1665
	Ratisbon	23 Feb 1665
	Ratisbon	26 Feb 1665
	Hamburg	7/17 Mar 1665
336rv	Newsletter	
	2pp (17x22)	
	Warsaw	25 Oct 1665
	Danzig	31 Oct 1665

337rv	Extracts of letters	5 Apr 1665
	Lubienietzki at Hamburg	
	Heinsius at Stockholm	11 Apr 1665 OS
	$1\frac{2}{3}$pp (17x22) Latin	[21 Apr 1665]
338r-339r	Extract of a letter	14 Apr 1665
	From Jona [Tornelii?] at Upsala	
	To Heinsius	
	$2\frac{1}{3}$pp (17x22) Latin	
339rv	Extract of a letter	26 Apr 1665
	[From Heinsius at Stockholm	
	To Tornelii?]	
	$\frac{2}{3}$p & $\frac{1}{2}$p (17x22) Latin	
340r-341v	News accounts	
	3pp (17x22) Latin	
	Danzig	2 Jun 1665
	Hamburg	11/21 Mar 1665
342rv	Copy of a letter [in L's hand]	28 Apr 1665
	From J. B. Riccioli at Bologna	
	To [Lubienietzki at Hamburg]	
	$1\frac{3}{4}$pp (17x22) Latin	
342v-343v	Copy of a letter	9 May 1665
	From Lubienietzki at Hamburg-Danzig	
	To Riccioli at [Bologna]	
	$\frac{1}{4}$p & $1\frac{2}{3}$pp (17x22) Latin	
344rv	From Lubienietzki at Hamburg	28 Aug 1666
	1p (17x22) Latin	
345r-347v	News accounts	
	6pp (17x22)	
	Warsaw	10 Aug 1666
	Warsaw	13 Aug 1666
	Danzig	11 Sep 1666
	Berlin	14 Sep 1666
	Warsaw	21 Aug 1666
348rv	Extract of a letter*	27 Aug 1666
	From Anonymous to Anonymous	
	1p (17x22)	
	*[French translation from Prussian]	
349r-358v	News accounts/extracts	
	18pp (17x22) French/Latin	
	Danzig	8 May 1666

	Danzig (2)	8 May 1666
	The Frontiers	10 May 1666
	Warsaw	30 May 1666
	Frontiers of Litvanie	30 May 1666
	Königsberg	4 Jun 1666
	Extract of a letter, "d'un homme de qualité touchant les affairs de Pologne..."	4 Jun 1666
	Frontiers of Poland	12 Jun 1666
	Danzig	12 Jun 1666
	Warsaw	11 Jun 1666
	Warsaw	25 Jun 1666
	Extract of a letter from Prussia	28 Jul 1666
	Extract of a letter from Prussia	30 Jul 1666
	Extract of a letter from Prussia	2 Aug 1666
	Warsaw	23 Jul 1666
	Königsberg	25 May 1666
	Warsaw	25 May 1666
359rv	From Lubienietzki at Hamburg 1¾pp (17x22) Latin	31 Jul 1666
360r-361v	News accounts 4pp (17x22)	
	Frontiers of Grand Poland	26 Jul 1666
	Warsaw	9 Jul 1666
	Frontiers of Grand Poland	14 Jul 1666
362rv	Newsletter Frontiers of Grand Poland 1½pp (17x22)	23 Jul 1666
363rv	From Lubienietzki at Hamburg 1½pp (17x22) Latin	5 Jun 1666
364rv	Latin verse 1¾pp (17x22) Latin	Kal Jun 1666 [1 Jun 1666]
365r-366v	From Lubienietzki at Hamburg 4pp (17x22) Latin	24 Jul 1666
367rv	From Lubienietzki at Hamburg To M. de Broussel at Paris ½p + ad (17x22) Latin	15 May 1666
368r-369v	From [Pieter Pels] at The Hague 2½pp + ad (17x22)	27 May 1666
370r-379r	News accounts/extracts 19pp (17x22) French/Latin [w/PS, 379r]	

	Hamburg, (14x16)	2 Jul 1666
	Vienna	3 Jun 1666
	Warsaw	23 Apr 1666
	Extract of a letter of Warsaw	23 Apr 1666
	Breslau [Wroclaw]	28 Apr 1666
	Danzig-Hamburg	13 Apr 1666
	Warsaw	19 Mar 1666
	Danzig	27 Mar 1666
	Danzig	27 Mar 1666
	Frontiers of Grand Poland	27 Mar 1666
	Warsaw	26 Mar 1666
	Frontiers of Grand Poland	2 Apr 1666
	Danzig	3 Apr 1666
379rv	From [Lubienietzki?] at Hamburg 7 lines (17x22) Latin	10 Apr 1666
380r-381v	News accounts/extracts 4pp (17x22)	
	Warsaw	16 Apr 1666
	Extract of a letter of Warsaw	undated
	Stockholm	21 Apr 1666
382r-383v	From Lubienietzki at Hamburg 3⅓pp (10x15) Latin	27 Feb 1666
384rv	News accounts 2pp (17x22) Latin	
	Frontiers of Grand Poland	25 Jan 1666
	Königsberg	26 Jan 1666
	Hamburg	6 Feb 1666
385rv	From Anonymous at Hamburg To Anonymous 1p (17x22) Latin	6 Feb 1666
386rv	From Lubienietzki at Hamburg 1½pp (13x16) Latin	23 Jan 1666
387rv	News account	
	Ratisbon ½p (13x16) Latin	14 Jan 1666
388r-389v	Brief dissertation 3½pp (17x22) Latin "Dissertatio de Statu Orbis Christiani A°/1666..."	undated

390r-391v	News accounts 2pp (13x16) Cracow Warsaw	 29 Dec 16[65] 7 Jan 1666
392r-393v	From Lubienietzki at Hamburg 2pp & 4 lines + ad (17x22) Latin	5 Feb 1667
394rv&397rv	From Lubienietzki at Hamburg 1p + ad (15x20) Latin	9 Jul 1667
395r-396v	From Lubienietzki at Amsterdam 3pp + ad (18x21)&(14x21) Latin	28 Nov 1668
398rv&401rv	From Lubienietzki at [Altenaviae] 2pp + ad (17x22) Latin	27 Apr 1669
399r-400v	From Lubienietzki at Hamburg 1½pp (13x16) Latin fol 400, blank	17 May 1672

▼ ▼ ▼

COLLECTION BOULLIAU, VOLUME XIV, FONDS FRANÇAIS 13032

1rv	From Ph Pels at Danzig 1p (15x20)	10 May 1661
2rv	From [Ph Pels] at Danzig 1p (15x20)	24 May 1661
3rv	From Ph Pels at Danzig 2pp (15x20)	31 May 1661
4rv	From [Ph Pels] at Danzig 1p (15x20)	14 Jun 1661
5rv	From Ph Pels at Danzig 1p (15x20)	29 Jun 1661
6rv	From Ph Pels at Danzig 1p (15x20)	12 Jul 1661
7rv	From Ph Pels at Danzig 1p (15x20)	19 Jul 1661
8rv&11rv	From Ph Pels at Danzig 1½pp + ad (15x20)	3 Dec 1661
9r-10v	From Ph Pels at Danzig B at The Hague 2pp & 2 lines + ad (15x20)	24 Dec 1661
12rv	From [Ph Pels] at Danzig B at The Hague 1p + ad (15x20)	28 Jan 1662
13r-14v	From [Ph Pels] at Danzig B at The Hague 2½pp + ad (15x20)	25 Feb 1662
15rv	Newsletter Danzig 1p (15x20)	11 Mar 1662
16r-17v	From [Ph Pels] at Danzig B at The Hague 3pp + ad (15x20)	18 Mar 1662
18rv	From [Ph Pels] at Danzig 1½pp (15x20)	1 Apr 1662

19rv	Anonymous note 9 lines (8x16)	undated
20rv	From [Ph Pels] at Danzig 2pp (15x20)	8 Apr 1662
21rv	From [Ph Pels at Danzig] 2pp (15x20)	6 May 1662
22rv	From [Pels] at The Hague 1½pp + ad (20x31)	17 Aug 1662
23rv	From [Ph Pels] at Danzig 1p + ad (15x20)	16 Sep 1662
24rv&26rv	From [Pels] at The Hague 3pp + ad (15x20)	21 Sep 1662
25rv	From [Pels] at The Hague 1p + ad (20x20)	28 Sep 1662
27rv	From [Pels] at The Hague 1p + ad (20x31)	30 Nov 1662
28r-29v	From [Pels] at The Hague 3pp + ad (15x20)	2 Nov 1662
30rv&35rv	From [Ph Pels] at Danzig 2pp & 3 lines + ad (15x20)	11 Nov 1662
31rv	From [Pels] at The Hague 1½pp + ad (15x20)	23 Nov 1662
32rv	From [Ph Pels] at Danzig 2pp (15x20)	18 Nov 1662
33rv	From [Ph Pels] at Danzig 1½pp (15x20)	24 Dec 1662
34rv	From O Homanne at Königsberg To Anonymous 2pp (15x17)	[10?] Dec 1662
36rv	From [Ph Pels] at Danzig 2pp (15x20)	29 Dec 1662
37r-38v	From [Pels] at The Hague 3pp + ad (15x20)	28 Dec 1662

39rv	From [Ph Pels at Danzig] 6 line fragment (15x20)	undated
40r-41v	From [Pels] at The Hague 3pp + ad (17x22)	3 Jan 1663
42rv	From [Pels] at The Hague 1p + ad (20x31)	11 Jan 1663
43rv	Epitaph 14 lines (15x18) Latin "Epitaphium in caedem Dni. Gonzewsky"	undated
44rv	From [Pels] at The Hague 1p + ad (20x31)	8 Feb 1663
45rv	From [Ph Pels] at Danzig 1p + ad (15x20)	3 Mar 1663
46r-47v	From [Pels] at The Hague 2½pp + ad (15x20)	15 Mar 1663
48rv	From [Pels] at The Hague 1p + ad (20x20)	3 Apr 1663
49rv	News accounts 1p & 6 lines (20x31) Léopol Warsaw Worborz	 2 Mar 1663 9 Mar 1663 2 Mar 1663
50r-51v	From [Pels] at The Hague 3pp + ad (15x20)	14 Jun 1663
52r-53v	From [Pels] at The Hague 3pp & 5 lines + ad (15x20)	5 Jul 1663
54r-55v	From [Pels] at The Hague 3pp + ad (15x20)	19 Jul 1663
56r-57v	From [Pels] at The Hague 3pp + ad (15x20)	2 Aug 1663
58rv	From [Pels] at The Hague 1p + ad (15x20)	30 Aug 1663
59r-60v	From [Ph Pels] at Danzig 3pp + ad (15x20)	18 Aug 1663

61rv	From [Pels] at The Hague 1p + ad (15x20)	1 Nov 1663
62rv	From [Ph Pels] at Danzig 1p + ad (15x20)	20 Oct 1663
63r-64v	From [Pels] at The Hague 4pp (15x20)	17 Jan 1664
65r-67v	From [Pels] at The Hague 5pp + ad (15x20) fol 66, news account, 30 Jan 1664, etc.	28 Feb 1664
68r-69v	From [Pels] at The Hague 3pp + ad (15x20)	7 Feb 1664
70r-71v	From [Pels] at The Hague 4pp (15x20)	20 Mar 1664
72r-73v	From [Pels] at The Hague 3pp + ad (15x20)	27 Mar 1664
74r-76v	From [Pels] at The Hague 5pp & 6 lines + ad (15x20) [fol 75, unsigned to Pels, 28 Mar 1664, Königsberg]	9 Apr 1664
77r-78v	From [Pels] at The Hague 4pp (15x20)	23 Apr 1664
79r-80v	From [Pels] at The Hague 3pp + ad (15x20)	8 May 1664
81r-82v	From [Pels] at The Hague 3pp + ad (15x20)	22 May 1664
83rv&86rv	From [Pels] at The Hague 2pp + ad (15x20)	29 May 1664
84r-85v	From [Pels] at The Hague 2½pp + ad (15x20)	18 Jun 1664
87r-88v	From [Pels] at The Hague 3pp + ad (15x20)	9 Jul 1664
89rv	From [Pels] at The Hague 1p & 5 lines (15x20)	17 Jul 1664

90r-91v	From [Ph Pels] at Danzig 3pp + ad (15x18)	19 Jul 1664
92rv	From [Pels] at The Hague 2pp (15x20)	31 Jul 1664
93rv	Text, untitled 1p (20x30) Latin	undated
94rv	From [Pels] at The Hague 1⅓pp + ad (15x20)	21 Aug 1664
95rv	From [Pels] at The Hague 1p + ad (15x20)	28 Aug 1664
96rv	From Ph Pels at Danzig 2pp (15x20)	6 Sep 1664
97rv	From [Pels] at The Hague 1p + ad (15x20)	11 Sep 1664
98rv&103rv	From [Pels] at The Hague 2pp + ad (15x20)	25 Sep 1664
99rv	From [Ph Pels] at Danzig 1⅓pp (15x20)	11 Oct 1664
100rv	From [Pels] at The Hague 1p + ad (15x20)	23 Oct 1664
101r-102v	From [Ph Pels] at Danzig 3pp + ad (15x20)	29 Nov 1664
104r-105v	From [Ph Pels] at Warsaw 3½pp + ad (15x20)	12 Dec [1664]
106rv	From [Pels] at The Hague 2pp (15x20)	1 Jan 1665
107rv	From [Pels] at The Hague 2pp (15x20)	29 Jan 1665
108rv	Text, "Decretũm Finale" 1p & 8 lines (18x28) Latin	undated
109rv	News accounts 2pp (13x18) Latin	
	Bononiae [Vidin]	23 Dec 1664
	Warsaw	14 Dec 1664
	?	8 Jan 1665

110r-111v	From [Ph Pels] at Danzig 3pp + ad (15x20)	21 Feb 1665
112rv&114rv	From [Pels] at The Hague 2pp + ad (15x20)	5 Mar 1665
113rv	From [Pels] at The Hague 1p + ad (15x20)	12 Mar 1665
115rv	From [Pels] at The Hague 2pp (15x20)	26 Mar 1665
116rv	From [Pels] at The Hague 2pp (15x20)	16 Apr 1665
117r-118v	From [Pels] at The Hague $2\frac{1}{2}$pp + ad (15x20)	23 Apr 1665
119r-120v	From [Ph Pels] at Danzig $2\frac{1}{2}$pp + ad (15x20)	25 Apr 1665
121r-122v	From [Pels] at The Hague $2\frac{1}{2}$pp + ad (15x20)	30 Apr 1665
123rv	From [Pels] at The Hague 1p + ad (15x20)	7 May 1665
124rv	From [Pels] at The Hague 1p & 6 lines + ad (15x20)	14 May 1665
125rv	From [Pels] at The Hague $1\frac{2}{3}$pp (15x20)	21 May 1665
126rv	From [Pels] at The Hague $1\frac{2}{3}$pp (15x20)	9 Jul 1665
127r-128v	Text $2\frac{1}{2}$pp (15x20) Latin "Johannes de Samoyocio Palatinus Sendomiriensis/Magnorum Cancellariorum Ari et Parentis..."	undated
129r-130v	Copy of a letter in Pels' hand From Brahe, Eleonora, Stenbock, Bonde, Gardie, etc. at Holmensi $3\frac{1}{3}$pp (20x30) Latin	20 Jun 1665
131rv	From [Pels] at The Hague 1p + ad (15x20)	16 Jul 1665

132r-133v	From [Pels] at The Hague 2½pp + ad (15x20)	23 Jul 1665
134r-135v	From [Pels] at The Hague 3pp + ad (15x20)	13 Aug 1665
136rv	From [Pels] at The Hague 1½pp (15x20) *[closing date 18 Aug]	20 Aug 1665*
137r-138v	From [Pels at The Hague] 2½pp + ad (15x20)	5 Aug 1665
139r-140v	Brief tract in Pels' hand 4pp (15x20) [Concerns the history of the Providence of Guelre]	undated
141r-142v	From [Pels] at The Hague 2½pp + ad (15x20)	27 Aug 1665
143r-144v	From [Pels] at The Hague 4pp (15x20)	3 Sep 1665
145r-146v	From [Pels] at The Hague 3pp + ad (15x20)	10 Sep 1665
147rv	Newsletter [in Ph Pels' hand] Warsaw 1p (15x20)	11 Sep 1665
148r-149v	From [Pels] at The Hague 3pp + ad (15x20)	17 Sep 1665
150r-151v	From [Pels] at The Hague 3½pp (15x20)	24 Sep 1665
152r-153v	From [Pels] at The Hague 4pp (15x20)	8 Oct 1665
154r-155v	From [Pels] at The Hague 4pp (15x20)	29 Oct 1665
156r-157v	From [Pels] at The Hague 4pp (15x20)	5 Nov 1665
158r-159v	From [Pels at The Hague] 3pp + ad (15x20)	12 Nov 1665

160r-161v	From [Pels] at The Hague 3⅓pp (15x20)	19 Nov 1665
162r-163v	From [Pels] at The Hague 3¾pp (15x20)	3 Dec 1665
164r-165v	From [Pels] at The Hague 3pp + ad (15x20)	10 Dec 1665
166r-167v	From [Pels] at The Hague 3½pp (15x20)	24 Dec 1665
168r-170v	From [Pels] at The Hague 4¾pp + ad (15x20)	7 Jan 1666
171r-175v	Copy of a document in Pels' hand 9¼pp (15x20) Latin "Anatomica Inspectio et indicium/ de/Corpore Regni Poloniae"	undated
176r-177v	From [Pels] at The Hague 4pp (15x20)	21 Jan 1666
178rv	From [Pels] at The Hague 2pp (15x20)	28 Jan 1666
179r-181v	From [Pels] at The Hague 5pp + ad (15x20)	23 Feb 1666
182r-183v	From [Pels] at The Hague 4pp (15x20)	11 Mar 1666
184r-185v	From [Pels] at The Hague 2½pp + ad (15x20)	18 Mar 1666
186r-187v	From [Pels] at The Hague 3pp + ad (15x20)	25 Mar 1666
188r-189v	From [Pels] at The Hague 2pp & 5 lines + ad (15x20)	1 Apr 1666
190r-191v	From [Pels] at The Hague 3pp & 6 lines + ad (15x20)	8 Apr 1666
192r-193v	From [Pels] at The Hague 3pp & 4 lines + ad (15x20)	15 Apr 1666
194r-195v	From [Pels] at The Hague 3pp + ad (15x20)	22 Apr 1666

196r-197v	From [Pels] at The Hague 3¼pp + ad (15x20)	29 Apr 1666
198r-200v	From [Pels] at The Hague 4pp + ad (15x20) fol 199, Latin encl	6 May 1666
201rv&204rv	From [Pels] at The Hague 2pp + ad (15x20)	13 May 1666
202r-203v	From [Pels] at The Hague 2½pp + ad (15x20)	20 May 1666
205r-206v	From [Pels] at The Hague 2⅔pp + ad (15x20)	17 Jun 1666
207r-208v	From [Pels] at The Hague 3pp + ad (15x20)	24 Jun 1666
209r-210v	From [Pels] at The Hague 2⅓pp + ad (15x20)	1 Jul 1666
211rv	From [Pels] at The Hague 1p + ad (15x20)	8 Jul 1666
212r-213v	From [Pels] at The Hague 2⅔pp + ad (15x20)	22 Jul 1666
214r-215v	From [Pels] at The Hague 4pp (15x20)	5 Aug 1666
216r-217v	From [Pels] at The Hague 2½pp + ad (15x20)	12 Aug 1666
218rv&221rv	From [Pels] at The Hague 2pp + ad (15x20)	19 Aug 1666
219r-220v	From [Pels] at The Hague 2⅔pp + ad (15x20)	2 Sep 1666
222rv	From [Pels] at The Hague 2pp (15x20)	9 Sep 1666
223rv	From [Pels] at The Hague 2pp (15x20)	23 Sep 1666
224r-225v	From [Pels] at The Hague 3pp + ad (15x20)	7 Oct 1666

226rv	From [Pels] at The Hague 1p & 5 lines + ad (15x20)	22 Oct 1666
227r-228v	From [Pels] at The Hague $2\frac{1}{2}$pp + ad (15x20)	14 Oct 1666
229r-230v	From [Pels] at The Hague $2\frac{1}{2}$pp + ad (15x20)	28 Oct 1666
231r-232v	From [Pels at The Hague] 3pp + ad (15x20) *dated by B	[4 Nov 1666]*
233r-234v	From [Pels] at The Hague 3pp + ad (15x20)	18 Nov 1666
235r-236v	From [Pels] at The Hague $2\frac{1}{2}$pp + ad (15x20)	2 Dec 1666
237r-238v	From [Pels] at The Hague $3\frac{1}{2}$pp (15x20)	16 Dec 1666
239r-240v	From [Pels] at The Hague $2\frac{1}{4}$pp + ad (15x20)	23 Dec 1666
241r-242v	From [Pels] at The Hague 3pp + ad (15x20)	6 Jan 1667
243rv&246rv	From [Pels] at The Hague $1\frac{1}{2}$pp + ad (15x20)	20 Jan 1667
244r-245v	From [Pels] at The Hague $2\frac{1}{2}$pp + ad (15x20)	3 Feb 1667
247r-248v	From [Pels] at The Hague 3pp + ad (15x20)	10 Feb 1667
249r-250v	From [Pels] at The Hague 3pp + ad (15x20)	3 Mar [1667]
251rv	From [Pels] at The Hague $1\frac{2}{3}$pp (15x20)	10 Mar 1667
252r-253v	From [Pels] at The Hague $2\frac{2}{3}$pp + ad (15x20)	17 Mar 1667
254rv	From [Pels] at The Hague 2pp (15x20)	7 Apr 1667

255rv	From [Pels] at The Hague 2pp (15x20)	14 Apr 1667
256rv	From [Pels] at The Hague 2pp (15x20)	31 Mar 1667
257rv	From [Pels] at The Hague 2pp (15x20)	28 Apr 1667
258r-259v	From [Pels] at The Hague 2½pp + ad (15x20)	5 May 1667
260rv	From [Pels] at The Hague 2pp (15x20)	12 May 1667
261r-262v	From [Pels] at The Hague 2½pp + ad (15x20)	19 May 1667
263r-264v	From [Pels] at The Hague 2¼pp + ad (15x20)	25 May 1667
265rv	From [Pels] at The Hague 1p + ad (15x20)	2 Jun 1667
266r-267v	From [Pels] at The Hague 3pp + ad (15x20)	16 Jun 1667
268rv	From [Pels] at The Hague 1p + ad (15x20)	23 Jun 1667
269r-270v	From [Pels at The Hague] 3¾pp (15x20)	29 Jun 1667
271r-272v	From [Pels] at The Hague 4pp (15x20)	7 Jul 1667
273r-274v	From [Pels at The Hague] 4pp (15x20)	14 Jul 1667
275r-277v	From [Pels at The Hague] 5pp + ad (15x20)	21 Jul 1667
278r-279v	From [Pels] at The Hague 2½pp + ad (15x20)	28 Jul 1667
280r-281v	From [Pels at The Hague] 3½pp (15x20)	12 Aug 1667

282r-283v	From [Pels at The Hague] 2½pp + ad (15x20)	18 Aug 1667
284r-285v	From [Pels] at The Hague 2½pp + ad (15x20)	25 Aug 1667
286rv	From [Pels at The Hague] 2pp (15x20)	[1] Sep 1667
287rv	From [Pels at The Hague] 2pp (15x20)	8 Sep 1667
288r-289v	From [Pels at The Hague] 2½pp + ad (15x20)	22 Sep 1667
290r-291v	From [Pels] at The Hague 3pp + ad (15x20)	[29 Sep] 1667
292rv	From [Pels at The Hague] 1½pp (15x20)	1 Oct 1667
293r-294v	From [Pels at The Hague] 3pp + ad (15x20)	19 Oct 1667
295rv	From [Pels] at The Hague 1½pp (15x20)	27 Oct 1667
296rv	Copy in Pels' hand 1p + ad (15x20) Latin "Magnae Britanniae/et/Foederati Belgii Ordinum/Leones reconciliati." Concludes: "F.K.V.H."	undated
297r-298v	From [Pels at The Hague] 2½pp + ad (15x20)	3 Nov 1667
299r-300v	From [Pels at The Hague] 4pp (15x20) Latin	17 Nov 1667
301rv	From [Pels] at The Hague 1¾pp (15x20)	24 Nov 1667
302rv	From [Pels] at The Hague 1½pp (15x20)	1 Dec 1667
303r-304v	From [Pels at The Hague] 2½pp + ad (15x20)	8 Dec 1667

305r-306v	From [Pels] at The Hague 3pp + ad (15x20)	15 Dec 1667
307r-308v	From [Pels at The Hague] 2pp & 4 lines + ad (15x20)	22 Dec 1667

▼ ▼ ▼

COLLECTION BOULLIAU, VOLUME XV, FONDS FRANÇAIS 13033

1rv&4rv	From [Pels at The Hague] 2⅓pp + ad (15x20)	5 Jan 1668
2r-3v	From [Pels at The Hague] 3pp + ad (15x20)	12 Jan 1668
5rv&8rv	From [Pels at The Hague] 2pp + ad (15x20)	19 Jan 1668
6r-7v	From [Pels at The Hague] 3pp + ad (15x20)	2 Feb 1668
9rv&13rv	From [Pels at The Hague] 2pp + ad (15x20)	9 Feb 1668
10rv	Newsletter Warsaw 1½pp (15x20) Latin	16 Jan 1668
11r-12v	From [Pels at The Hague] 2⅔pp + ad (15x20)	16 Feb 1668
14rv	From [Pels at The Hague] 1½pp (15x20)	23 Feb [1668]
15r-17v	From [Pels at The Hague] 3pp + ad (15x20)	1 Mar 1668
18r-19v	Newsletter Warsaw 2⅔pp + ad (15x20) French/Latin	11 Feb NY
20r-21v	From [Pels at The Hague] 3⅓pp* (15x20) *w/enclosure	[8 Mar 1668]
22rv&25rv	From [Pels at The Hague] 1½pp + ad (15x20)	8 Mar 1668
23r-24v&26rv	From [Pels at The Hague] 4⅓pp + ad (15x20)	15 Mar 1668
27r-28v	From [Pels] at The Hague 3⅔pp (15x20)	22 Mar 1668

29r-30v	From [Pels] at The Hague 2½pp + ad (15x20)	28 Mar 1668
31rv&36rv	From [Pels] at The Hague 1⅔pp + ad (15x20)	5 Apr 1668
32rv&35rv	From [Pels] at The Hague 2pp & 3 lines + ad (15x20)	19 Apr 1668
33r-34v	From [Pels] at The Hague 3pp + ad (15x20)	12 Apr 1668
37r-38v	From [Pels at The Hague] 2½pp + ad (15x20)	26 Apr 1668
39rv&44rv	From [Pels at The Hague] 2pp + ad (15x20)	3 May 1668
40rv&43rv	From [Pels at The Hague] 2pp & 4 lines + ad (15x20)	17 May 1668
41r-42v	From [Pels] at The Hague 2½pp + ad (15x20)	10 Apr 1668
45r-46v	From [Pels at The Hague] 3pp + ad (15x20)	31 May 1668
47rv	From [Pels at The Hague] 2pp (15x20)	7 Jun 1668
48rv	Latin verse in Pels' hand 1p + ad (15x20) Latin "Quelqu'ûn de nos poëtes a fait les vers/ suivants sûr les conjointures présents."	undated
49rv	From [Pels at The Hague] 1½pp (15x20)	21 Jun 1668
50rv	From [Pels at The Hague] 2pp (15x20)	28 Jun 1668
51rv	From [Pels at The Hague] 2pp (15x20)	5 Jul 1668
52r-53v	From [Pels at The Hague] 3pp + ad (15x20)	19 Jul 1668
54r-55v	From [Pels at The Hague] 2½pp + ad (15x20)	12 Jul 1668

56r-57v	From [Pels at The Hague] 3pp + ad (15x20)	[26 Jul 1668]
58r-59v	From [Pels at The Hague] 3pp + ad (15x20)	2 Aug 1668
60r-61v	From [Pels at The Hague] 3pp + ad (15x20)	9 Aug 1668
62r-64v	From [Pels at The Hague] 4$\frac{1}{2}$pp + ad (15x20)	16 Aug 1668
65rv	From [Pels at The Hague] 2pp (15x20)	23 Aug 1668
66rv	From [Pels at The Hague] 2pp (15x20)	30 Aug 1668
67r-68v	From [Pels] at The Hague 3$\frac{2}{3}$pp + ad (15x20)	5 Sep 1668
69r-70v	From [Pels] at The Hague 2$\frac{2}{3}$pp + ad (15x20)	13 Sep 1668
71r-72v	From [Pels] at The Hague 3$\frac{2}{3}$pp (15x20)	20 Sep 1668
73r-74v	From [Pels] at The Hague 4pp (15x20)	27 Sep 1668
75r-77v	From [Pels] at The Hague 5pp + ad (15x20)	4 Oct 1668
78rv	From [Pels] at The Hague 1p + ad (15x20)	11 Oct 1668
79r-80v	From [Pels] at The Hague 3pp + ad (15x20)	18 Oct 1668
81r-82v	From [Pels] at The Hague 4pp (15x20)	25 Oct 1668
83r-84v	From [Pels] at The Hague 3pp + ad (15x20)	1 Nov 1668
85r-86v	From [Pels at The Hague] 3pp + ad (15x20) Latin	8 Nov 1668

87rv	From [Pels] at The Hague 2pp (15x20) Latin	15 Nov 1668
88rv	From [Pels at The Hague] 2pp (15x20)	22 Nov 1668
89r-90v	From [Pels] at The Hague 3pp + ad (15x20)	29 Nov 1668
91r-92v	From [Pels] at The Hague 3pp + ad (15x20)	6 Dec 1668
93r-94v	From [Pels] at The Hague 2½pp + ad (15x20)	13 Dec 1668
95r-97v	From [Pels at The Hague] 5pp + ad (15x20)	20 Dec 1668
98r-99v	From [Pels] at The Hague 3pp & 3 lines + ad (15x20)	[27] Dec 1668
100rv&105rv	From [Pels at The Hague] 1½pp + ad (15x20) *[corrected from 1668]	3 Jan 1669*
101rv&104rv	From [Pels] at The Hague 2pp & 5 lines + ad (15x20)	17 Jan 1669
102r-103v	From [Pels at The Hague] 3pp + ad (15x20)	24 Jan 1669
106r-108v	From [Pels at The Hague] 4⅓pp + ad (15x20)	31 Jan 1669
109r-110v	From [Pels] at The Hague 2⅔pp + ad (15x20)	14 Feb 1669
111r-112v	From [Pels] at The Hague 3pp + ad (15x20)	21 Feb 1669
113r-114v	From [Pels at The Hague] 3pp + ad (15x20)	28 Feb 1669
115rv	From [Pels] at The Hague 1p + ad (15x20)	7 Mar 1669
116rv&121rv	From [Pels] at The Hague 2pp & 4 lines + ad (15x20)	13 Mar 1669

117rv	From [Pels] at The Hague 2pp (15x20)	21 Mar 1669
118rv	From [Pels] at The Hague 1p + ad (15x20)	28 Mar 1669
119r-120v	From [Pels] at The Hague 1½pp + ad (15x20) *B corrects from March	4 [Apr]* 1669
122rv	From [Pels at The Hague] 2pp (15x20)	11 Apr 1669
123r-124v	From [Pels] at The Hague 2⅔pp + ad (15x20)	18 Apr 1669
125r-126v	From [Pels at The Hague] 3pp + ad (15x20)	25 Apr 1669
127r-128v	From [Pels at The Hague] 2⅔pp + ad (15x20)	2 May 1669
129r-130v	From [Pels at The Hague] 3pp + ad (15x20)	16 May 1669
131rv	From [Pels at The Hague] 2pp (15x20)	23 May 1669
132r-133v	From [Pels at The Hague] 3pp + ad (15x20)	30 May 1669
134r-135v	From [Pels] at The Hague 4pp (15x20)	6 Jun 1669
136r-138v	From [Pels at The Hague] 4⅔pp + ad (15x20) fol 138, fragment	20 Jun 1669
139r-140v	From [Pels] at The Hague 4pp (15x20)	27 Jun 1669
141r-142v	From [Pels at The Hague] 3⅔pp (15x20)	4 Jul 1669
143r-144v	From [Pels] at The Hague 3½pp (15x20)	11 Jul 1669
145r-146v	From [Pels] at The Hague 3pp + ad (15x20)	18 Jul 1669

147rv&152rv	From [Pels] at The Hague 2pp & 1 line + ad (15x20)	25 Jul 1669
148r-151v	From [Pels] at The Hague 5½pp + ad (15x20)	26 Aug 1669
153r-154v	From [Pels] at The Hague 2¼pp + ad (15x20)	16 Oct 1669
155rv	From [Pels] at The Hague 1⅓pp (15x20)	31 Oct 1669
156r-157v	From [Pels] at The Hague 2⅔pp + ad (15x20)	7 Nov 1669
158r-160v	From [Pels] at The Hague 4½pp + ad (15x20)	14 Nov 1669
161r-163v	From [Pels at The Hague] 4½pp + ad (15x20)	21 Nov 1669
164r-165v	From [Pels] at The Hague 4pp (15x20)	28 Nov 1669
166r-168v	From [Pels] at The Hague 5pp + ad (15x20)	5 Dec 1669
169rv	From [Pels] at The Hague 2pp (15x20)	12 Dec 1669
170rv	From [Pels] at The Hague 1p & 8 lines + ad (15x20)	19 Dec 1669
171r-172v	From [Pels at The Hague] 3pp + ad (15x20)	2 Jan 1670
173r-174v	From [Pels] at The Hague 3pp + ad (15x20)	9 Jan 1670
175rv	From [Pels] at The Hague 1p (15x20)	16 Jan 1670
176rv	From [Pels] at The Hague 1½pp (15x20)	23 Jan 1670
177r-178v	From [Pels] at The Hague 2½pp + ad (15x20)	30 Jan 1670

179rv	From [Pels] at The Hague 2pp (15x20)	13 Feb 1670
180rv&182rv	From [Pels] at The Hague 2pp & 2 lines + ad (15x20)	19 Feb 1670
181rv	From [Pels at The Hague] 1½pp (15x20)	27 Feb 1670
183rv	From [Pels] at The Hague 1p & 4 lines (15x20)	13 Mar 1670
184r-185v	From [Pels at The Hague] 2⅔pp + ad (15x20)	27 Mar 1670
186r-188v	From [Pels at The Hague] 4⅔pp + ad (15x20)	3 Apr 1670
189rv&194rv	From [Pels] at The Hague 2pp & 1 line + ad (15x20)	10 Apr 1670
190rv	From [Pels] at The Hague 1⅔pp (15x20)	17 Apr 1670
191rv	From [Pels] at The Hague 1p + ad (15x20)	24 Apr 1670
192r-193v	From [Pels] at The Hague 4pp (15x20)	1 May 1670
195r-196v	From [Pels at The Hague] 2½pp + ad (15x20)	8 May 1670
197r-198v	From [Pels at The Hague] 2½pp + ad (15x20)	22 May [1670]
199r-200v	From [Pels] at The Hague 3pp + ad (15x20)	29 May 1670
201rv	From [Pels] at The Hague 2pp (15x20)	5 Jun 1670
202rv	Latin verse, untitled fragment 5 lines (6x15) Latin	undated
203rv&205rv	From [Pels] at The Hague 3pp + ad (15x20)	19 Jun 1670

204rv	Latin verse, fragment 9 lines (8x13) Latin "Antiphone"	undated
206rv	Printed list 1p (17x27) "Generale/CARGA/Of Landinge/Van vyftien Ooft-Indifche Retour-Schepen..."	[16-17 Nov 1669]
207rv	From [Pels at The Hague] 2pp (15x20)	26 Jun 1670
208rv	From [Pels] at The Hague 2pp (15x20)	3 Jul 1670
209rv	From [Pels] at The Hague 2pp (15x20)	10 Jul 1670
210rv	From [Pels] at The Hague 1½pp (15x20)	17 Jul 1670
211r-212v	From [Pels] at The Hague 2¾pp + ad (15x20)	24 Jul 1670
213r-214v	From [Pels] at The Hague 3pp + ad (15x20) *dated 31 Jul 1670 by B	[7 Aug 1670]*
215rv	From [Pels] at The Hague 1⅔pp (15x20) *dated 7 Aug 1670 by B	[31 Jul 1670]*
216r-217v	From [Pels] at The Hague 2pp & 5 lines + ad (15x20)	21 Aug 1670
218rv	From [Pels] at The Hague 1½pp (15x20)	28 Aug 1670
219rv	From [Pels] at The Hague ⅔p + ad (15x20)	4 Sep 1670
220rv	From [Pels at The Hague] 1p + ad (15x20)	11 Sep 1670
221r-222v	From [Pels] at The Hague 3pp + ad (15x20)	25 Sep 1670

223rv	From [Pels] at The Hague 2pp (15x20)	9 Oct 1670
224r-225v	From [Pels at The Hague] 3⅓pp (15x20)	15 Oct 1670
226rv	From [Pels at The Hague]* 2pp (15x20) *in B's hand	[23 Oct 1670]*
227r-228v	From [Pels at The Hague]* 2pp & 7 lines + ad (15x20) *in B's hand	[30 Oct 1670]*
229rv	From [Pels] at The Hague 1½pp (15x20)	6 Nov 1670
230rv	From [Pels] at The Hague 2pp (15x20)	13 Nov 1670
231r-232v	From [Pels] at The Hague 3pp & 4 lines + ad (15x20)	20 Nov 1670
233r-234v	From [Pels] at The Hague 2⅓pp + ad (15x20)	4 Dec 1670
235rv	From [Pels] at The Hague ¾p (15x20)	11 Dec 1670
236r-237v	From [Pels at The Hague] 2⅔pp + ad (15x20)	18 Dec 1670
238r-239v	From [Pels] at The Hague 2⅔pp + ad (15x20)	25 Dec 1670
240r-241v	From [Pels] at The Hague 2½pp + ad (15x20)	1 Jan 1671
242r-243v	From [Pels at The Hague] 2½pp + ad (15x20)	15 Jan 1671
244r-245v	From [Pels] at The Hague 2⅔pp + ad (15x20)	22 Jan 1671
246rv	From [Pels at The Hague] 1p + ad (20x31) *in B's hand	[29 Jan 1671]*

247rv	From [Pels] at The Hague 2pp (15x20)	5 Feb 1671
248r-249v	From [Pels] at The Hague $2\frac{2}{3}$pp + ad (15x20)	12 Feb 1671
250r-251v	From [Pels] at The Hague 3pp + ad (15x20)	26 Feb 1671
252rv	From [Pels] at The Hague 2pp (15x20)	5 Mar 1671
253r-254v	From [Pels at The Hague] $2\frac{2}{3}$pp + ad (15x20)	12 Mar 1671
255rv	From [Pels] at The Hague 1p + ad (15x20)	19 Mar 1671
256rv	From [Pels] at The Hague 2pp (15x20)	26 Mar 1671
257rv	From [Pels at The Hague] $1\frac{1}{3}$pp + ad (20x31) *in B's hand	[9 Apr 1671]*
258r-259v	From [Pels] at The Hague $2\frac{1}{2}$pp + ad (15x20)	16 Apr 1671
260r-261v	From [Pels at The Hague] 3pp + ad (15x20) *in B's hand	[30 Apr 1671]*
262rv&268rv	From [Pels] at The Hague 2pp + ad (15x20)	7 May 1671
263rv	Latin verse $1\frac{1}{2}$pp (15x20) Latin "Catastrophe Ducis Lotharingiae/ ante funus nuper sepulti"	undated
264rv&267rv	From [Pels] at The Hague 2pp & 6 lines + ad (15x20)	14 May 1671
265r-266v	From [Pels at The Hague] 3pp + ad (15x20) *in B's hand	[28 May 1671]*

269r-271v	From [Pels at The Hague] $2\frac{2}{3}$pp & 13 lines + ad (15x20) fol 271 (8x20)	4 Jun 1671
272rv	From [Pels] at Utrecht 1p & 7 lines + ad (20x35)	10 Jun 1671
273rv&275rv	From [Pels] at The Hague 3pp + ad (15x20)	25 Jun 1671
274rv	Latin copy 1p (15x20) Latin	undated
276rv&279rv	From [Pels] at The Hague 2pp & 3 lines + ad (15x20)	2 Jul 1671
277r-278v	From [Pels] at The Hague $2\frac{2}{3}$pp + ad (15x20)	9 Jul 1671
280rv	From [Pels] at The Hague $1\frac{1}{3}$pp (15x20)	16 Jul 1671
281rv	From [Pels] at The Hague $1\frac{2}{3}$pp (15x20)	23 Jul 1671
282rv	From [Pels] at The Hague $1\frac{2}{3}$pp (15x20)	6 Aug 1671
283rv	From [Pels] at The Hague $1\frac{2}{3}$pp (15x20)	13 Aug 1671
284rv&288rv	From [Pels] at The Hague 2pp + ad (15x20)	20 Aug 1671
285rv&287rv	From [Pels] at The Hague 2pp & 3 lines + ad (15x20)	10 Sep 1671
286rv	From [Pels] at The Hague 1p + ad (15x20)	17 Sep 1671
289r-290v	From [Pels] at The Hague $2\frac{3}{4}$pp + ad (15x20)	24 Sep 1671
291rv	From [Pels] at The Hague 1p + ad (15x20)	1 Oct 1671
292rv&297rv	From [Pels at The Hague] 2pp + ad (15x20)	8 Oct 1671

293rv	From [Pels] at The Hague $\frac{3}{4}$p + ad (15x20)	15 Oct 1671
294rv	From [Pels] at The Hague $\frac{3}{4}$p + ad (15x20)	22 Oct 1671
295rv	From [Pels] at The Hague 8 lines + ad (15x20)	29 Oct 1671
296rv	From [Pels] at The Hague 1p + ad (15x20)	5 Nov 1671
298rv	From [Pels] at The Hague 1p + ad (15x20)	26 Nov 1671
299rv&301rv	From [Pels] at The Hague 2pp & 3 lines + ad (15x20)	10 Dec 1671
300rv	From [Pels] at The Hague 1p + ad (20x31)	17 Dec 1671
302rv	From [Pels] at The Hague 1p & 10 lines + ad (20x31)	24 Dec 1671
303r-304v	From [Ph] Pels at Danzig 4pp (17x22)	12 Dec 1671
305rv	From [Pels] at The Hague $\frac{3}{4}$p + ad (15x20)	31 Dec 1671
306rv	From [Pels at The Hague] $\frac{3}{4}$p + ad (14x18)	8 Dec 1672
307rv	From [Pels at The Hague] $\frac{1}{2}$p + ad (20x20)	18 May 1672
308rv	From [Pels] at The Hague 1$\frac{2}{3}$pp (15x20)	7 Jan 1672
309rv	From [Pels] at The Hague 1p + ad (15x20)	14 Jan 1672
310rv	From [Pels at The Hague] 1$\frac{2}{3}$pp (15x20)	22 Jan 1672
311rv	From [Pels] at The Hague 1$\frac{3}{4}$pp (15x20)	28 Jan 1672

312rv	From [Pels] at The Hague 2pp (15x20)	4 Feb 1672
313r-315v	From [Pels] at The Hague 3pp & 6 lines + ad (15x20) fol 314, 6 line note (5x12)	10 Feb 1672
316r-317v	From [Pels at The Hague] 3pp + ad (15x20)	18 Feb 1672
318rv&321rv	From [Pels] at The Hague 2pp & 1 line + ad (15x20)	25 Feb 1672
319r-320v	From [Pels] at The Hague 3pp + ad (15x20)	3 Mar 1672
322rv&325rv	From [Pels] at The Hague 1½pp + ad (15x20)	10 Mar 1672
323r-324v	From Ph Pels at Danzig 3pp + ad (15x20)	2 Mar 1672
326rv&329rv	From [Pels] at The Hague 1p & 4 lines + ad (15x20)	7 Apr 1672
327rv	From [Pels at The Hague] ½p + ad (15x20)	6 Jul 1673
328rv	From [Pels at The Hague] 1p + ad (15x20)	3 May 1673
330r-331v	From [Pels] at The Hague 3pp & 7 lines + ad (15x20)	17 Mar 1672
332r-333v	From [Pels] at The Hague 2¾pp + ad (15x20)	24 Mar 1672
334r-335v	From [Pels] at The Hague 4pp (15x20)	31 Mar 1672
336rv&339rv	From Pels at The Hague 2pp + ad (15x20)	$\frac{14}{24}$ Jan 1680
337rv	From [Pels at The Hague] 1p + ad (20x31)	26 Apr 1674
338rv	From [Pels] at The Hague ⅔p + ad (17x17)	20 Mar 1675

▼ ▼ ▼

COLLECTION BOULLIAU, VOLUME XVI, FONDS FRANÇAIS 13034

Folio	Description	Date
1r-2v	Newsletter 2⅔pp (14x20) Italian *[internal date, 20 Jun 1670]	undated*
3rv	From Anonymous at Pera 2pp (13x20)	20 Dec [1666?]
4rv	Anonymous note ½p (13x19) Italian *[dated by Des Noyers]	6 Oct 1668*
5rv	Newsletter Rome 1½pp (13x19) Italian	28 Dec 1669
6r-7v	Newsletter Rome 3pp (13x29) Italian *[ends w/31 Jan 1669 Ab Inc, 31 Jan 1670]	25 Jan [1670]*
8rv	Newsletter Rome 1⅔pp (13x19) Italian	1 Feb 1670
9rv	Newsletter Rome 1½pp (13x19) Italian	15 Feb 1670
10rv	Newsletter Rome 1⅓pp (13x19) Italian	11 Feb 1670
11rv	Copies of letters 1⅔pp (19x25) Italian	8 Feb 1670 & 14 Feb [1670]
12rv	Newsletter Florence 2pp (13x19) Italian	28 Feb 1670
13r-14v	Newsletter Florence 3pp (15x20) Italian	27 Feb 1670

15r-16v	Newsletter Rome 3pp & 2 lines (13x19) Italian	1 Mar 1670
17r-18v	Newsletter Rome 3pp (13x19) Italian	1 Mar 1670
19r-20v	Newsletter Rome 2½pp + ad (13x19) Italian [PS, Rome, 10 Mar 1670]	8 Mar 1670
21r-22v	Newsletter Rome 3pp + ad (13x19) Italian	15 Mar 1670
23r-24v	Newsletter Florence 2⅔pp + ad (13x19) Italian	28 Mar 1670
25r-26v	Newsletter Rome 3pp (13x19) Italian	12 Apr 1670
27r-30v	Newsletter Florence 5½pp (15x20) Italian/French fol 30, blank	25 Apr 1670
31r-32v	Newsletter Florence 3pp (15x20) Italian	20 Jun 1670
33r-34v	Newsletter Florence 1p (15x20) Italian fol 34, blank	11 Jul 1670
35r-36v	Newsletter Florence 2⅔pp (15x20) Italian	26 Dec 1670
37r-38v	Newsletter 3pp & 5 lines (13x19) Italian	
39rv	Newsletter Florence 2pp (13x19) Italian	18 Jul 1670

Folios	Description	Date
40r-41v	Newsletter Constantinople $1\frac{2}{3}$pp & 7 lines (13x19) Italian/French	28 Aug 1670 21 Sep 1670
42rv	Newsletter [Smirna?] 1p & 4 lines (15x20) Italian	1 Dec 1670
43rv	Newsletter Constantinople 2pp (13x19) Italian	13 Dec [1670]
44rv	Newsletter Florence 1p (13x19) Italian	20 Mar 1670 Ab Inc [20 Mar 1671]
45rv	Newsletter Florence $1\frac{1}{2}$pp (15x20) Italian	29 Aug 1670
46rv	Newsletter Florence $\frac{2}{3}$p (13x19) Italian	5 Sep 1670
47rv	Newsletter Florence 1p (13x19) Italian	31 Oct 1670
48r-49v	Newsletter Constantinople $2\frac{1}{2}$pp (13x18) Italian	12 Jan 1671
50rv	Copy Florence 2pp (15x20) Italian	26 Feb 1671 Ab Inc [26 Feb 1672]
51r-52v	Newsletter Constantinople $2\frac{1}{2}$pp (13x19) Italian	4 Mar 1672
53r-54v	Newsletter Florence 4pp (15x20) Italian/French	28 Apr 1672
55rv	Newsletter Genoa 1p & 2 lines (13x19) Italian	25 [Jun] NY

Folio	Description	Date
56rv	Newsletter Florence 9 lines (13x19) Italian	1 Jul 1672
57r-58v	Newsletter Rome 4pp (13x19) Italian	29 Oct 1672
59rv	Newsletter Genoa 2pp (13x19) Italian	30 Oct 1672
60rv	Newsletter Genoa 1½pp (13x19) Italian	3 Nov 1672
61r-62v	Newsletter 3¾pp (13x19) Italian	5 Nov 1672
63r-64v	Newsletter 4pp (13x19) Italian Vienna Warsaw	 23 Oct 1672 12 [Oct 1672]
65r-66v	From Marucelli at Florence 2pp + ad (13x19)	28 Feb 1666
67r-68v	From Marucelli as Pisa 3pp (13x19) Italian	26 Jan 1674
69rv	From Marucelli at Florence To Anonymous 1p (13x19) Italian	16 Mar 1674
70rv	From Marucelli at Pisa To [B?] 1p & 3 lines (15x20) Italian	8? [?] 1675
71r-72v	From Marucelli at Florence 2½pp (15x20) Italian/French	7 Oct 1675
73rv	From Marucelli at Florence 1p (15x20) Italian	20 Oct 1675
74rv	From Marucelli at Pisa 2pp (15x20) Italian	22 May 1676
75rv	From Marucelli at Pisa 1p (15x20) Italian	2 May 1677

76rv	From Marucelli at Florence 1p (13x18) Italian	16 Jun 1679
77r-78v	From Marucelli at Florence 2pp & 4 lines (15x20) Italian/French	3 Feb 1677
79rv	From Marucelli at Rome 1p (13x19) Italian	9 Nov 1677
80r-81v	From Marucelli at Rome 2pp & 7 lines* Italian *fol 80 (13x19) [fol 81, 7 line undated note from TLB (Burattini) (9x19)]	[8?] Jan 1678
82rv	From Marucelli at Pisa 1p (13x19) Italian	21 [Apr?] 1679
83rv	From Marucelli at Rome 1p (15x20) Italian	23 Jul [1680?]
84rv	From Marucelli at Rome 1½pp (13x19) Italian	9 Jun 1681
85r-86v	From Marucelli at Rome 2pp & 5 lines (13x19) Italian	31 Dec 1686
87rv	From Marucelli at Rome 2pp (20x23) Italian	25 Feb 1681
88rv	Newsletter 2pp (17x22) Italian Constantinople Florence	 30 Dec 1671 11 Mar 1671 Ab Inc [11 Mar 1672]
89r-90v	From Marucelli at Rome 3pp (13x19) Italian	[25?] Jan 1686
91rv	From Oldenburg at London 1½pp (15x20) Publ, OC, Vol III, pp 398-400, No 633	22 Apr 1667 OS [2 May 1667]
92rv	From Oldenburg at London 2pp (13x19) Publ, OC, Vol IX, pp 66-68, No 1978	20 May 1672 OS [30 May 1672]

93rv	From Oldenburg at London 2pp (15x20) Publ, OC, Vol XI, pp 96-97, No 2549	5 Oct 1674 OS [15 Oct 1674]
94rv	From Oldenburg at London 1p + ad (15x20) Publ, OC, Vol XI, pp 137-138, No 2574	7 Dec 1674 OS [17 Dec 1674]
95rv	From Oldenburg at London 1p + ad (13x19) Publ, OC, Vol XI, pp 160-161, No 2594	18 Jan 1675 OS [28 Jan 1675]
96rv	From Oldenburg at London 2pp (15x20) French/Latin Publ, OC, Vol XI, pp 198-200, No 2613	18 Feb 1675 OS [28 Feb 1675]
97rv	Copy in Oldenburg's hand 1½pp (15x20) Latin "joh. Flamstedii Observationes Lunae/ jun 26. 75." See OC, Vol XI, p 414, fn 1	
98rv	From Oldenburg at London ½p + ad (15x20) Latin Publ, OC, Vol XI, pp 413-414, No 2707	22 Jul 1675 OS [1 Aug 1675]
99rv	From Oldenburg at London 1p (13x19) Publ, OC [forthcoming]	17 Nov 1676 OS [27 Nov 1676]
100rv	Copy of a letter From Oldenburg at London To Hevelius at Danzig 1½pp (22x34) Latin Publ, OC, Vol III, pp 72-76, No 503	30 Mar 1666 OS [9 Apr 1666]
101r-102v	Copy (enclosed w/fol 100) 1⅔pp + ad (22x30) Latin "Erotemata/Celeberrimo Dno Hevelio commendata" Publ, OC, Vol III, pp 73-75; 77-78	
103rv	Copy of a letter in Hevelius' hand From Oldenburg at London To Hevelius at Danzig 1⅔pp (15x20) Latin Publ, OC, Vol III, pp 215-221, No 562 (complete)	[24 Aug 1666 OS] [3 Sep 1666]

104rv	Extract of a letter in B's hand 1⅓pp (15x20) "De la Lettre de M. Bigot à M. Heinsius de Rome le 27 Avril 1660"	
105rv	From Bigot at [Rouen] 1p (13x18)	7 Sep 1675
106rv	From Bigot at [Rouen] 1½pp (10x15)	8 Feb 1677
107r-109v	From [Portner* at Ratisbon] To Anonymous 3pp & 4 lines + ad (15x20) fol 109r, 4 lines in B's hand + ad fol 109v, figures (8x16) *[identified by B]	22 Mar 1656
110r-111v	From Portner at Ratisbon To J-A de Thou at Paris 2pp + ad (13x19)	10 Jan 1657
112r-113v	From Portner at Ratisbon 4pp (15x20) Latin	V Id Jun 1659 [9 Jun 1659]
114r-115v	From [Portner at Ratisbon] 4pp (15x20) Latin/French fol 115v, list of books supplied to B	Non Oct 1659 [7 Oct 1659]
116r-117v	From [Portner at Ratisbon] 2⅔pp + ad (15x20) Latin	Non Nov 1659 [5 Nov 1659]
118rv	From [Portner at Ratisbon] 2pp (15x20) Latin	VII Kal Dec 1659 [25 Nov 1659]
119r-122v	From [Portner at Ratisbon] 4pp + ad + ad (15x20) Latin/French fol 120, insert w/fol 119, French Messrs. Faesch, Rychiner et Socin to Portner	Proprid Kal Jan 1660 [30 Dec 1659]
123rv	From [Portner at Ratisbon] 2pp (15x20) Latin	Proprid Non Feb [1660] [3 Feb 1660]
124rv	From [Portner at Ratisbon] 1⅔pp (15x20) Latin	VIII Kal Apr [1660] [25 Mar 1660]
125r-126v	From [Portner at Ratisbon] 3½pp (15x20) Latin	Idib Jun 1660 [13 Jun 1660]

127r-128v	From Portner at Ratisbon B at The Hague 2½pp + ad (15x20) Latin	Non Oct 1660 [7 Oct 1660]
129r-130v	From [Portner] at Ratisbon 2pp + ad (15x20) Latin	VI Id Jul 1661 [10 Jul 1661]
131r-132v	From [Portner] at Ratisbon 4pp (15x20) Latin	Non Oct 1661 [7 Oct 1661]
133r-134v	From [Portner at Ratisbon] 4pp (15x20) Latin	VII Id Sep [1661?] [7 Sep 1661]
135r-136v	From Portner at Ratisbon B at The Hague 3pp + ad (15x20) Latin	IV Kal Jan 1662 [29 Dec 1661]
137rv	From Portner at Ratisbon 2pp (15x20) Latin	VII Kal Feb 1662 [26 Jan 1662]
138r-139v	From [Portner at Ratisbon] 3pp + ad (15x20) Latin	XVI Kal Mar 1662 [14 Feb 1662]
140r-141v	From Portner at Ratisbon B at The Hague 3pp & 3 lines + ad (15x20) Latin	XIII Kal May 1662 [19 Apr 1662]
142r-143v	From [Portner at Ratisbon] 3½pp (15x20) Latin	VII Kal Dec 1664 [26 Dec 1664]
144r-145v	From Portner at [Ratisbon] 3pp + ad (15x20) Latin	Proprid Kal Feb 1663 [30 Jan 1663]
146r-147v	From Portner at [Ratisbon] 3pp (15x20) Latin	Non Jun 1663 [5 Jun 1663]
148r-149v	From Portner at [Ratisbon] 4pp (15x20) Latin	Prid Kal Nov 1663 [31 Oct 1663]
150r-151v	From [Portner at Ratisbon] 3pp + ad (15x20) Latin	V Kal Nov 1664 [31 Oct 1664]
152rv	From Portner at Ratisbon ¾p (15x20) Latin	IV Id Mar 1665 [12 Mar 1665]
153r-154v	From [Portner at Ratisbon] 3pp + ad (15x20) Latin	VIII Kal Apr 1665 [25 Mar 1665]

155rv	From Portner at Ratisbon 1½pp (15x20) Latin	Non Apr 1665 [5 Apr 1665]
156r-157v	From [Portner at Ratisbon] 3pp (15x20) Latin	IV Id Nov 1665 [10 Nov 1665]
158r-159v	From [Portner at Ratisbon] 3pp (15x20) Latin	Kal Sep 1665 [1 Sep 1665]
160r-161v	From Portner at [Ratisbon] 3½pp (15x20) Latin	Proprid Kal Oct 1665 [29 Sep 1665]
162r-163v	From Portner at Ratisbon 3⅔pp (15x20) Latin	VII Kal Feb 1666 [26 Jan 1666]
164rv	From Portner at Ratisbon ½p + ad (15x20)	17 Nov 1665
165rv	From Portner at Ratisbon 1½pp (15x20) Latin	XIII Kal Apr 1666 [20 Mar 1666]
166rv	From Portner at Ratisbon 1p & 3 lines (15x20) Latin	VII Kal May 1666 [25 Apr 1666]
167r-168v	From Portner at Ratisbon 3½pp (15x20) Latin	X Kal Apr 1666 [23 Mar 1666]
169r-170v	From Portner at [Ratisbon] 3⅓pp (15x20) Latin	[21?] Dec 1666
171r-172v	From Portner at Ratisbon 4pp (15x20) Latin/French	VIII Kal Sep 1666 [25 Aug 1666]
173rv&175rv	From Portner at Ratisbon 4pp (15x20) Latin	Prid Non Apr 1667 [4 Apr 1667]
174rv	Note in Portner's hand 9 lines (13x13) [concerns books purchased]	undated
176rv	From Portner at Ratisbon 2pp (15x20) Latin	VIII Kal Jun 1667 [25 May 1667]
177rv	Note in Portner's hand 10 lines (10x13)	undated
178rv	From Portner at [Ratisbon?] 1p (15x20) Latin	VI Kal Jul 1667 [26 Jun 1667]

179r-180v	From Portner at Ratisbon 4pp (15x20) Latin fol 180v, PS	Postrid Id Jul 1667 [15 Jul 1667]
181rv	From Portner at Ratisbon 1p + ad (15x20) Latin	VII Kal ultil. 1667 [26 Jul 1667]
182r-183v	From Portner at [Ratisbon] 3pp (15x20) Latin	Proprid Kal Sep 1667 [30 Aug 1667]
184rv	Book list in Portner's hand 1p (10x10)	undated
185rv	Book list in Portner's hand/signed 1p (10x12)	undated
186rv	From Faesch, Rychiner & Socin at Basle 3 lines + ad (19x19)	12 May 1667
187r-188v	From Portner at [Ratisbon] 3pp + ad (15x20) Latin	V Kal Sep 1667 [23 Aug 1667]
189rv	From Portner at Vienna 1½pp (15x20) Latin	IV Id Apr 1674 [10 Apr 1674]
190rv	From Portner at Ratisbon 1p (15x20) Latin	XIII Kal Jul 1669 [19 Jun 1669]
191r-192v	From Portner at Ratisbon 3pp + ad (15x20) Latin	XIV Kal Oct 1669 [18 Sep 1669]
193rv&195rv	From Rautenstein at Frankfurt B at The Hague 3pp + ad (15x20) Latin	7 Jan 1661
194rv	Latin verse (insert w/fol 193) 8 lines (15x16) Latin "In Delphiniēm Galliæ, natūm i Nov. i66i"	
196rv	Newsletter in Rautenstein's hand Mereciâ in Lithuaniâ 2pp (19x20) Latin	26 Oct 1661

197rv	From [Rautenstein]* at Frankfurt B at The Hague ½p + ad (15x20) Latin *[signed "M.N."]	6 Dec 1661
198r-199v	From Rautenstein at Frankfurt B at The Hague 2⅔pp + ad (15x20) Latin	4 Dec 1661
200rv	From Rautenstein at Frankfurt 2pp (15x20) Latin	11 Dec 1661
201rv&203rv	From Rautenstein at Frankfurt B at The Hague 3pp + ad (15x20) Latin	24 Dec 1661
202rv	Latin transcription 2pp (15x20) Latin [Rautenstein notes: "Inscriptiēm Sarcophago stanneo"]	[Dec 1661]
204rv	Note, date for horoscopes 1p (12x16) Latin	
205rv	From [Rautenstein] at Frankfurt 2pp (15x20) Latin	14 Feb 1662
206rv	From Rautenstein at Frankfurt 1½pp (15x20) Latin	4 Mar 1662
207r-208v	From [Rautenstein] at Frankfurt B at The Hague 3pp + ad (15x20) Latin	11 Mar 1662
209rv&214rv	From [Rautenstein] at Frankfurt 2pp & 2 lines + ad (15x20) Latin	4 May 1662
210rv&213rv	From [Rautenstein] at Frankfurt 2½pp + ad (15x20) Latin	Kal Jun 1662 [1 Jun 1662]
211rv	Verse 1½pp (15x20) Latin "Ad D. Pastorium de Scriptis/D. Venatoris"	15 May 1662
212rv	Verse, same hand as fol 211* ¾p (15x20) Latin *fol 212r, signed "B. Venator"	undated

215r-216v	From Rautenstein at Frankfurt 3pp + ad (18x26) Latin *B writes "Receüe le 24. Octob. 1662."	17 Aug 1662*
217r-218v	From [Rautenstein] at Frankfurt 4pp (18x26) Latin	3 Dec 1662
219rv	From [Rautenstein] at Frankfurt 2pp (15x20) Latin	8 Mar 1663
220r-222v	From [Rautenstein] at Ratisbon 3⅓pp + ad (15x20) Latin	14 May 1663
223r-224v	From [Rautenstein] at Ratisbon 3pp + ad (15x20) Latin	30 Jul 1663
225r-226v	From [Rautenstein] at Ratisbon 3pp + ad (15x20) Latin	Kal Nov 1663 [1 Nov 1663]
227r-228v	From Rautenstein at Frankfurt 2½pp + ad (20x31) Latin	8 Feb 1663
229r-230v	From Rautenstein at Ratisbon 2pp & 4 lines + ad (15x20) Latin	24 [Mar?] 1664
231rv	From Rautenstein at Ratisbon 1p (15x20) Latin	12 May 1664
232rv	From Rautenstein at Ratisbon 2pp (15x20) Latin fol 232v, extract of a letter: "Ex literis P. Courtii [Albert Curtz]/ Neob. 13 Jun. 1664"	16 Jun 1664
233rv	From Rautenstein at Ratisbon 2pp (15x20) Latin	22 Sep 1664
234r-235v	From Rautenstein at Ratisbon 2½pp + ad (15x20) Latin	20 Nov 1664
236rv	From Rautenstein at [Ratisbon] 1p (15x20) Latin	[22?] Dec 1664
237rv	From Rautenstein at Ratisbon 1p (15x20) Latin	5 Feb 1665
238rv&241rv	From Rautenstein at Ratisbon 1¼pp + ad (15x20) Latin	18 May 1665

239r-240v	From Rautenstein at Ratisbon 2pp & 5 lines + ad (15x20) Latin	13 Jul 1665

▼ ▼ ▼

COLLECTION BOULLIAU, VOLUME XVII, FONDS FRANÇAIS 13035

1rv	From Gratta at Danzig To Anonymous 1p (12x24) Italian	16 Nov 1660
2r-3v	From Gratta at Danzig 3pp + ad (11x17) Italian	30 Dec 1660
4rv	From Gratta at Danzig To Anonymous 1p (11x17) Italian	10 Nov 1661
5rv	From Gratta at Danzig 2pp (11x17) Italian	17 Dec 1661
6rv	From Gratta at [Danzig] 2pp (11x17) Italian	14 Jan 1662
7r-8v	From Gratta at Danzig B at The Hague 2⅔pp + ad (11x17) Italian	7 Jan 1662
9r-10v	From Gratta at Danzig B at The Hague 2pp & 2 lines + ad (11x17) Italian	28 Jan 1662
11rv&14rv	From Gratta at Danzig B at The Hague 2pp + ad (11x17) Italian	4 Feb 1662
12r-13v	From Gratta at Danzig B at The Hague 3pp + ad (11x17) Italian	18 Feb 1662
15r-16v	From Gratta at Danzig B at The Hague 3pp + ad (11x17) Italian	25 Feb 1662
17rv	From Gratta at Danzig 2pp (11x17) Italian	4 Mar 1662
18rv	From Gratta at Danzig 2pp (11x17) Italian	11 Mar 1662
19rv	From Gratta at Danzig 1½pp (11x17) Italian	18 Mar 1662

20r-21v	From Gratta at Danzig B at The Hague 3pp + ad (11x17) Italian	25 Mar 1662
22rv	From Gratta at Danzig 2pp (11x17) Italian	1 Apr 1662
23rv	From Gratta at Danzig 2pp (11x17) Italian	8 Apr 1662
24rv	From Gratta at Danzig B at The Hague 2pp (11x17) Italian	15 Apr 1662
25rv	From Gratta at Danzig 1½pp (16x22) Italian	29 Apr 1662
26r-27v	From Gratta at Danzig B at The Hague 3pp + ad (16x22) Italian	6 May 1662
28r-29v	From Gratta at Danzig 2½pp + ad (11x17) Italian	27 May 1662
30rv	From Gratta at Danzig 2pp (11x17) Italian	3 Jun 1662
31rv	From Gratta at Danzig 1½pp (11x17) Italian	10 Jun 1662
32rv	From Gratta at Danzig 1½pp (11x17) Italian	17 Jun 1662
33rv	From Gratta at Danzig 1½pp (11x17) Italian	24 Jun 1662
34rv&37rv	From Gratta at Danzig 1p + ad (11x17) Italian	1 Jul 1662
35r-36v	From Gratta at Danzig 2½pp + ad (11x17) Italian	5 Aug 1662
38rv&41rv	From Gratta at Danzig 1p & 1 line + ad (11x17) Italian	19 Aug 1662
39r-40v	From Gratta at Danzig 2pp & 4 lines + ad (11x17) Italian	26 Aug 1662

42rv	From Gratta at Danzig 2pp (11x17) Italian	2 Sep 1662
43rv	From Gratta at Danzig 1⅔pp (11x17) Italian	9 Sep 1662
44r-45v	From Gratta at Danzig 3pp + ad (11x17) Italian	23 Sep 1662
46r-47v	From Gratta at Danzig 3pp + ad (11x17) Italian	30 Sep 1662
48r-49v	From Gratta at Danzig 3pp + ad (11x17) Italian	21 Oct 1662
50rv&53rv	From Gratta at Danzig 2pp + ad (11x17) Italian	28 Oct 1662
51r-52v	From Gratta at Danzig 3pp + ad (11x17) Italian	4 Nov 1662
54r-55v	From Gratta at Danzig 2pp & 3 lines + ad (11x17) Italian	11 Nov 1662
56r-57v	From Gratta at Danzig 3pp + ad (16x22) Italian	2 Dec 1662
58rv&61rv	From Gratta at Danzig 2pp + ad (11x17) Italian	9 Dec 1662
59r-60v	From Gratta at Danzig 3pp + ad (11x17) Italian	23 Dec 1662
62rv	From Gratta at Danzig 2pp (16x22) Italian	10 Feb 1663
63r-64v	From Gratta at Danzig 2⅓pp + ad (11x17) Italian	8 Dec 1663
65r-66v	From Gratta at Danzig 3pp + ad (11x17) Italian	15 Dec 1663
67r-68v	From Gratta at Danzig 3pp + ad (11x17) Italian	1 Dec 1663
69r-70v	From Gratta at Danzig 3pp + ad (11x17) Italian	6 Jan 1663

71r-72v	From Gratta at Danzig 3pp + ad (11x17) Italian	13 Jan 1663
73r-74v	From Gratta at Danzig 3pp + ad (11x17) Italian	20 Jan 1663
75r-76v	From Gratta at Danzig 2½pp + ad (11x17) Italian	27 Jan 1663
77r-78v	From Gratta at Danzig 2½pp + ad (11x17) Italian	3 Feb 1663
79r-80v	From Gratta at Danzig 3pp + ad (11x17) Italian	17 Feb 1663
81r-82v	From Gratta at Danzig 3pp + ad (11x17) Italian	24 Feb 1663
83rv	From Gratta at Danzig 2pp (11x17) Italian	3 Mar 1663
84rv	From Gratta at Danzig 2pp (11x17) Italian	10 Mar 1663
85r-86v	From Gratta at Danzig 3pp + ad (11x17) Italian	17 Mar 1663
87r-88v	From Gratta at Danzig 2⅔pp + ad (11x17) Italian	24 Mar 1663
89r-90v	From Gratta at Danzig 2⅔pp + ad (11x17) Italian	31 Mar 1663
91r-92v	From Gratta at Danzig 3pp + ad (11x17) Italian	7 Apr 1663
93r-94v	From Gratta at Danzig 3pp + ad (11x17) Italian	14 Apr 1663
95rv&98rv	From Gratta at Danzig 2pp + ad (11x17) Italian	21 Apr 1663
96r-97v	From Gratta at Danzig 3pp + ad (11x17) Italian	28 Apr 1663
99r-100v	From Gratta at Danzig 2⅔pp + ad (11x17) Italian	5 May 1663

101r-102v	From Gratta at Danzig 3pp + ad (11x17) Italian	12 May 1663
103r-104v	From Gratta at Danzig 2$\frac{1}{3}$pp + ad (11x17) Italian	19 May 1663
105r-106v	From Gratta at Danzig 2$\frac{2}{3}$pp + ad (11x17) Italian	2 Jun 1663
107rv	From Gratta at Danzig 2pp (11x17) Italian	16 Jun 1663
108rv	From Gratta at Danzig 1$\frac{2}{3}$pp (11x17) Italian	23 Jun 1663
109r-110v	From Gratta at Danzig 2$\frac{1}{2}$pp + ad (11x17) Italian	30 Jun 1663
111r-112v	From Gratta at Danzig 2pp & 1 line + ad (11x17) Italian	9 Jul 1663
113rv	From Gratta at Danzig 2pp (11x17) Italian	14 Jul 1663
114rv	From Gratta at Danzig 1$\frac{1}{2}$pp (11x17) Italian	21 Jul 1663
115r-116v	From Gratta at Danzig 2$\frac{2}{3}$pp + ad (11x17) Italian	7 Jul 1663
117r-118v	From Gratta at Danzig 2$\frac{1}{2}$pp + ad (11x17) Italian	4 Aug 1663
119r-120v	From Gratta at Danzig 2$\frac{1}{2}$pp + ad (11x17) Italian	11 Aug 1663
121rv	From Gratta at Danzig 1$\frac{2}{3}$pp (11x17) Italian	18 Aug 1663
122rv	From Gratta at Danzig 2pp (11x17) Italian	25 Aug 1663
123r-124v	From Gratta at Danzig 2pp & 1 line + ad (11x17) Italian	1 Sep 1663
125rv	From Gratta at Danzig 1$\frac{1}{2}$pp (11x17) Italian	15 Sep 1663

126rv	From Gratta at Danzig 1½pp (11x17) Italian	29 Sep 1663
127rv	From Gratta at Danzig 1p & 1 line (16x22) Italian	22 Sep 1663
128rv	From Gratta at Danzig 2pp (11x17) Italian	20 Oct 1663
129rv	From Gratta at Danzig 2pp (11x17) Italian	6 Oct 1663
130r-131v	From Gratta at Danzig 3pp + ad (11x17) Italian	2 Aug 1664
132r-133v	From Gratta at Danzig 4pp (11x17) Italian	9 Aug 1664
134r-135v	From Gratta at Danzig 3pp + ad (11x17) Italian	26 Aug 1664
136r-137v	From Gratta at Danzig 2pp & 3 lines + ad (11x17) Italian	7 Feb 1665
138r-140v	From Gratta at Danzig 4¾pp + ad (11x17) Italian	23 Aug 1664
141r-142v	From Gratta at Danzig 3pp + ad (11x17) Italian	30 Aug 1664
143r-144v	From Gratta at Danzig 3pp + ad (11x17) Italian	6 Sep 1664
145r-146v	From Gratta at Danzig 2½pp + ad (11x17) Italian	13 Sep 1664
147rv&150rv	From Gratta at Danzig 2pp + ad (11x17) Italian	21 Sep 1664
148r-149v	From Gratta at Danzig 2pp & 3 lines + ad (11x17) Italian	4 Oct 1664
151r-152v	From Gratta at Danzig 2⅔pp + ad (11x17) Italian	25 Oct 1664
153r-154v	From Gratta at Danzig 2½pp + ad (11x17) Italian	8 Nov 1664

155rv&158rv	From Gratta at Danzig 1½pp + ad (11x17) Italian	15 Nov 1664
156r-157v	From Gratta at Danzig 2½pp + ad (11x17) Italian	29 Nov 1664
159r-160v	From Gratta at Danzig 2⅔pp + ad (11x17) Italian	6 Dec 1664
161rv&164rv	From Gratta at Danzig 1½pp + ad (11x17) Italian	13 Dec 1664
162r-163v	From Gratta at Danzig 3pp + ad (11x17) Italian	20 Dec 1664
165r-166v	From Gratta at Danzig 3pp + ad (11x17) Italian	27 Dec 1664
167rv&170rv	From Gratta at Danzig 2pp + ad (11x17) Italian	5 Jan 1664
168r-169v	From Gratta at Danzig 3pp + ad (11x17) Italian	26 Jan 1664
171r-172v	From Gratta at Danzig 3pp + ad (11x17) Italian	1 Feb 1664
173r-174v	From Gratta at Danzig 2⅓pp + ad (11x17) Italian	9 Feb 1664
175r-176v	From Gratta at Danzig 2½pp + ad (11x17) Italian	16 Feb 1664
177rv&180rv	From Gratta at Danzig 2pp + ad (11x17) Italian	23 Feb 1664
178r-179v	From Gratta at Danzig 2pp & 3 lines + ad (11x17) Italian	7 Mar 1664
181r-182v	From Gratta at Danzig 2pp & 4 lines + ad (11x17) Italian	15 Mar 1664
183r-184v	From Gratta at Danzig 2⅔pp + ad (11x17) Italian	22 Mar 1664
185r-186v	From Gratta at Danzig 3pp + ad (11x17) Italian	29 Mar 1664

187r-188v	From Gratta at Danzig 3½pp (11x17) Italian	19 Apr 1664
189r-190v	From Gratta at Danzig 2½pp + ad (11x17) Italian	10 May 1664
191r-192v	From Gratta at Danzig 2½pp (11x17) Italian	17 May 1664
193r-194v	From Gratta at Danzig 2½pp + ad (11x17) Italian	24 May 1664
195r-196v	From Gratta at Danzig 3pp + ad (11x17) Italian	7 Jun 1664
197r-198v	From Gratta at Danzig 3pp + ad (11x17) Italian	27 Jun 1664
199r-200v	From Gratta at Danzig 2½pp + ad (11x17) Italian	21 Jun 1664
201r-202v	From Gratta at Danzig 2pp & 1 line + ad (11x17) Italian	14 Jun 1664
203r-204v	From Gratta at Danzig 3pp + ad (11x17) Italian	5 Jul 1664
205r-206v	From Gratta at Danzig 3pp + ad (11x17) Italian	12 Jul 1664
207r-208v	From Gratta at Danzig 3pp + ad (11x17) Italian	19 Jul 1664
209r-210v	From Gratta at Danzig 3pp + ad (11x17) Italian	26 Jul 1664
211r-212v	From Gratta at Danzig 2pp & 2 lines + ad (11x17) Italian	12 Jan 1664
213r-214v	From Gratta at Danzig 3pp + ad (11x17) Italian	10 Jan 1665
215r-216v	From Gratta at Danzig 3pp + ad (11x17) Italian	17 Jan 1665
217r-218v	From Gratta at Danzig 2½pp + ad (11x17) Italian	24 Jan 1665

219r-220v	From Gratta at Danzig 2pp & 2 lines + ad (11x17) Italian *corrected from 1664 to 1665 by B	3 Jan [1665]*
221r-222v	From Gratta at Danzig 3pp + ad (11x17) Italian	14 Feb 1665
223r-224v	From Gratta at Danzig 2pp & 2 lines + ad (11x17) Italian	21 Feb 1665
225rv&228rv	From Gratta at Danzig 1½pp + ad (11x17) Italian	28 Feb 1665
226r-227v	From Gratta at Danzig 2⅔pp + ad (11x17) Italian	7 Mar 1665
229r-230v	From Gratta at Danzig 2½pp + ad (11x17) Italian	14 Mar 1665
231r-232v	From Gratta at Danzig 2½pp + ad (11x17) Italian	21 Mar 1665
233rv	From Gratta at Danzig 2pp (11x17) Italian	28 Mar 1665
234rv	From Gratta at Danzig 1⅔pp (11x17) Italian	4 Apr 1665
235r-236v	From Gratta at Danzig 2½pp + ad (11x17) Italian	11 Apr 1665
237r-238v	From Gratta at Danzig 3pp + ad (11x17) Italian	18 Apr 1665
239r-240v	From Gratta at Danzig 2½pp + ad (11x17) Italian	25 Apr 1665
241r-242v	From Gratta at Danzig 2pp & 3 lines + ad (11x17) Italian	2 May 1665
243r-244v	From Gratta at Danzig 2⅔pp + ad (11x17) Italian	9 May 1665
245rv	From Gratta at Danzig 1p & 2 lines (11x17) Italian	16 May 1665
246rv	From Gratta at Danzig 1p & 1 line (11x17) Italian	23 May 1665

247rv	From Gratta at Danzig 1p (11x17) Italian	30 May 1665
248r-249v	From Gratta at Danzig 2½pp + ad (11x17) Italian	6 Jun 1665
250rv	From Gratta at Danzig 2pp (11x17) Italian	20 Jun 1665
251rv	From Gratta at Danzig 1⅔pp (11x17) Italian	27 Jun 1665
252rv	From Gratta at Danzig 2pp (11x17) Italian	4 Jul 1665
253rv	From Gratta at Danzig 1⅔pp (11x17) Italian	11 Jul 1665
254r-255v	From Gratta at Danzig 2½pp + ad (11x17) Italian	18 Jul 1665
256rv&259rv	From Gratta at Danzig 1½pp + ad (11x17) Italian	25 Jul 1665
257r-258v	From Gratta at Danzig 3pp + ad (11x17) Italian	1 Aug 1665
260rv	From Gratta at Danzig 1½pp (11x17) Italian	8 Aug 1665
261rv	From Gratta at Danzig 1⅔pp (11x17) Italian	15 Aug 1665
262r-263v	From Gratta at Danzig 2pp & 5 lines + ad (11x17) Italian	22 Aug 1665
264r-265v	From Gratta at Danzig 3pp + ad (11x17) Italian	29 Aug 1665
266r-267v	From Gratta at Danzig 2½pp + ad (11x17) Italian	5 Sep 1665
268r-269v	From Gratta at Danzig 3pp + ad (11x17) Italian	12 Sep 1665
270r-271v	From Gratta at Danzig 3pp + ad (11x17) Italian	19 Sep 1665

272r-273v	From Gratta at Danzig 2pp & 5 lines + ad (11x17) Italian	26 Sep 1665
274r-275v	From Gratta at Danzig 3pp + ad (11x17) Italian	3 Oct 1665
276r-277v	From Gratta at Danzig 2⅔pp (11x17) Italian	10 Oct 1665
278rv	From Gratta at Danzig 2pp (11x17) Italian	17 Oct 1665
279r-281v	From Gratta at Danzig 2pp & 3 lines + ad (11x17) Italian fol 281, blank	24 Oct 1665
282r-283v	From Gratta at Danzig 2½pp + ad (11x17) Italian	31 Oct 1665
284r-285v	From Gratta at Danzig 2pp & 2 lines + ad (11x17) Italian	7 Nov 1665
286rv	From Gratta at Danzig 1p & 5 lines (11x17) Italian	14 Nov 1665
287rv	From Gratta at Danzig 1p (11x17) Italian	21 Nov 1665
288rv	From Gratta at Danzig 1½pp (11x17) Italian	28 Nov 1665
289r-290v	From Gratta at Danzig 3pp + ad (11x17) Italian	5 Dec 1665
291rv&294rv	From Gratta at Danzig 2pp + ad (11x17) Italian	12 Dec 1665
292r-293v	From Gratta at Danzig 1p + ad (11x17) Italian	19 Dec 1665

▼ ▼ ▼

COLLECTION BOULLIAU, VOLUME XVIII, FONDS FRANÇAIS 13036

1rv	From Gratta at Danzig 1p & 3 lines (11x17) Italian	9 Jan 1666
2rv	From Gratta at Danzig 2pp (11x17) Italian	16 Jan 1666
3rv	From Gratta at Danzig 1p (11x17) Italian	23 Jan 1666
4rv	From Gratta at Danzig 1p (11x17) Italian	6 Feb 1666
5rv	From Gratta at Danzig 2pp (11x17) Italian	13 Feb 1666
6rv	From Gratta at Danzig 1p & 5 lines (11x17) Italian	20 Feb 1666
7rv	From Gratta at Danzig 2pp (11x17) Italian	27 Feb 1666
8rv	From Gratta at Danzig 1½pp (11x17) Italian	6 Mar 1666
9rv&12rv	From Gratta at Danzig 2pp + ad (11x17) Italian	20 Mar 1666
10r-11v	From Gratta at Danzig 2½pp + ad (11x17) Italian	27 Mar 1666
13rv&16rv	From Gratta at Danzig 1½pp + ad (11x17) Italian	3 Apr 1666
14r-15v	From Gratta at Danzig 2pp & 3 lines + ad (11x17) Italian	10 Apr 1666
17rv	From Gratta at Danzig 1p (11x17) Italian	17 Apr 1666
18rv	From Gratta at Danzig 1½pp (11x17) Italian	24 Apr 1666
19rv	From Gratta at Danzig 2pp (11x17) Italian	1 May 1666

20rv	From Gratta at Danzig 1p (11x17) Italian	8 May 1666
21rv&24rv	From Gratta at Danzig 1½pp + ad (11x17) Italian	15 May 1666
22r-23v	From Gratta at Danzig 2pp & 5 lines + ad (11x17) Italian	29 May 1666
25rv&28rv	From Gratta at Danzig 1½pp + ad (11x17) Italian	5 Jun 1666
26r-27v	From Gratta at Danzig 2½pp + ad (11x17) Italian	26 Jun 1666
29rv&32rv	From Gratta at Danzig 1¾pp + ad (11x17) Italian	3 Jul 1666
30r-31v	From Gratta at Danzig 2pp & 3 lines + ad (11x17) Italian	10 Jul 1666
33r-34v	From Gratta at Danzig 3pp + ad (11x17) Italian	17 Jul 1666
35rv	From Gratta at Danzig 1¾pp (11x17) Italian	31 Jul 1666
36rv	From Gratta at Danzig 1p & 4 lines (11x17) Italian	14 Aug 1666
37rv	From Gratta at Danzig 1½pp (11x17) Italian	28 Aug 1666
38r-39v	From Gratta at Danzig 2pp & 2 lines + ad (11x17) Italian	4 Sep 1666
40r-41v	From Gratta at Danzig 2½pp + ad (11x17) Italian	11 Sep 1666
42rv	From Gratta at Danzig 1½pp (11x17) Italian	18 Sep 1666
43rv	From Gratta at Danzig 2pp (11x17) Italian	25 Sep 1666
44r-45v	From Gratta at Danzig 2pp & 2 lines + ad (11x17) Italian	2 Oct 1666

46r-47v	From Gratta at Danzig 2½pp + ad (11x17) Italian	9 Oct 1666
48rv	From Gratta at Danzig 2pp (11x17) Italian	16 Oct 1666
49rv	From Gratta at Danzig 1½pp (11x17) Italian	23 Oct 1666
50rv	From Gratta at Danzig 1½pp (11x17) Italian	30 Oct 1666
51rv	From Gratta at Danzig 1p & 3 lines (11x17) Italian	13 Nov 1666
52rv	From Gratta at Danzig 1¾pp (11x17) Italian	27 Nov 1666
53rv	From Gratta at Danzig 1p & 5 lines (11x17) Italian	6 Nov 1666
54rv	From Gratta at Danzig 1p & 4 lines (11x17) Italian	20 Nov 1666
55rv	From Gratta at Danzig 1p & 3 lines (11x17) Italian	4 Dec 1666
56rv	From Gratta at Danzig 2pp (11x17) Italian	11 Dec 1666
57rv&60rv	From Gratta at Danzig 2pp + ad (11x17) Italian	18 Dec 1666
58r-59v	From Gratta at Danzig 2pp + ad (11x17) Italian	22 Jan 1667
61r-62v	From Gratta at Danzig 2½pp + ad (11x17) Italian	8 Jan 1667
63r-64v	From Gratta at Danzig 3pp + ad (11x17) Italian	15 Jan 1667
65r-66v	From Gratta at Danzig 2pp & 2 lines + ad (11x17) Italian	16 Apr 1667
67rv	From Gratta at Danzig 2pp (11x17) Italian	23 Apr 1667

68rv	From Gratta at Danzig 1p & 3 lines (11x17) Italian	30 Apr 1667
69rv	From Gratta at Danzig 1⅔pp (11x17) Italian	1 Jan 1667
70rv	From Gratta at Danzig 2pp (11x17) Italian	18 Feb 1667
71rv	From Gratta at Danzig 2pp (11x17) Italian	21 Feb 1667
72rv	From Gratta at Danzig 2pp (11x17) Italian	29 Jan 1667
73rv	From Gratta at Danzig 2pp (11x17) Italian	2 Apr 1667
74r-75v	From Gratta at Danzig 1p + ad (11x17) Italian	9 Apr 1667
76rv	From Gratta at Danzig 2pp (11x17) Italian	7 May 1667
77rv	From Gratta at Danzig 2pp (11x17) Italian	19 Mar 1667
78rv	From Gratta at Danzig 1p & 2 lines (11x17) Italian	26 Mar 1667
79r-80v	From Gratta at Danzig 2pp & 2 lines + ad (11x17) Italian	21 May 1667
81rv&86rv	From Gratta at Danzig 1½pp + ad (11x17) Italian	5 Mar 1667
82rv&87rv	From Gratta at Danzig 1p & 2 lines + ad (11x17) Italian	12 Mar 1667
83rv&88rv	From Gratta at Danzig 1½pp + ad (11x17) Italian	14 May 1667
84r-85v	From Gratta at Danzig 2pp + ad (11x17) Italian	28 May 1667
89rv	From Gratta at Danzig 2pp (11x17) Italian	11 Jun 1667

90rv	From Gratta at Danzig 1½pp (11x17) Italian	25 Jun 1667
91rv	From Gratta at Danzig 1⅔pp (11x17) Italian	2 Jul 1667
92rv	From Gratta at Danzig 1p & 5 lines (11x17) Italian	9 Jul 1667
93rv&96rv	From Gratta at Danzig 2pp + ad (11x17) Italian	23 Jul 1667
94r-95v	From Gratta at Danzig 2pp & 2 lines + ad (11x17) Italian	30 Jul 1667
97r-98v	From Gratta at Danzig 2½pp + ad (11x17) Italian	6 Aug 1667
99rv&102rv	From Gratta at Danzig 2pp + ad (11x17) Italian	13 Aug 1667
100r-101v	From Gratta at Danzig 2pp & 2 lines + ad (11x17) Italian	20 Aug 1667
103r-104v	From Gratta at Danzig 2pp & 4 lines + ad (11x17) Italian	3 Sep 1667
105r-106v	From Gratta at Danzig 2pp & 4 lines + ad (11x17) Italian	10 Sep 1667
107r-108v	From Gratta at Danzig 2pp & 6 lines + ad (11x17) Italian	17 Sep 1667
109r-110v	From Gratta at Danzig 2pp & 1 line + ad (11x17)	24 Sep 1667
111rv	From Gratta at Danzig 2pp (11x17) Italian	1 Oct 1667
112rv	From Gratta at Danzig 1p & 2 lines (11x17) Italian	8 Oct 1667
113rv	From Gratta at Danzig 2pp (11x17) Italian	22 Oct 1667
114r-116v	From Gratta at Danzig 4pp & 7 lines + ad (11x17) Italian	5 Nov 1667

117rv&122rv	From Gratta at Danzig 1p + ad (11x17) Italian	12 Nov 1667
118rv&121rv	From Gratta at Danzig 1⅔pp + ad (11x17) Italian	26 Nov 1667
119r-120v	From Gratta at Danzig 2⅔pp + ad (11x17) Italian	3 Dec 1667
123rv	From Gratta at Danzig 2pp (11x17) Italian	10 Dec 1667
124rv	From Gratta at Danzig 1p & 3 lines (11x17) Italian	17 Dec 1667
125rv	From Gratta at Danzig 1p (11x17) Italian	24 Dec 1667
126r-127v	From Gratta at Danzig 2⅔pp + ad (11x17) Italian	31 Dec 1667
128r-129v	From Gratta at Danzig 3pp + ad (11x17) Italian	7 Jan 1668
130r-131v	From Gratta at Danzig 2pp & 2 lines + ad (11x17) Italian	21 Jan 1668
132rv	From Gratta at Danzig 1p & 1 line (11x17) Italian	28 Jan 1668
133rv	From Gratta at Danzig 1p (11x17) Italian	4 Feb 1668
134r-135v	From Gratta at Danzig 2pp & 2 lines + ad (11x17) Italian	11 Feb 1668
136rv	From Gratta at Danzig 2pp (11x17) Italian	18 Feb 1668
137rv	From Gratta at Danzig 1½pp (11x17) Italian	25 Feb 1668
138r-139v	From Gratta at [Danzig]* 2½pp + ad (11x17) Italian *in B's hand	[10 Mar 1668]*
140rv	From Gratta at Danzig 2pp (11x17) Italian	3 Mar 1668

141r-142v	From Gratta at Danzig 3pp + ad (11x17) Italian	31 Mar 1668
143rv	From Gratta at Danzig 2pp (11x17) Italian	7 Apr 1668
144rv	From Gratta at Danzig 1½pp (11x17) Italian	14 Apr 1668
145rv	From Gratta at Danzig 1½pp (11x17) Italian	21 Apr 1668
146r-147v	From Gratta at Danzig 3pp + ad (11x17) Italian	28 Apr 1668
148r-149v	From Gratta at Danzig 2½pp + ad (11x17) Italian	5 May 1668
150rv&153rv	From Gratta at Danzig 1p + ad (11x17) Italian	12 May 1668
151r-152v	From Gratta at Danzig 2½pp + ad (11x17) Italian	19 May 1668
154rv	From Gratta at Danzig 1p & 5 lines (11x17) Italian	26 May 1668
155rv	From Gratta at Danzig 1½pp (11x17) Italian	2 Jun 1668
156r-157v	From Gratta at Danzig 2pp & 4 lines + ad (11x17) Italian	8 Jun 1668
158r-159v	From Gratta at Danzig 3½pp (11x17) Italian	16 Jun 1668
160r-161v	From Gratta at Danzig 2½pp + ad (11x17) Italian	23 Jun 1668
162r-163v	From Gratta at Danzig 2pp & 2 lines + ad (11x17) Italian	30 Jun 1668
164r-165v	From Gratta at Danzig 2pp & 2 lines + ad (11x17) Italian	7 Jul 1668
166rv	From Gratta at Danzig 2pp (11x17) Italian	14 Jul 1668

167rv	From Gratta at Danzig $1\frac{3}{4}$pp (11x17) Italian	21 Jul 1668
168r-169v	From Gratta at Danzig 2pp & 6 lines + ad (11x17) Italian	28 Jul 1668
170rv&173rv	From Gratta at Danzig 2pp + ad (11x17) Italian *corrected from Jul by B	4 [Aug]* 1668
171r-172v	From Gratta at Danzig 2pp & 3 lines + ad (11x17) Italian	18 Aug 1668
174r-175v	From Gratta at Danzig 2pp & 2 lines + ad (11x17) Italian	25 Aug 1668
176rv&180rv	From Gratta at Danzig $1\frac{1}{2}$pp + ad (11x17) Italian	1 Sep 1668
177rv&181rv	From Gratta at Danzig 2pp + ad (11x17) Italian	8 Sep 1668
178r-179v	From Gratta at Danzig 2pp & 5 lines + ad (11x17) Italian	15 Sep 1668
182r-183v	From Gratta at Danzig $2\frac{1}{2}$pp + ad (11x17) Italian	10 Nov 1668
184rv	From Gratta at Danzig $\frac{2}{3}$p + ad (16x24) Italian	22 Sep 1668
185r-186v	From Gratta at Danzig 2pp & 6 lines + ad (11x17) Italian	29 Sep 1668
187r-188v	From Gratta at Danzig $2\frac{1}{2}$pp + ad (11x17) Italian	13 Oct 1668
189rv&192rv	From Gratta at Danzig $1\frac{1}{2}$pp + ad (11x17) Italian	20 Oct 1668
190r-191v	From Gratta at Danzig 2pp & 4 lines + ad (11x17) Italian	3 Nov 1668
193rv	From Gratta at Danzig 2pp (11x17) Italian	16 Nov 1668
194rv	From Gratta at Danzig $1\frac{1}{2}$pp (11x17) Italian	24 Nov 1668

195rv	From Gratta at Danzig 1½pp (11x17) Italian	1 Dec 1668
196r-197v	From Gratta at Danzig 3pp + ad (11x17) Italian	8 Dec 1668
198rv	From Gratta at Danzig 1½pp (11x17) Italian	15 Dec 1668
199rv	From Gratta at Danzig 1½pp (11x17) Italian	22 Dec 1668
200r-201v	From Gratta at Danzig 2pp & 2 lines + ad (11x17) Italian	29 Dec 1668
202rv&205rv	From Gratta at Danzig 1½pp + ad (11x17) Italian *corrected from 1668 by B	5 Jan [1669]*
203r-204v	From Gratta at Danzig 2pp & 6 lines + ad (11x17) Italian *[1668 is written]	19 Jan [1669]*
206r-207v	From Gratta at Danzig 3pp + ad (11x17) Italian	9 Mar 1669
208rv&211rv	From Gratta at Danzig 2pp (11x17) Italian	22 Jun 1669
209r-210v	From Gratta at Danzig 2pp & 3 lines + ad (11x17) Italian	13 Apr 1669
212r-213v	From Gratta at Danzig 2pp + ad (11x17) Italian	20 Apr 1669
214r-215v	From Gratta at Danzig 2⅔pp + ad (11x17) Italian	4 May 1669
216r-217v	From Gratta at Danzig 2pp & 5 lines + ad (11x17) Italian	11 May 1669
218r-219v	From Gratta at Danzig 3pp + ad (11x17) Italian	18 May 1669
220r-221v	From Gratta at Danzig 1p & 6 lines + ad (11x17) Italian	8 Jun 1669

222r-223v	From Gratta at Danzig 3pp + ad (11x17) Italian	25 May 1669
224r-225v	From Gratta at Danzig 2⅔pp + ad (11x17) Italian	1 Jun 1669
226r-227v	From Gratta at Danzig 2pp + ad (11x17) Italian	15 Jun 1669
228rv	From Gratta at Danzig 1½pp (11x17) Italian	13 Jul 1669
229rv	From Gratta at Danzig 1½pp (11x17) Italian	20 Jul 1669
230r-231v	From Gratta at Danzig 3pp + ad (11x17) Italian	29 Jun 1669
232rv&235rv	From Gratta at Danzig 2pp + ad (11x17) Italian	6 Jul 1669
233r-234v	From Gratta at Danzig 3pp + ad (11x17) Italian	16 Mar 1669
236r-237v	From Gratta at Danzig 2½pp + ad (11x17) Italian	23 Mar 1669
238r-239v	From Gratta at Danzig 3pp + ad (11x17) Italian	30 Mar 1669
240r-241v	From Gratta at Danzig 2½pp + ad (11x17) Italian	6 Apr 1669
242r-243v	From Gratta at Danzig 2½pp + ad (11x17) Italian	26 Jan 1669
244r-245v	From Gratta at Danzig 2pp & 2 lines + ad (11x17) Italian	16 Feb 1669
246r-247v	From Gratta at Danzig 4pp (11x17) Italian	2 Mar 1669
248r-249v	From Gratta at Danzig 3pp + ad (11x17) Italian	23 Feb 1669
250rv	From Gratta at Danzig 1⅔pp (11x17) Italian	12 Jan 1669

251rv	From Gratta at Danzig 1½pp (11x17) Italian	27 Apr 1669
252rv	From Gratta at Danzig 1½pp (11x17) Italian	7 Sep 1669
253r-254v	From Gratta at Danzig 4pp (11x17) Italian	14 Dec 1669
255r-256v	From Gratta at Danzig 2pp + ad (11x17) Italian	27 Jul 1669

▼ ▼ ▼

COLLECTION BOULLIAU, VOLUME XIX, FONDS FRANÇAIS 13037

Folio	Description	Date
1r-2v	From Longomontanus at Copenhagen 3pp + ad (21x30) Latin [B writes: "Redditae 20 Maii 1633"]	Prid Kal Aug 1632 OS [10 Aug 1632]
3r-5v	Copy of a letter in B's hand From Longomontanus at Copenhagen 1½pp + ad (20x29) Latin fol 5, blank	Ult Id Quint 1635 [OS] [25 Jul 1635]
6r-7v	From Longomontanus at Copenhagen 3pp + ad (17x20) Latin	Kal Oct 1645 OS [11 Oct 1645]
8rv	From Ogier at Stockholm 1p + ad (19x31)	16 [Apr?] 1634
9r-10v	From Schickard at Tübingen To Gassendi at Paris 3pp + ad (20x31) Latin	4 Oct 1630 [OS] [14 Oct 1630]
11rv	Astronomical data and drawings in Schickard's hand 1p (20x31) Latin "Eclipsis ☽ observata Tubingae/A: 1634 d: 4. Mart" [enclosed w/fol 9]	
12rv	From Schickard at Tübingen 1p (17x18) Latin Publ, MC, IV, p 277, No 366	[end of Jul 1634?]
13rv	From Schickard at Tübingen 1p + ad (19x34) Latin See MC, IV, p 184	25 Mar 1634 OS [4 Apr 1634]
14rv	From Gassendi at Aix 1p + ad (21x30) Publ, MC, III, pp 400-402, No 250	17 May 1633
15rv	From Gassendi at Tanaron 1p + ad (21x34) Latin Publ, MC, III, pp 464-465, No 266 Publ extract, GGO, XVIII, p 431, No 2639bis Publ, GOO, VI, p 58	Eidibus Aug 1633 [13 Aug 1633]

16rv	From Gassendi at Digne 1p + ad (19x27) Publ, MC, IV, pp 11-12, No 304	24 Jan 1634
17rv	From Gassendi at Aix 1p + ad (20x29) Publ, MC, IV, pp 324-326, No 372	22 Aug 1634
18r-22v	To Gassendi at [Digne] 5pp & 6 lines (18x29) Latin fols 21 & 22, blank Publ, MC, V, pp 75-78, No 410	4 Kal Mar 1635 [26 Feb 1635]
23r-27v	To [Peiresc] at Aix 8¾pp (21x30) Publ, MC, V, pp 474-476, No 509	15 Nov 1635
28rv	From Gassendi at Aix 1p + ad (20x29)	6 May 1636
29rv	From Gassendi at Aix 1p + ad (14x20)	10 Feb 1637
30rv	From Gassendi at Aix 1p + ad (20x29) Incl astronomical information in B's hand	3 Mar 1637
31rv	From Gassendi at Aix 1p + ad (13x20)	31 Mar 1637
32r-33v	Astronomical data 3pp (20x28) Latin "Agrus-sextiis/Anno 1637, mense Martio/ Die 29 vesperi...*Occultatio, Quadrilateri* *Pleiadum a* ☾..."	
34rv	From Gassendi at Aix 1p + ad (20x29)	25 Aug 1637
35r-36v	From Gassendi at Aix 2¾pp + ad (21x31) Publ, MC, VI, pp 308-310, No 626	ce jor St Michel 1637 [29 Sep 1637]
37r-38v	From Gassendi at Aix 3pp + ad (15x21) Publ, MC, VI, p 320, No 630	3 Nov 1637

39r-41v	From Gassendi at Aix 4pp + ad (20x30)	3 Jan 1638
42r-43v	From Gassendi at Aix 3pp + ad (17x20)	16 Mar 1638
44rv	From Gassendi at Digne $\frac{3}{4}$p + ad (14x20)	18 Jul 1638
45r-46v	From Gassendi at Digne 3pp + ad (20x28) Incl astronomical data in B's hand	[4 Dec] 1638
47r-48v	From Gassendi at Digne $3\frac{1}{3}$pp + ad (20x28) French/Latin Incl tables of ☾ eclipse, Dec 1638	25 Dec 1638
49rv	From Gassendi at ? 1p (14x18) *in B's hand	[13 Feb 1639]*
50rv	From Gassendi at Aix 1p + ad (18x27)	15 Mar 1639
51r-52v	From Gassendi at Aix $\frac{3}{4}$p + ad (13x19)	24 May 1639
53r-54v	From Gassendi at Aix $1\frac{2}{3}$p + ad (15x21)	16 Aug 1639
55r-56v	From Gassendi at Paris B at Bonneval 3pp + ad (15x21)	30 Jul 1642
57r-58v	From Gassendi at [Paris] B at Bonneval 3pp + ad (14x20)	undated
59r-60v	From Gassendi at Paris 3pp (14x20) French/Latin	6 Feb 1646
61r-62v	From Gassendi at Paris 2pp + ad (14x20) French/Latin	6 Nov 1646
63r-64v	From Gassendi at Toulon 1p + ad (15x21)	6 Dec 1649

65r-66v	From Gassendi at Toulon 2pp + ad (14x20) *in B's hand	17 Jan [1650]*
67rv	From Gassendi at Digne 1¾pp (15x20)	28 Sep 1652
68rv	Chart, lunar eclipse 1p + ad (15x20)	undated
69r-70v	From Gassendi at Digne 1p + ad (16x22)	[25?] Oct 1652
71rv	Eclipse chart (phase, altitude, time) in Gassendi's hand 1p + ad (16x22)	undated
72r-73v	From Gassendi at Digne ⅔p + ad (16x22)	4 Dec 1652
74r-75v	From Gassendi at Digne ¾p + ad (16x22) *[1652 is written]	7 Jan [1653]*
76r-77v	From Gassendi at Digne 2⅔pp + ad (16x21)	16 Jan 1653
78r-79v	From Gassendi at Digne 1p + ad (16x33)	31 Jan 1653
80rv	From Anonymous* 7 lines (10x15) *[probably Jacques de Valois] **in B's hand	[Aug 1645]**
81r-82v	From Gassendi at Menil-St-Denis 1p (16x21) fol 82, blank	11 Sep 1654
83r-86v	Copy of a letter in B's hand From Wendelin at Brussels To Gassendi & B 6pp (20x30) Latin fol 86, blank	29 Nov 1647

Folio	Description	Date
87r-88v	From Mersenne at Rome 2⅔pp + ad (20x27) Publ, MC, XIII, pp 315-322, No 1333	16 Jan 1645
89r-90v	From Mersenne at Paris B at Venice 3pp + ad (15x20) Publ, MC, XIV, pp 136-142, No 1445	16 Mar 1646
91r-92v	From Mersenne at Paris B at Venice 2pp + ad (16x21) Publ, MC, XIV, pp 559-562, No 1541	26 Oct 1646
93rv	From Ravius at Berlin 1p + ad (20x32) Latin	2 Jan 1671
94r-95v	From Hortensius at Amsterdam 2¼pp + ad (19x30) Latin	10 Mar 1637
96rv	From Hortensius at Amsterdam ½p + ad (20x30) Latin	19 Jul 1638
97rv	From Hortensius at Amsterdam 1p + ad (16x18) Latin	17 Jan 1639
98rv	From Blaeu at Amsterdam ½p (19x21)	6 Feb 1638
99r-100v	From Galileo at Florence 1p (20x28) Latin fol 100, blank Publ, GGO, XVII, pp 245-246, No 3633	Kal Jan 1638 [1 Jan 1638]
101r-102v	From Galileo at Florence ½p (19x27) Latin fol 102, blank Publ, GGO, XVIII, p 134, No 3954	30 Dec 1639
103rv	"Epitaphe" for Galileo 1p (20x27) Latin See MC, XI, p 32, No 2	undated
104rv	Extract of a letter 1p (19x28) Italian "Extraict de la Lettre de Monsieur Galilei à Monsieur Diodati" [Arcetri, 11 Dec 1639] Publ, GGO, XVIII, pp 128-129, No 3949	

105r-106v	From Bardi at Florence 1⅓pp + ad (19x25) Italian	10 [Oct?] 1643
107r-108v	To Heydon at [London] 3¼pp (20x30) Latin	19 Jul 1640
109r-110v	From Heydon at London 3¼pp (25x36) Latin	27° sextelei 1640 OS [6 Aug 1640]
111r-112v	To Heydon at [London] 4pp (21x30) Latin	Prid Kal Sep 1640 [31 Aug 1640]
113r-114v	From Heydon at [London] 1p + ad (19x30)	23 Feb [1640] OS [5 Mar 1641]
115rv	From Langius at Amsterdam 1p & 5 lines + ad (22x35) Latin	IV Non Quintiliis 1652 [OS] [14 Jul 1652]
116r-117v	To Langius at [Amsterdam?] 2¾pp (18x25) Latin	12 Jul 1652
118r-119v	From Langius at Padua 2⅓pp + ad (20x30) Latin/Italian	Prid Kal Jan [NY] [OS] [10 Jan]
120rv	From Langius at Copenhagen 1p + ad (20x30) *[actually 1653]	XV Kal Jan 1652* [OS] [28 Dec 1652]
121r-122v	To Langius at Copenhagen 1½pp (18x23) Latin fol 122, blank	25 Jan 1653
123rv	From Langius at Copenhagen 1¾pp (21x30) Latin	VII Kal Apr 1653 OS [5 Apr 1653]
124r-125v	To Langius at Copenhagen 2pp (17x21) Latin fol 125, blank	6 Jun 1653
126rv	From Langius at Copenhagen 1½pp + ad (20x30) Latin	Prid Id Nov 1653 OS [22 Nov 1653]
127r-128v	To Langius at Copenhagen 2⅔pp (20x30) Latin	23 Sep 1654

129r-130v	To Langius at Copenhagen B at The Hague 1⅓pp (17x22) Latin fol 130, blank	8 May 1657
131r-132v	To Langius at Copenhagen B at The Hague 1p (17x22) fol 132, blank	9 Jul 1657
133r-134v	To Golius at Leiden 2⅔pp (17x21) Latin	17 Jan 1653
135r-136v	To Golius at Leiden 2⅓pp (17x22) Latin	7 Feb 1653
137rv	From [Golius at Leiden] 1¼pp (19x21) Latin/Greek	undated
138r-139v	From Antonini at [Venice?] ¾p* (20x29) Italian *3 initials on fol 139v	27 Jun 1646
140r-143v	Copy of a letter From Galileo at Arcetri To Anonymous 7pp (20x29) Italian	20 Feb 1638
144r-145v	To Antonini at Venice B at Venice 1⅓pp (18x28) Italian fol 145, blank	23 Jun 1646
146r-148v	To Antonini at [Venice] B at Venice 3pp & 5 lines (20x30) Latin fol 148, blank	30 Jun 1646
149r-150v	From Santini at Rome B at Florence 1⅔pp + ad (20x26) Italian Publ, MC, XIV, pp 542-543, No 1534	23 Oct 1646
151r-152v	From Van Langren at Brussels 1p (15x20) fol 152, blank Publ, MC, XIII, pp 449-450, No 1373	3 Jun 1645

153r-154v	To Van Langren at [Brussels] B at Venice 2⅔pp (15x20) Latin	18 Nov 1645
155r-156v	From Van Langren at Brussels 1¾pp (15x19) fol 156, blank	2 Jun 1652
157r-158v	From Van Langren at Brussels 1¼pp (15x19) fol 158, blank	Jour de Solstice [NY] [22 Jun]
159r-160v	From Van Langren at Brussels 2⅔pp + ad (15x19)	7 Sep 1652
161r-162v	From Van Langren at Brussels 1½pp (15x19) fol 162, blank	26 Oct 1652
163rv	Map, designed and annotated by Van Langren 1p (13x16) Map of Oostende [Ostend, Belgium]	undated
164rv	Four printed maps [Ostend] 1p (15x19) [proposed plan of Van Langren]	undated
165rv	Two printed maps [Ostend] 1p (19x32) [proposed plan of Van Langren]	undated
166r-167v	From Van Langren at Brussels 1⅔pp + ad (15x18)	27 [Dec?] 1652
168r-169v	From Van Langren at Brussels 1p + ad (15x18)	Jour de St Paul 1653 [25 Jan?] 1653
170rv	From Van Langren at Brussels ½p (17x18)	15 Feb 1653
171r-172v	From Van Langren at Brussels 2⅔pp + ad (15x18)	21 Mar 1653

173r-174v	From Van Langren at Brussels 3pp (15x18) [174r, drawing of ☾ in Pleides]	5 Apr 1653
175r-176v	To J. Wallis at Oxford $1\frac{2}{3}$pp (17x21) Latin fol 176, blank	24 May 1655
177rv&180r-181v	From S. Ward at Oxford B at The Hague 2pp + ad + ad (15x20) Latin	27 Jul 1657 OS [6 Aug 1657]
178r-179v	To S. Ward at [Oxford] B at The Hague $\frac{3}{4}$p (17x22) Latin fol 179, blank	9 Jul 1657
182r-183v	From Huet at Caen 2pp + ad (16x23) Latin	Eid Oct 1654 [15 Oct 1654]
184rv	To Huet at [Caen] $1\frac{1}{4}$pp (20x30) Latin	30 Dec 1654
185rv	Missing	
186r-187v	Charts, etc. in Huet's hand 4pp (11x34) Latin "Observationes Poli Cadomensis per altitudines meridianas Solis" and others [1619-25]	
188r-189v	Charts in Huet's hand 4pp (11x34) Latin [Eclipses of Dec 1619 & Nov 1631]	
190r-191v	To Huet at [Caen] $1\frac{2}{3}$pp & 1 line* (15x20) Latin *fol 191v, 1 line in B's hand	31 Jan 1655
192r-193v	From G. della Pax at Smyrna [Izmir] $\frac{2}{3}$p + ad (21x30) Italian	8 Jun 1648
194r-195v	From G. della Pax at Smyrna $1\frac{2}{3}$pp (21x33) Latin fol 195, blank	Kal Oct [1649?] [1 Oct]
196r-197v	From G. della Pax at Smyrna $2\frac{2}{3}$pp (20x30) Italian	24 Jan 1654

198rv	From Torrino at Turin 1p (20x30) Italian	15 Jul 1656
199rv	To Torrino at Turin 1½pp (15x20) Latin	15 Sep 1656
200r-203v	From Torrino at Turin 2pp & 3 lines + ad (15x20) Latin fol 202, blank	8 [Nov?] 1656
204rv&207rv	From Wangnereckio at Vienna 1⅓pp + ad (20x30) Latin/Green	XIV Kal Apr 1656 [19 Mar 1656]
205r-206v	To Wangnereckio at [Vienna] 3¾pp (15x22) Latin fol 206, list of books	28 Apr 1656
208r-209v	To S. Ward at Oxford 2pp (14x20) Latin fol 209, blank	24 Jan 1657
210r-211v	From Vincent at Genoa 2⅔pp + ad (16x22) "Anno 1656 Jan. die 26. Solis eclipsii observata..." with data	5 Sep 1656
212r-213v	From de Venasque at Monaco 3pp (21x30)	1 Nov [1643?]
214r-215v	From Deriennes at La Fleche 1p + ad (13x18)	8 Mar 1653
216r-217v	From Debeaune at Blois 1½pp + ad (20x30) Publ, MC, VIII, pp 441-446, No 745	18 Jun 1639
218rv&221rv	From A. Morretti at Venice 1p + ad (18x27) Italian	7 Dec 1657
219r-220v	To A. Morretti at Venice 2pp (17x22) Latin fol 220, blank	29 Jul 1658
222r-223v	From Santini at [Carnaro?] To Anonymous at Rome ½p + ad (20x27) Italian	2 Jan 1653

224r-225v	Astronomical data and text in J. de Valois' hand 2½pp (19x27) Latin [Solar eclipse of 12 Aug 1654; sent with fols 228&229]	undated
226r-227v	From [de Valois] at Grenoble 3pp & 4 lines + ad (12x16)	13 Sep 1654
228r-229v	From [de Valois] at Grenoble 3pp + ad (12x16)	30 Aug 1654
230r-232v	From Léotaud at Embrun 3½pp + ad* (18x25) fol 232, blank *[unnumbered fol between 230&231]	12 Jan 1655
233r-239v	"Solis Ecclipsis observata Ebreduni die Augusti 12/ad longitudinem 27 Gr. A latitudinem 43 G. 50 M." 13pp (18x28) Latin, unsigned fol 239v, 3 geometrical diagrams	undated
240r-243v	From Saint-Légier* at Avignon To Peiresc at Aix 4pp + ad (19x28) fol 242, blank *[Tondut de Saint-Légier, Pierre François]	26 Mar 1634
244rv	From Saint-Légier at Avignon 1p + ad (18x27)	22 Aug 1640
245rv	From Saint-Légier at Avignon ⅔p (18x27)	18 Sep 1640
246r-247v	From Saint-Légier at Avignon 2pp + ad (18x27)	12 Dec 1640
248rv	From Saint-Légier at [Avignon] 1p + ad (18x27)	undated
249r-250v	From Saint-Légier at Avignon 2½pp + ad (14x20)	11 Nov 1641
251r-252v	From Saint-Légier at Avignon 3pp + ad (14x20)	1 Jan 1647

253rv	Chart: Inclinatio Grad. 1.0' 1p (17x22) Latin	undated
254rv	Text, unsigned 1p (14x20) Latin [Concerns chart (fol 253); planetary latitudes & Kepler's hypothesis]	undated
255r-256v	From Saint-Légier at Avignon 2pp + ad (15x22)	9 Jun 1652
257r-258v	From Saint-Légier at Avignon 1½pp + ad (14x19)	28 Apr 1653

▼ ▼ ▼

COLLECTION BOULLIAU, VOLUME XX, FONDS FRANÇAIS 13038

1r-2v	From Grémonville* at Grémonville 1½pp + ad (17x22) *[Bretel de Grémonville, Nicolas]	5 Oct 1638
3r-4v	From Grémonville at Turin 1p + ad (17x22)	1 Oct 1643
5r-6v	From Grémonville at Venice 3½pp (20x30)	29 Sep 1646
7r-8v	From Grémonville at Venice 3pp (20x30)	6 Oct 1646
9r-10v	From Grémonville at Venice 2pp & 4 lines (20x30)	13 Oct 1646
11r-12v	From Grémonville at Venice 3pp (20x30)	20 Oct 1646
13r-14v	From Grémonville at Venice 4pp (20x30)	27 Oct 1646
15r-16v	From Matharel at Venice 4pp (15x20)	29 Sep 1646
17r-18v	From Matharel at Venice 2pp (15x20) fol 18, blank	7/8 Oct 1646
19r-20v	From Matharel at [Venice] 1⅔pp & 11 lines (15x20)	13 Oct 1646
21r-22v	From Matharel at Venice 2⅔pp (15x20) Italian	20 Oct 1646
23r-24v	From Grémonville at Venice 4pp (20x30)	9 [Feb?] 1647
25r-26v	From Matharel at [Venice] 4pp (20x30)	4 Apr 1648
27r-28v	Printed text 4pp (17x23) Italian	

	"Vittoria/Ottenuta dalle Armifelicissme/ Della Sereniss.^{ma} Repub.^{ca}./Veneta/Nell' impressa dell'inespugnabile Fortezza/ DICLISSA...Venetia...[1648]"	
29r-30v	From Matharel at Venice 2pp (15x20) fol 30, blank	11 Apr 1648
31r-32v	From Matharel at Venice 4pp (20x30)	30 May 1648
33r-34v	From Matharel at Venice 2¼pp (20x30)	13 Jun 1648
35r-36v	From Matharel at Venice 3½pp (20x30)	20 Jun 1648
37r-38v	From Matharel at Venice 2½pp (20x30)	27 Jun 1648
39r-40v	From Matharel at Venice 2pp (20x30) fol 40, blank	4 [Jul?] 1648
41r-42v	From Matharel at Venice 1p (20x30) fol 42, blank	11 Jul 1648
43r-44v	From Matharel at Venice 2pp (20x30) fol 44, blank	18 Jul 1648
45r-46v	From Matharel at Venice 2½pp (15x20)	25 Jul 1648
47r-48v	Newsletter Venice 2pp (20x30) fol 48, blank	25 Jul 1648
49r-50v	From Matharel at Venice 2pp & 5 lines (20x30)	1 Aug 1648
51r-52v	From [Matharel] at Venice 4pp (20x30)	15 Aug 1648

53r-54v	From Matharel at Venice 3pp (20x30)	22 Aug 1648
55r-56v	From Matharel at Venice 3½pp (20x30)	29 Aug 1648
57r-58v	From Matharel at Venice 2⅔pp (29x30)	5 Sep 1648
59r-60v	From Matharel at Venice 2pp (20x30) fol 60, blank	12 Sep 1648
61r-62v	Newsletter Venice 2pp (20x30) fol 62, blank	19 Sep 1648
63r-64v	From [Matharel at Venice] 4pp (20x30)	undated
65r-66v	From Matharel at Venice 3pp (20x30)	3 Oct 1648
67r-68v	From Matharel at Venice 4pp (20x30)	10 Oct 1648
69r-70v	From Matharel at Venice 2pp & 4 lines (20x30)	17 Oct 1648
71r-72v	From [Matharel] at Venice 2⅔pp (20x30)	24 Oct 1648
73r-74v	From Matharel at Venice 2½pp (20x30)	31 Oct 1648
75r-76v	From Matharel at Venice 2⅔pp (20x30)	7 Nov 1648
77r-78v	From [Matharel] at Venice 2pp (20x30) fol 78, blank	14 Nov 1648
79r-80v	From Matharel at Venice 2½pp (20x30)	5 Dec 1648

81r-82v	Newsletter Venice 1⅔pp (20x30) fol 82, blank	12 Dec [1648]
83rv	List 2pp (20x30) Italian "Lista delle famiglie fatte/nobili perdenaro nell'occasione/della guerra del Turco...,/Giorno 12 Xbre 1648"	undated
84r-85v	From Matharel at Venice 3½pp (20x30)	19 Dec 1648
86r-87v	From [Matharel] at Venice 4pp (20x30)	2 Jan 1649
88r-89v	From [Matharel] at Venice 3pp (20x30)	9 Jan [1649]
90r-91v	From [Matharel] at Venice 3pp (20x30)	16 Jan 1649
92r-93v	From [Matharel] at Venice 3pp (20x30)	23 Jan 1649
94r-95v	From [Matharel] at Venice 3⅔pp (15x20)	30 Jan 1649
96r-97v	From Matharel at Venice 2pp (20x30) fol 97, blank	6 Feb 1649
98r-99v	From [Matharel] at Venice 3½pp (15x20)	13 Feb 1649
100r-101v	From [Matharel] at Venice 3pp (20x30)	8 May 1649
102r-103v	From [Matharel] at Venice 3pp (20x30)	15 May 1649
104rv&107rv	From [Matharel] at Venice 2pp (20x30) fol 107, blank	22 May 1649

105r-106v	Newsletter Candia 1p & 6 lines (15x20) fol 106, blank	3 Apr 1649
108r-109v	From Matharel at Venice 3pp (20x30)	29 May 1649
110r-111v	From Matharel at Venice 2½pp (20x30)	5 Jun 1649
112r-113v	From Matharel at Venice 3pp & 3 lines (20x30)	19 Jun 1649
114r-117v	Printed text 7pp (15x21) Italian "Relatione della/VITTORIA/OTTENVTA/ Dalle Armi della Serenissima/Republica di Venetia/...Venetia [1649]"	
118r-119v	From Matharel at Venice 2pp (20x30) fol 119, blank	26 Jun 1649
120r-121v	From Matharel at Venice 3pp (20x30)	3 Jul 1649
122r-123v	From [Matharel] at Venice 2½pp (20x30)	17 Jul 1649
124r-125v	From Matharel at Venice 2½pp (20x30)	10 Jul 1649
126r-127v	From Matharel at Venice 3½pp (20x30)	24 Jul 1649
128r-129v	From [Matharel] at Venice 3pp (20x30)	31 Jul 1649
130r-131v	From [Matharel] at Venice 1⅔pp (20x30) fol 131, blank	7 Aug 1649
132r-133v	From [Matharel] at Venice 3pp (20x30)	14 Aug 1649
134r-135v	From [Matharel] at Venice 3pp (20x30)	21 Aug 1649

136r-137v	From Matharel at Venice 1p (20x30) fol 137, blank	28 Aug 1649
138r-139v	Newsletter Bologna 2pp & 8 lines (20x30)* Italian *fol 139 (15x20)	13 Aug 1649
140rv	Newsletter Reggio $\frac{2}{3}$p (20x30) Italian	12 Aug 1649
141r-142v	Newsletter Venice 3$\frac{2}{3}$pp (20x30) Italian	28 Aug 1649
143r-144v	Newsletter Venice 3$\frac{1}{2}$pp (20x30) Italian	4 Sep 1649
145r-146v	From [Matharel?] at Venice 2$\frac{2}{3}$pp (20x30)	4 Sep 1649
147r-148v	Newsletter Venice 3$\frac{3}{4}$pp (20x30) Italian	10 Sep 1649
149r-150v	From [Matharel at Venice?] 2$\frac{1}{2}$pp (20x30)	[11?] Sep 1649
151r-152v	Newsletter Venice 3$\frac{1}{2}$pp (20x30) Italian	17 Sep 1649
153r-154v	From [Matharel?] at Venice 1$\frac{2}{3}$pp (20x30) fol 154, blank	18 Sep 1649
155r-157v	Newsletter Venice 4$\frac{2}{3}$pp (20x30) Italian	24 Sep 1649
158r-159v	From Matharel at Venice 2$\frac{2}{3}$pp (20x30)	25 Sep 1649

160r-161v	Newsletter Venice 3½pp (20x30) Italian	 2 Oct 1649
162r-163v	From Matharel at Venice 1⅔pp (20x30) fol 163v, astronomical notes in B's hand	2 Oct 1649
164rv	From [Matharel] at Venice ⅔p (20x30)	9 Oct 1649
165r-166v	Newsletter Venice 3pp + ad (20x30) Italian	 9 Oct 1649
167r-168v	From [Matharel] at Venice B at Meslay 2⅔pp + ad (20x30)	16 Oct 1649
169r-170v	Newsletter Venice 3pp & 5 lines (20x30) Italian	 23 Oct 1649
171r-172v	From [Matharel at Venice] 2⅓pp (20x30)	23 Oct 1649
173r-174v	Newsletter Venice 4pp (20x30) Italian	 29 Oct 1649
175r-176v	Newsletter Venice 2½pp (29x30)	 30 Oct 1649
177r-178v	Newsletter Venice 3pp & 4 lines (20x30) Italian	 5 Nov 1649
179r-180v	From Matharel at Venice 3pp (20x30)	6 Nov 1649
181r-182v	From Matharel at Venice 3pp (20x30)	13 Nov 1649
183r-184v	Newsletter Venice 3½pp (20x30) Italian	 [12?] Nov 1649

185r-186v	Newsletter Venice 3pp & 3 lines (20x30) Italian	19 Nov 1649
187r-188v	From Matharel at Venice 1½pp (20x30) fol 188, blank	20 Nov 1649
189r-190v	Newsletter Venice 3pp & 5 lines (20x30) Italian	3 Dec 1649
191r-192v	From [Matharel] at Venice 2pp & 4 lines (20x30)	4 Dec 1649
193r-194v	Newsletter Venice 3pp & 6 lines (20x30) Italian	10 Dec 1649
195r-196v	From Anonymous at Venice 2⅓pp (20x30)	11 Dec 1649
197r-198v	From [Matharel] at Venice 3pp & 4 lines (20x30)	18 Dec 1649
199r-200v	From Matharel at Venice 2pp & 5 lines (20x30)	25 Dec 1649

▼ ▼ ▼

COLLECTION BOULLIAU, VOLUME XXI, FONDS FRANÇAIS 13039

1r-2v	To Gaston d'Orléans at [Blois] ½p (17x22) fol 2, blank	28 Mar 1657
3r-4v	From d'Orléans at Blois B at The Hague ½p + ad (17x22)	15 May 1657
5r-6v	To Leopoldo at Florence 1⅓pp (20x31) Latin fol 6, blank	[21] May 1649
7r-8v	From Leopoldo at Florence 1½pp + ad (20x31) Italian	17 Jun 1649
9r-10v	To Leopoldo at [Florence] B at [Paris] ½p (20x31) Latin fol 10, blank	undated
11r-35v	To Ferdinando 45⅔pp (20x31)* Latin [copy of a mémoire concerning B's trip to Italy and Constantinople; 1651] fol 35, blank *fol 29 (15x25)	
36rv&40rv	From Leopoldo at Pisa 1p + ad (20x31) Italian	4 Feb 1651 Ab Inc [4 Feb 1652]
37r	To Leopoldo at Florence 1p (20x31) Latin	13 Jan 1651
37v	To Leopoldo at Florence 1p (20x31) Latin	28 Apr 1651
38r-39v	From Leopoldo at Florence ½p + ad (20x31) Italian	30 Mar 1651
41r-42v	From Leopoldo at Florence 1p + ad (20x31) Italian	23 May 1652
43r-44v	To Leopoldo at Florence 1½pp (17x22) Latin fol 44, blank	28 Jun 1652

45r-46v	From Leopoldo at Florence ½p + ad (20x31) Italian	13 Sep 1652
47r-48v	To Leopoldo at Florence 1⅓pp (18x25) Latin fol 48, blank	11 Oct 1652
49r-50v	From Leopoldo at Florence ⅔p + ad (20x31) Italian	7 Nov 1652
51r-52v	From Leopoldo at Florence ½p (20x31) Italian fol 52, blank	27 Oct 1652
53r-54v	To Leopoldo at Florence 1p & 5 lines (18x25) Latin fol 54, blank	17 Jan 1653
55r-56v	From Leopoldo at Livorno [Leghorn] 6 lines + ad (20x31) Italian	17 Jan 1652 Ab Inc [17 Jan 1653]
57r-58v	From Leopoldo at Florence ½p + ad (20x31) Italian	13 Mar 1652 Ab Inc [13 Mar 1653]
59r-60v	To Leopoldo at Florence 2pp (18x25) Latin fol 60, blank Publ, LIF, II, pp 197-198	17 Sep 1655
61r-62v	From Leopoldo at Florence ⅔p & 3 lines* (20x31) Italian *fol 62v, in B's hand	9 Jun [1657?]*
63r-64v	From Leopoldo at Florence ⅔p (20x31) Italian fol 64, blank	21 Oct 1655
65r-66v	To Leopoldo at Florence 1p (15x20) Latin fol 66, blank	4 Feb 1656
67r-68v	To Leopoldo at Florence ⅔p (15x20) Latin fol 68, blank	31 Mar 1657
69r-70v	To Leopoldo at Florence 1½pp (17x22) Latin fol 70, blank	14 Sep 1657

71r-72v	From Leopoldo at Florence ⅔p (20x31) Italian fol 72, blank	13 Oct 1657
73r-74v	To Leopoldo at Florence 2pp (17x22) Latin fol 74, blank	30 Aug 1658
75rv	From Leopoldo at Florence 1p (18x28) Italian	21 Sep 1658
76r-78v	Text, unsigned ⅔p (20x31) Latin fols 77 & 78, blank	undated
79rv&82rv	From Leopoldo at Pisa ⅔p (20x31) Italian fol 82, blank	23 Dec 1658
80r-81v	To Leopoldo at Florence 2⅓pp (15x20) Latin Publ, HO, III, pp 459-461, No 592[a]	28 Feb 1659
83r-84v	To Leopoldo at Florence 2¾pp (17x22) Latin [fol 84v, drawing]	4 Apr 1659
85rv&88rv	From Leopoldo at Florence 1p & 6 lines (18x27) Italian Publ, HO, III, pp 461-462, No 604[a] fol 88, blank	31 Mar 1659
86r-87v	To Leopoldo at Florence 1⅓pp (17x22) Latin fol 87, blank Publ, HO, III, pp 462-463, No 609[a]	2 May 1659
89r-92v	From Leopoldo at Florence 2⅓pp (20x27) Italian [fol 91r, drawing of physics instrument] fol 92, blank Publ, HO, III, pp 464-465, No 621[a]	22 May 1659
93r-94v	From Leopoldo at [Podio Caiano] 3pp (20x27) Italian	24 Apr 1659

95r-96v	To Leopoldo at Florence 2⅔pp (17x22) Latin Publ, HO, III, pp 465-467, No 637[a]	13 Jun 1659
97r-98v	From Rinaldini at Florence ⅓p + ad (18x27)	7 Aug 1659
99r-100v	From Rinaldini at [Pisa] To [Leopoldo at Florence] 1p Italian fol 100, blank (19x27) [copy sent w/following treatise]	[7 Aug 1659]
101r-117v	"Traitté des expériences sur le vif-argent" [by Carlo Rinaldini] 11p & 3 lines (19x27) Italian/French fols 116 & 117, blank	undated
118r-119v	To Leopoldo at Florence 2¾pp (17x22) Latin Publ, LIF, II, pp 198-202	19 Dec 1659
120r-125v	To Rinaldini at Pisa 10½pp (17x22) Latin [fols 101-115, B's response to Rinaldini's treatise]	14 Dec 1659
126r-141v	Copy of a treatise in B's hand From Paolo del Buono To Leopoldo at [Florence] [fol 126r, B writes "Cet escrit m'a esté envoyé/de Pologne par Mr Des Noyers/dont j'ay renvoyé l'original"] Publ, WEK Middleton, AHES 6(1969):1-28. 30pp (17x22) Italian fol 141, blank	[Aug/Sep 1657?]
142r-143v	From Leopoldo at Florence 2pp (19x27) Italian Publ, HO, III, pp 467-469, No 655[a] fol 143, blank	21 Aug 1659
144r-146v	From Leopoldo at Florence ⅔p (19x27) Italian fols 145 & 146, blank Publ, HO, III, pp 469-470, No 473[a]	9 Oct 1659

147rv	Copy of a letter From Viviani at [DiCasa] To Leopoldo at [Florence] 1p (20x31) Italian Publ, HO, III, pp 470-472, No 673[b] see also, Publ, GGO, XIX, p 647	[20 Aug 1659]
148r-159v	Treatise, "histoire de l'invention d'un horologe/et pendule par Galileé, Ecrite/ par vincentio viviani" 16½pp (21x30) Italian fols 156 & 157, blank; 158 (30x41) & 159 (26x38) Drawings fol 155r, DiCasa, 20 Aug 1659, signed Vincenzio Viviani Publ, HO, III, pp 472-484, No 673[b]	
160r-161v	To Leopoldo at [Florence?] 1p (17x22) Latin fol 161, blank	7 Nov 1659
162r-163v	From Leopoldo at Pisa ½p (19x27) Italian fol 163, blank	15 Dec 1659
164r-165v	To Viviani at [Florence?] 1p (17x22) Latin fol 165, blank	26 Mar 1660
166r-167v	From Leopoldo at Florence ½p (19x27) Italian fol 167, blank	1 Jul 1660
168r-169v	From Leopoldo at Florence 1½pp (19x27) Italian fol 169, blank	19 Aug 1660
170r-171v	Text & drawing* in B's hand 2pp (20x29) Italian fol 171r, drawing of experimental apparatus [*cf. Saggi di naturali esperienze nell' Academia del Cimento, LXXXXIII]	undated
172r-173v	To Leopoldo at Florence 1⅓pp (17x22) Latin fol 173, blank Publ, LIF, II, pp 203-205	30 Jul 1660

▼ ▼ ▼

COLLECTION BOULLIAU, VOLUME XXII, FONDS FRANÇAIS 13040

1r-2v	Anonymous Tract 3pp (20x31) Latin	undated
3r-5v	From Dufresne at Würzburg To B? 6pp (18x25)	9 Jan 1669
6r-7v	Copy of letters in B's hand From A. Magliabecchi at Florence To E. Menagio; To E. Bigot 2½pp (18x25) Italian	30 Jul 1678
8rv	From Morin at Paris 1p (17x25)	27 Dec 1650
9r-10v	Printed letter concerning geometry 3pp (18x24) Latin [inscribed: "Clarissimo viro D.K.D."]	undated
11r-12v	From [?] at Florence 2½pp + ad (20x31)	12 Jun 1672
13r-15v	From Anonymous To Hevelius at [Danzig] 4pp & 4 lines (18x25) Latin	undated
16rv&18rv	From Butsner at Danzig B at Warsaw 2pp & 6 lines + ad (20x31)	9 May 1661
17rv	Note in B's hand concerning Père Jesuit Tremelius 1p & 6 lines (10x15)	undated
19rv	From Butsner at Danzig 1⅓pp (20x31) Latin	20 Apr 1661
20r-23v	Copies of letters 9pp (20x31) "Copie des Lettres escrittes au S^{r}. de Castelfranc par Monsieur de l'Escale" Hollande, 2⅓pp Leiden, 1⅔pp Leiden, 1⅓pp & 2 lines Leiden, 1⅔pp	 19 Oct 1604 7 Apr 1605 16 Feb 1606 15 Mar 1618

Folio	Description	Date
24rv	From Chappuis at [Châlon?] To Lantin at Paris 1p (20x31)	1 [Sep] 1677
25r-26v	From Lantin at Dijon $2\frac{2}{3}$pp + ad (17x22)	29 Jan 1662
27rv&32rv	From U. Boulliau at Marseille $\frac{1}{2}$p & $\frac{1}{3}$p + ad (17x22)	22 Sep 1688
28r-31v	From U. Boulliau at Douvres 6pp & 3 lines + ad (17x22)	18 Jul 1690
33rv	Copy of a letter in B's hand Concerning Comte Nicolas de Serin $\frac{2}{3}$p (18x25) Latin	30 Jun 1664
34rv	Copy of a letter in B's hand "Lettre du Comte de Serin a l'Evesque de Zagrabia 28 Mars 1664" $\frac{2}{3}$p (18x25) Latin	28 Mar 1664
35r-37v	Copy of a letter From Chiaramanci at Bursa in Asia To J. Conte di Rabatta $4\frac{1}{3}$pp (18x25) Italian	22 Aug 1659
38r-39v	From Lanquet at [Paris?] 3pp (11x17) Latin/French	undated
40r-41v	From Anonymous To the sister of de Thou* 2pp (17x22) *[Louise-Gabrielle de Thou]	21 Dec 1659
42r-44v	From Siri at Vienna To Des Noyers 5pp (20x31) Italian	11 Sep 1663
45r-47v	From [Siri] at Vienna To Burattini 6pp (20x31) Italian	1 Oct 1663
48r-50v	Copy of a letter From Adam Comte Forgae at Comoro To the Emperor at Vienna $2\frac{2}{3}$pp & 9 lines (20x31) Italian	27 Sep 1663

51r-52v	From Siri at Vienna To Anonymous 3pp + ad (20x31) Italian	19 Sep 1663
53r-54v	From [Siri at Vienna?] To Des Noyers 4pp (20x31) Italian	undated
55rv	Copy of a letter in B's hand "Lettre de l'Evesque de Luceorie/a Mr de Harlay Procur^r^. Gnal./6 Octob. 1685" 1⅓pp (18x25) Latin	6 Oct 1685
56rv	Note, Anonymous 1p (12x17) [concerns fol 55]	undated
57r-58v	From Chanut at ? To [J-A de Thou?] 3pp + ad (17x22)	undated
59r-60v	From [L. S. Mark d'Anglie?] at Rome To Anonymous 2½pp (20x31) Italian	15 Jun 1630
61r-62v	Tract, unsigned 3⅓pp (20x31) Italian	undated
63rv	Copy of a letter From Pépin [Pipinus] at Bologna To Mazaud [Mazaldum?] 1p (20x31) Latin	[Ascension, 1550] [15 May 1550]
64rv	From Montesson at St-Jean-de-Luz 1½pp (17x22)	13 Oct [1658?]
65r-67v	From Montesson at St-Jean-de-Luz 2pp & 2 lines (17x22) fol 67, blank	2 Oct [1658?]
68r-69v	Copy of a letter From Neerkassel at Utrecht To [Louis XIV at Paris] 1½pp & ⅔p (20x31)	3 Jan 1662
70r-72v	From Tervilliers* at ? To Rivière at [Paris] 2pp + ad (17x22) fol 72, blank *[Viole de Tervilliers de Saint-Maure]	undated

73r-74v	From Joly at London 3pp + ad (17x22)	4 Oct 1671
75rv	From [de Thou?] To Pontcarré, conseiller au Parlement 1p (11x17)	25 Sep 1677
76r-78v	From [de Thou?] To Pontcarré, conseiller au Parlement 3pp + ad + ad (11x17)	24 Sep 1677
79rv&81rv	From Chappuis at Châlon 1p + ad (12x17)	2 Oct 1677
80rv	From Pentzenau at ? To Rivière at [Paris] ⅔p + ad + ad (18x25) Latin	Jan 1661
82rv	From Deuringius at Groningae To Anonymous 2pp (20x31) Latin	21 Jan 1660
83rv	Extract in B's hand "Extraict d'une lettre de Mr Boritzius de Lion le 18 Juillet 1665." ½p (17x22)	18 Jul 1665
84rv	From Wagenseil at Altdorf 1½pp (17x22) Latin	18 Aug 1682
85r-86v	From Blot at Amsterdam [To J-A de Thou?] 1⅔pp + ad (18x25)	16 Jun 1667
87r-88v	From Marchais at Blois 1½pp + ad (18x25)	9 Mar 1656
89rv	From Deneuré at Caen To Anonymous 1p (18x25)	1 Nov 1654
90rv	From Deneuré at Caen To Anonymous 1p (18x25)	30 Apr 1655
91rv	Extracts of letters in B's hand 1p (18x25) Latin [Julii Reichelo? on comets]	

92r-93v	From Justel at [London] 1½pp + ad (14x19)	3 Mar 1691
94rv	From Justel at [London] 1p + ad (14x19)	11 Feb 1686
95rv	From Justel at [London] 1p (14x19) *[B indicates rec'd end of Jul 1685]	undated*
96r-98v	From [Nijolanti?] at Paris To [Marechal d'Estrèes?] 6pp (20x31) Italian	2 May 1684
99rv	Four astrological squares in B's hand 1p (20x31) [Concerns fols 96-98]	
100r-101v	From C-H Muller at Leipzig 2pp + ad (16x19) Latin Includes astrological chart	24 Sep 1661
102rv&104rv	From Christophorus [?] at Leipzig To [Hevelius at Danzig] 2⅓pp (17x22) Latin	8 Nov 1684
103rv	Astronomical data 1p (20x31) "Eclipsis Solaris,/GEDANI/Anno à nato Christo, 1666, Die 2 Julii, s.n. ante m. observata,/à Johanne Hevelio" [w/diagrams of phases]	
105r-106v	From Kirchmayer at Wittenberg 2pp & 6 lines + ad (16x19) Latin	XIII Kal Oct 1661 [19 Sep 1661]
107r-108v	From Saint-Remy at Bonneval 2pp & 3 lines + ad (12x18) *[rec'd 4 Feb 1669]	23 Jan 1669*
109rv	From Bruchet at The Hague To Anonymous 1p (16x20)	21 Apr 1661
110rv	From Beverlant at London 1½pp (20x31) Greek/Latin [fol 110v, Greek passage from Pappus]	Jan 1684

111r-113v	From Moscheni at Ancone To Anonymous [Monseigneur] 6pp (20x31)	13 Jun 1665
114rv	From [Cölero at Almâ Viadtinâ?]* 1p (16x20) Latin *[Placentinus?]	31 Mar 1665
115rv	From Cölero at [Stettin] B at [Warsaw] 1p (17x22) Latin	Pridie Kal Jun 1661 [31 May 1661]
116rv	To Cölero at [Stettin] B at Warsaw 2pp (11x15) Latin	4 Jul 1661
117rv	Fragment (list) 1p (19x19) Latin	23 Sep 1662
118rv	From Dilgerus at ? ½p + ad (18x29) Latin	undated
119rv&121rv	Note [of Basile Titel] 2½pp (15x19) Latin w/fol 120	undated
120rv	From Titel at Leipzig 1½pp (15x19)	$\frac{12}{22}$ Feb 1669
122r-123v	From Findegueller at Leipzig 3pp (11x16)	10 May NY
124rv	List in B's hand 1p (10x13) Latin	undated
125rv	From ? at Liège 1p & 4 lines (17x22)	18 Mar 1667
126rv	From Anonymous at Poitiers 2pp (17x22)	Pentecoste 1666 [13 Jun 1666]
127rv	From Priandi at Paris 1p (18x20)	25 May 1657
128rv	From Priandi at Paris 1p (18x20)	4 May 1657
129rv	From De Sacqueville at La Flèche 1p (17x22)	30 [Oct?] NY

130r-131v	From Wigan at London 1p & 4 lines (17x22)* Latin/French *fol 131 (9x17)	Kal Nov 1685 [11 Nov 1685]
132rv	From De Méru at [Nevers?] 1p (18x25)	18 Jul 1649
133rv	From De Méru at [Nevers?] 1⅔pp (18x25)	2 Aug 1659
134rv	From Bury at Orléans 1⅔pp (18x25)	20 Oct 1655
135rv	From Bury at Blois 2pp (18x25)	28 Feb 1658
136rv	From Bury at Blois 1p (18x25)	12 Apr 1658
137r-138v	From Van Beuninghen at Amsterdam 2⅓pp + ad (18x25)	28 Feb 1669
139r-140v	Extract of a letter in B's hand 3½pp (20x31) "Extrait d'une lettre de Mr de Fermat a Mr Carcavi" See FO, I, pp XXV-XXVI	undated
141r-142v	Copy of a letter 3½pp (20x31) "Lettre de Jean du Belley...a Monsr. l'Evesque d'Auxerre Ambassadeur du Roy a Rome." [from Fontainbleau, 28 Aug 1532]	
143rv	From J. Martin at Hamburg 1p (17x22)	7 Mar 1662
144r-145v	Copy of a letter in B's hand "Othonis Sperling, ad David Blondellum/ epistola." 3⅓pp (15x20) Latin/Greek	[Jan 1654]
146rv&153rv	Copy of a letter in B's hand [From V. Fabricius at Danzig] "Lettre escrite a M. Sarau Con.er au Parlement" 2⅓pp (20x31) Latin	VII Eid. Oct [1648?] [9 Oct 1648?]

147r-148v	Copy of a letter in B's hand "De la lettre de Mr de Boissieu a Mr Nublé/du 18 Mars 1648. de Vienne" 1⅓pp (12x18)	18 Mar 1648
149r-152v	Copy of a letter in B's hand "Lettre de Monsieur de Saumaise/a/ Monsieur le President de Thou." 6⅔pp (10x15)	1 Apr 1615
154rv	Copy of a letter in B's hand "V.C./Dno. Claudio Salmazio/SD/ Anna Maria à Schurman" 1⅓pp (18x25) Latin	Prid. Id Dec 1639 [?] [12 Dec 1639]
155r-156v	Extract of a letter in B's hand "Extraict d'une lettre de M. Saumaize a Monsieur de/Sarau/18. Octobre 1642" 2⅓pp (17x22)	18 Oct 1642
157rv	Extracts of letters in B's hand 1⅔pp (20x31) Latin "Cujacius Johanni Ambrosio Brassicano..." etc.	undated
158r-161v	Copy of a letter 7½pp (15x20) Latin Nichol. Cragius ad Tychonem Dabantur Haffniae pridie Nonas Decembres anno 1599 Publ, BOO, VIII, pp 211-215	Prid Non Dec 1599 [OS] [14 Dec 1599]
162rv	Extract of a letter in B's hand 1½pp (18x25) Latin "Fragmentum epistolae scriptae a viro Nobili/Tychone Brahe Dano ad virum clarissm./Dominum D Casparum Peucerum" Publ, BOO, VI, pp 127-141	13 Sep 1688 [OS] [23 Sep 1588]
162v-167v	Extract of a letter in B's hand 9½pp & ⅔p (18x25) Latin/Greek "Ad Dominum Tychonem Brahe Danum/ Mathematicum Clarissimum, Virum/ Nobilissimum" [Casparus Peucerus ad Tychonem, Datae Dessae 10. die Maij Iuliani anno 1589] Publ, BOO, VI, pp 184-191	10 May 1589 [OS] 20 May 1589

168r-169v	Copy of a letter in B's hand "Georg. S. Sedlius [Prague] Casparo Peucero" 3⅓pp (20x31) Latin	23 Apr 1562
170r-171v	From Guillou at Loudun 2½pp + ad (17x22) fol 171, 4 horoscope squares	5 Sep 1652
172r-173v	From Lapeyrère at Paris 2½pp (17x22) Publ, LLI, pp 21-24 [cf.]	18 Feb 1661
174rv	From Lapeyrère at Paris 1p + ad (12x18) Publ, LLI, pp 24-26	6 May 1661
175rv	From Lapeyrère at Paris B at Warsaw 1p + ad (17x18) Publ, LLI, pp 26-29	29 Jul 1661
176rv	From Wachschlager at [Brunn?] 1p (20x31) Latin	20 Aug 1661
177rv	From Wachschlager To Anonymous 1p & 3 lines (16x20) Latin	2 Oct 1673
178r-179v	From Andreas Arnold at Nürnberg 2pp & 6 lines + ad	24 Jun 1686
180rv	From [J. Messinge?] at The Hague 1p (17x22)	16 Nov 1662
181rv	Extract of a letter in B's hand "Ex litteris P. Antonii Maria de Rheita Coloniae scriptis/9 Januarii 1643 ad Ericum Puteanum" 1p + ad (17x22) Latin	9 Jan 1643
182rv&186rv	From [Fantoni] at Paris To Gr. Chan.er of Lithuania 1p (17x22) Italian fol 186, blank	28 Sep 1660
183rv	From Morstin at Stockholm 2pp (20x31)	25 Apr 1655

Folios	Description	Date
184r-185v	From Morstin at Stockholm 3½pp (20x31)	20 Mar 1655
187r-188v	Copy of a letter From Radinewski at Hamburg To Fleury (Con. de la Reine de Pologne) 3pp & 5 lines (20x31)	17 Apr 1655
189rv	From ? at Vienna To Anonymous 1p (20x31)	17 May 1656
190r-191v	Copy of a letter From Opalinski at [Siracowicz] To Fleury 4pp (20x31) Latin	28 Feb 1649
192rv	From O. Worm at [Copenhagen] To Fleury 1p + ad (18x24)	Idib. Sep 1637 [13? Sep 1637]
193rv&196rv	From Flournoy at Amsterdam To B at The Hague 1p + ad (18x24)	1 Jun 1657
194r-195v	Copy of a letter in B's hand From [Godefroy?] To Anonymous 3½pp (18x25)	22 Jul 1653
197r-199v	From Gentilot at The Hague To J-A de Thou at Paris 2pp & 7 lines + ad (12x17)	2 Dec 1667
200r-201v	From Van Beuninghen at The Hague To [de la Volpiliere?] at Paris 2pp (18x25)	23 Mar 1673
202r-205v	Copies of letters in B's hand Between Théodore de Bèze and Désiré Érasme, etc. 6½pp (15x20) Latin fol 205, blank	27 Jan 1593; 1586; 29 Aug 1593
206rv	Copy of a letter From Rochevernay To Florent Chrestien 1p + ad (10x19) Greek/Latin/French	29 Dec 1569

Folio	Description	Date
207rv	Copy of a letter ? From Salmon Macrin To Florent Chrestien ½p + ad (13x20)	[?] Feb 1570
208r-209v	From ? 2½pp + ad (18x25) Incl horoscopes	undated
210r-211v	From Dufresne at Paris To J-A de Thou 1½pp + ad (20x31)	11 Oct 1671
212rv	From [Caron?] at Fort-Dauphin To Anonymous 2pp (20x31)	15 Oct 1667
213rv	From F. Burrius at Amsterdam To Anonymous 1p (20x31) Latin	2 Aug 1662
214r-215v	From [Caron at ?] To Anonymous 4pp (20x31)	3 Feb 1669
216rv	Copy of a letter? From François, évêque de Genève* at Annecy To Anonymous 1⅔pp (20x31) *[François de Sales]	3 May 1615
217rv	From l'évêque d'Orléans To Anonymous 1p (17x22)	17 Apr NY
218r-219v	From [Bertrand?] at Laon 2½pp + ad (17x22)	24 [Oct 1671]
220r-221v	From Anonymous at Adrianople To Anonymous 4pp (20x31)	13 Mar 1660
222r-223v	From [De Vreuin?] To Anonymous 1½pp & 2 lines (17x22)	16 Aug 1667
224rv	From [De Vreuin?] To Anonymous 1½pp (17x22)	14 Sep [1667?]

225rv	From [De Vreuin?] To Anonymous 1p (17x22)	14 Sep [1667?]
226r-227v	Fragment 4pp (17x22)	undated
228r-229v	From Anonymous at Paris To Anonymous 4pp (17x22)	14 Oct 1660
230rv	From Montesson at St-Jean-de-Luz To Anonymous 2pp (17x22)	18 Sep NY
231rv&240rv?	From Fantone at Paris To Comte de Lezno 1p + ad (17x22) Italian	28 Sep 1660
232rv&241rv?	From Fantone at Paris To Comte de Lezno 1p +ad? (17x22) Italian	28 Sep 1660
233rv	From Lubomirski To Anonymous ½p (17x22) Polish *dated in different hand	[30 Sep 1660]*
234rv	From Lubomirski To Anonymous ⅔p (17x22) Polish *dated in different hand	[30 Sep 1660]*
235rv	From Lubomirski To ? ⅔p (17x22) Polish *dated in different hand	[30 Sep 1660]
236rv&239rv	From Anonymous at Copenhagen* To Anonymous 2½pp (17x22) *fols 236-249, same hand	[22?] Oct 1664
237r-238v&240rv	From Anonymous at Copenhagen To Anonymous 2½pp + ad (17x22)	[25?] Oct 1664
242r-243v	From Anonymous* at [Copenhagen] To Anonymous 3pp (17x22) *[fols 242-247, feld-maréchal Vurr?]	29 Oct 1664

244rv — From Anonymous at Copenhagen — [1?] Nov 1664
To Anonymous
2pp (17x22)

245r-246v — From Anonymous at Copenhagen — 29 Nov 1664
To Anonymous
3pp (17x22)

247rv&249rv — From Anonymous at Copenhagen — [24?] Dec 1664
To ? at Hamburg
3pp + ad (17x22)

248rv — To [? at Paris] — 29 Jun 1651*
B at Basle
$1\frac{1}{2}$pp (12x17)
*w/PS, 4 Jul 1651

250rv&252rv — From Bernegger at Strasbourg — 16 Dec 1653
2pp & 2 lines (17x22)

251rv — From Bernegger at Strasbourg — 2 Jul 1660
2pp (17x22)

253rv — From [de Cailleur?] — 15 Aug 1675
1p (17x22)
fragment

254rv — From [de Cailleur?] — 31 Aug [1675]
1p (17x22)

255r-256v — From Snabelius at Haarlem — 13 Mar 1680
3pp + ad (17x22) Latin
[fol 256, extracts of letters]

257rv — From Williamson at London — 14 Jan $\frac{1674}{1675}$
1p (17x22) Latin
[24 Jan 1675]

258rv — Copy of a letter in B's hand — 13 Nov 1659
From Siri at St-Jean-de-Luz
$\frac{1}{2}$p (20x31) Italian

259rv&262rv — From [Milet?] at The Hague — 27 Jul 1662
1p + ad (17x22)

260r-261v — From [Milet?] at The Hague — 17 Aug 1662
3pp + ad (17x22)

▼ ▼ ▼

COLLECTION BOULLIAU, VOLUME XXIII, FONDS FRANÇAIS 13041

Folio	Description	Date
1r-2v	From Portner at Leiden 1½pp + ad (15x20) Latin	Non Sep 1652 [5 Sep 1652]
3r-4v	To Portner at [Ratisbon] 4pp (17x22) Latin	20 Dec 1652
5r-10v	From Portner at Ratisbon 11½pp + ad (17x22) Latin	Idib Jan 1653 [13 Jan 1653]
11r	To Portner at [Ratisbon] 1½pp (17x22) Latin	22 Feb 1653
11v	Copy of a letter in B's hand To Leopoldo at Florence ⅓p (17x22) Latin	21 Feb 1653
12r	Copy of a letter in B's hand To Princess Elizabeth 1p (17x22)	22 Feb 1653
12v	Copy of a letter in B's hand To Patri Gansero at Ratisbon 1p (17x22) Latin	22 Feb 1653
13rv	From Portner at Ratisbon ½p + ad (15x20) Latin	10 Mar 1653
14r-17v	From Portner at Ratisbon 7pp + ad (15x20) Latin	17 Mar 1653
18r-19v	To Portner at [Ratisbon] 4pp (18x23) Latin	25 Apr 1653
20rv	From Portner at Ratisbon 2pp (15x20) Latin	Postrid Kal May 1653 [2 May 1653]
21rv	From Portner at Ratisbon ½p (15x20) Latin	22 May 1653
22r-23v	From Portner at Ratisbon 3pp + ad (15x20) Latin	17 May 1653
24r-25v	Copy of a letter in B's hand From Maria Cunitz at Bicinii Silesiorum To Portner at [Ratisbon] 1½pp (16x21) Latin fol 25, blank	22 Mar 1653

26r-27v	From Portner at Ratisbon 1½pp + ad (17x22) Latin	Postrid Kal Jun 1653 [2 Jun 1653]
28r-29v	To Portner at [Ratisbon] 3⅔pp (18x25) Latin	27 Jun 1653
30r-31v	From Lenz at Ratisbon 2pp & 3 lines + ad (17x22) Latin	15 Jul 1653
32r-33v	From Portner at Ratisbon To Naudé at Paris 1½pp + ad (15x20) Latin	XVI Kal Aug 1653 [17 Jul 1653]
34r-35v	From Portner at Ratisbon 2½pp + ad (15x20) Latin	XV Kal Aug 1653 [18 Jul 1653]
36r-37v	From Portner at Ratisbon 3½pp (17x22) Latin	X Kal Sep 1653 [23 Aug 1653]
38rv	To Lenz at [Ratisbon] 2pp (15x20) Latin	12 Sep 1653
39rv	To Portner at Ratisbon 2pp (15x20) Latin	12 Sep 1653
40r-41v	From Portner at Ratisbon 1⅓pp + ad (17x22) Latin	Kal Oct 1653 [1 Oct 1653]
42r-43v	From Portner at Ratisbon 3½pp (17x22) Latin	IV Eid Oct 1653 [12 Oct 1653]
44rv	To Portner at [Ratisbon] 1p (20x31) Latin	23 Oct 1653
45r-47v	To Portner at [Ratisbon] 2½pp (20x31) Latin fol 47, blank	30 Oct 1653
48r-49v	From Lenz at Ratisbon 2½pp + ad (17x22) Latin	Kal Nov 1653
50r-51v	To Lenz at [Ratisbon] 2pp & 5 lines + ad (17x22) Latin	13 Feb 1654
52r-53v	From Portner at Ratisbon 2¾pp (20x31) Latin	VII Id Nov 1653 [7 Nov 1653]

54r-57v	From Portner at Ratisbon 7⅓pp (15x20) Latin	IV Kal Dec 1653 [28 Nov 1653]
58r-59v	To Portner at [Ratisbon] 3pp (15x20) Latin	13 Feb 1654
60r-61v	From Portner at Ratisbon 2⅓pp + ad (15x20) Latin	X Kal Feb 1654 [23 Jan 1654]
62r-63v	From Portner at Ratisbon 3⅔pp (17x22) Latin	5 Mar 1654
64r-65v	From Lenz at Ratisbon 1p + ad (17x22) Latin	24 Apr 1654
66r-67v	From Portner at Paris B at Meslay ¾p + ad (17x22) Latin	[16?] Nov 1654
68rv	To Portner at [Ratisbon] 1p (18x25) Latin	1 Jan 1655
69r-70v	From Portner at Ratisbon 2⅔pp + ad (17x22) Latin	Non Jan 1655 [5 Jan 1655]
71r-72v	From Portner at Ratisbon 2pp + ad (17x22) Latin	XIII Kal Feb 1655 [20 Jan 1655]
73r-74v	To Portner at [Ratisbon] 3½pp (15x20) Latin	26 Feb 1655
75r-76v	From Portner at Ratisbon 1⅓pp + ad (15x20) Latin	12 Feb 1655
77r-78v	From Portner at Ratisbon 1⅔pp + ad (15x20) Latin	Idib Feb 1655 [13 Feb 1655]
79r-80v	To Portner at [Ratisbon] 1¾pp (18x29) Latin fol 80, blank	12 Mar 1655
81r-82v	From Portner at Ratisbon 1⅓pp + ad (15x20) Latin	V Kal Mar 1655 [25 Feb 1655]
83r-84v	To Portner at [Ratisbon] 1½pp (15x20) Latin fol 84, blank	26 Mar 1655

Folios	Description	Date
85r-86v	From Portner at Ratisbon 3⅔pp (15x20) Latin	10 Mar 1655
87r-88v	To Portner at [Ratisbon] 2½pp (17x22) Latin	9 Apr 1655
89r-90v	From Portner at Ratisbon 1p + ad (15x20) Latin	[22?] Apr 1655
91r-92v	To Portner at [Ratisbon] 1⅔pp (17x22) Latin fol 92, blank	23 Apr 1655
93r-94v	From [Portner] at Ratisbon 4pp (15x20) Latin	IV Id Apr 1655 [10 Apr 1655]
95r-96v	From Portner at Ratisbon 3pp (15x20) Latin	Idib Apr 1655 [13 Apr 1655]
97r-98v	To Portner at [Ratisbon] 2¼pp (17x22) Latin	7 May 1655
99r-100v	From Lenz at Ratisbon 2pp & 1 line + ad (17x22) Latin	$\frac{18}{28}$ Apr 1655 [28 Apr 1655]
101r-102v	From Portner at Ratisbon 2⅔pp (15x20) Latin	28 Apr 1655
103rv	To Lenz at [Ratisbon] 1p (17x22) Latin [sent w/fol 104]	21 May 1655
104rv	To Portner at [Ratisbon] 2pp (17x22) Latin	21 May 1655
105r-106v	From [Portner] at Ratisbon 3⅔pp (15x20) Latin	III Id May 1655 [13 May 1655]
107r-108v	To Portner at [Ratisbon] 1½pp (17x22) Latin fol 108, blank	4 Jun 1655
109rv	From [Portner?] at Ratisbon 1¼pp (17x30) Latin	V Id May 1655 [11 May 1655]
110r-111v	From Portner at Ratisbon 1¾pp (15x20) Latin fol 111, blank	26 May 1655

112r-113v	To Portner at [Ratisbon] 2½pp (15x20) Latin	18 Jun 1655
114r-115v	From Portner at Ratisbon 2½pp (15x20) Latin	V Id Jun 1655 [9 Jun 1655]
116r-117v	To Portner at [Ratisbon] 3pp (15x20) Latin	2 Jul 1655
118r-119v	From Portner at Ratisbon 2¼pp (15x20) Latin *[B writes 30 June]	Prid Kal Jul 1655* [30 Jun 1655]
120r-121v	From Portner at [Ratisbon] 1p & 5 lines (15x20) Latin fol 121, blank *[B writes 24 Jun]	VIII Kal Jul 1655* [24 Jun 1655]
122rv	Note From Boecler at Strasbourg To Portner at [Ratisbon] 5 lines + ad (15x17) Latin Publ, URI, p 224	28 May 1655
123rv	From Boecler at Strasbourg To Portner at [Ratisbon] ⅔p (14x21) Latin see URI, p 225	7 Jun 1655
124r-125v	To Portner at [Ratisbon] 3pp (17x22) Latin	16 Jul 1655
126rv	From Portner at Ratisbon ⅔p + ad (15x22)	1 Jul 1655
127r-128v	From Portner at Ratisbon 2¾pp (15x20) Latin	Nonis Jul 1655 [7 Jul 1655]
129r-130v	To Portner at [Ratisbon] 2pp (17x22) Latin fol 130, blank	30 Jul 1655
131rv	From [Portner at Ratisbon] 1p + ad (17x22) Publ, URI, pp 225-226	XII Kal Aug 1655 [21 Jul 1655]

132r-133v	To Portner at [Ratisbon] 2pp (15x20) Latin fol 133, blank	6 Aug 1655
134r-135v	From [Portner at Ratisbon] 2¾pp + ad (15x20) Latin Publ extract, URI, p 226	V Kal Aug [1655] [28 Jul 1655]
136r-137v	To Portner at [Ratisbon] 2½pp (15x20) Latin	20 Aug 1655
138r-139v	From [Portner at Ratisbon] 1½pp + ad (17x22) Latin Publ extract, URI, p 227	Prid Id Aug 1655 [12 Aug 1655]
140r-141v	From [Portner at Ratisbon] 3⅓pp (15x22) Latin *[B writes 16 Aug 1655]	XVI Kal Sep 1655* [17 Aug 1655]
142r-143v	To Portner at [Ratisbon] 2½pp (15x20) Latin	3 Sep 1655
144r-145v	From [Portner at Ratisbon] 2⅔pp + ad (17x22) Latin	Kal Sep 1655 [1 Sep 1655]
146r-147v	From [Portner at Ratisbon] 4pp (15x20) Latin See URI, p 228	VI Id Sep 1655 [8 Sep 1655]
148r-149v	To Portner at [Ratisbon] 1½pp (15x20) Latin fol 149, blank	17 Sep 1655
150r-151v	To Portner at [Ratisbon] 2⅓pp (17x22) Latin	1 Oct 1655
152r-153v	To Portner at [Ratisbon] 2pp (17x22) Latin fol 153, blank	15 Oct 1655
154r-155v	From [Portner at Ratisbon] 4pp (15x20) Latin *[B writes 24 Sep 1655] Publ extract, URI, pp 228-229	VIII Kal Oct 1655 [24 Sep 1655]*
156r-157v	From [Portner at Ratisbon] 3⅓pp (15x20) Latin *[B writes 6 Oct 1655]	Prid Non Oct 1655* [6 Oct 1655]

158r-161v	To Portner at [Ratisbon] 5pp (17x22) Latin fol 161, blank Publ, Petro Gassendo Praeposito Diniensi...Paris, 1656	29 Oct 1655
162r-165v	From [Portner at Ratisbon] 6pp & 6 lines (15x20) Latin	XII Kal Nov 1655 [21 Oct 1655]
166r-167v	To Portner at [Ratisbon] 3¾pp (17x22) Latin	12 Nov 1655
168r-171v	From [Portner at Ratisbon] 5¾pp & 5 lines (15x20) Latin	III Non Nov 1655 [3 Nov 1655]
172r-173v	To Portner at [Ratisbon] 2pp (17x22) Latin fol 173, blank	19 Nov 1655
174r-177v	From [Portner at Ratisbon] 4pp (15x20) Latin fol 176, list, 1¼pp fol 177, Latin enclosure, ¾p see URI, pp 230-231	XV Kal Dec 1655 [17 Nov 1655] 10 Oct 1655 28 Sep 1655
178r-179v	To Portner at [Ratisbon] 3pp (15x20) Latin	3 Dec 1655
180r-181v	To Portner at [Ratisbon] 2pp (17x22) Latin fol 181, blank	17 Dec 1655
182r-183v	From [Portner at Ratisbon] 3½pp (15x20) Latin	Kal Dec 1655 [1 Dec 1655]
184r-185v	From Portner at [Ratisbon] 1⅓pp (15x20) Latin fol 185, blank	8 Dec 1655
186r-187v	To Portner at [Ratisbon] 1⅔pp (15x20) Latin fol 187, blank	24 Dec 1655
188r-189v	To Portner at [Ratisbon] 1p (15x20) Latin fol 189, blank	31 Dec 1655

190r-192v	From Portner at [Ratisbon] 4pp (15x20) Latin fol 192r, $\frac{1}{2}$p, copy of a letter From Queen Christine To Carol. Gustauven [Charles X]	X Kal Jan 1655 [23 Dec 1655] 8 Nov 1655
193r-194v	To Portner at [Ratisbon] $1\frac{1}{3}$pp (15x20) Latin fol 194, blank *probably 1656	7 Jan 1655*
195r-196v	To Portner at [Ratisbon] 2pp (15x20) Latin fol 196, blank	21 Jan 1656
197r-198v	From [Portner at Ratisbon] 2pp + ad (15x20) Latin	Nonis Jan 1656 [5 Jan 1656]
199r-200v	Latin text (not B's hand) 4pp (15x20) Latin "Cataplasma/Ducimutinensi [...] Apud Augustum Miradei Anno MDCLV"	
201r-202v	To Portner at [Ratisbon] $2\frac{1}{3}$pp (15x20) Latin	4 Feb 1656
203r-204v	From Lenz at Ratisbon 2pp & 4 lines + ad (15x20) Latin	Kal Jan 1656 [1 Jan 1656]
205r-206v	To Lenz at [Ratisbon 1p (15x20) Latin fol 206, blank	18 Feb 1656
207r-208v	From [Portner at Ratisbon] $3\frac{1}{2}$pp (15x20) Latin	XIII Kal Feb 1656 [20 Jan 1656]
209r-210v	To Portner at [Ratisbon] $2\frac{1}{3}$pp (15x20) Latin	18 Feb 1656
211r-212v	From [Portner at Ratisbon] 3pp + ad (15x20) Latin	IV Non Feb 1656 [2 Feb 1656]
213r-214v	From [Portner at Ratisbon] 2pp & 6 lines (15x20) Latin	Idib Feb 1656 [13 Feb 1656]
215rv	From Abdias Trew To Anonymous 2pp (20x31) Latin	24 Jan 1656

216r-217v	To Portner at [Ratisbon] 1½pp (15x20) Latin fol 217, blank	3 Mar 1656
218r-219v	From [Portner at Ratisbon] 3pp (15x20) Latin	Kal Mar 1656 [1 Mar 1656]
220r-221v	To Portner at [Ratisbon] 1½pp (15x20) Latin fol 221, blank	17 Mar 1656
222r-223v	From [Portner at Ratisbon] 2½pp (15x20) Latin	undated
224r-225v	From [Boecler at Strasbourg?] To Anonymous 3⅓pp (20x31) Latin	undated
226r-227v	To Portner at [Ratisbon] 1p (15x20) Latin fol 227, blank	31 Mar 1656
228r-229v	From [Portner at Ratisbon] 2pp & 7 lines + ad (18x25) Latin	IV Kal Apr 1656 [29 Mar 1656]
230r-231v	To Portner at [Ratisbon] 2pp & 6 lines (15x20) Latin	14 Apr 1656
232r-233v	To Portner at [Ratisbon] 2pp (15x20) Latin fol 233, blank	28 Apr 1656
234r-235v	From [Portner at Ratisbon] 4pp (15x20) Latin	III Id Apr 1656 [11 Apr 1656]
236r-237v	To Portner at [Ratisbon] 2pp (15x20) Latin fol 237, blank	12 Mar 1656
238r-239v	From [Portner at Ratisbon] 3½pp (15x20) Latin	VI Kal May 1656 [26 Apr 1656]
240r-241v	To Portner at [Ratisbon] 2pp (15x20) Latin fol 241, blank	19 May 1656

242r-243v	From [Portner at Ratisbon] 2⅓pp + ad (15x20) Latin	10 May 1656
244r-245v	From Portner at Ratisbon 2¼pp + ad (15x20) Latin	III Non Apr 1656 [3 Apr 1656]
246r-247v	From [Portner at Ratisbon] 2½pp + ad (15x20) Latin	IX Kal Jun 1656 [24 May 1656]
248rv	To Portner at [Ratisbon] 1⅔pp (15x20) Latin	26 May 1656
249rv	To [Portner at Ratisbon] 1¼pp (17x22) Latin	2 Jun 1656
250r-251v	To Portner at [Ratisbon] 1½pp (17x22) Latin fol 251, blank	16 Jun 1656
252r-253v	From [Portner at Ratisbon] 1⅓pp + ad (15x20) Latin	Prid Kal Jun 1656 [31 May 1656]
254r-255v	From [Portner at Ratisbon] 2pp & 4 lines (15x20) Latin	VII Id Jun 1656 [7 Jun 1656]
256r-257v	List of books 2pp (15x20) Latin fol 257, blank	undated
258r-259v	To Portner at [Ratisbon] 2pp (15x20) Latin fol 259, blank	23 Jun 1656
260r-261v	From Portner at [Ratisbon] 2⅓pp + ad (15x20) Latin	IX Kal Jul 1656 [23 Jun 1656]
262r-263v	To Portner at [Ratisbon] 2pp (17x22) Latin fol 263, blank	7 Jul 1656
264r-265v	To Portner at [Ratisbon] 2pp (15x20) Latin fol 265, blank	21 Jul 1656
266-267v	From [Portner at Ratisbon] 2⅔pp + ad (15x20) Latin *in B's hand	[5 Jul]*

268r-269v	From Portner at [Ratisbon] 3pp (15x20) Latin	$\frac{12}{2}$ Jul 1656 [12 Jul 1656]
270r-271v	To Portner at [Ratisbon] $1\frac{2}{3}$pp (17x22) Latin fol 271, blank	28 Jul 1656
272r-273v	From [Portner at Ratisbon] 4pp (15x20) Latin	VII Kal Aug 1656 [26 Jul 1656]
274r-275v	To Portner at [Ratisbon] 2pp (17x22) Latin fol 275, blank	11 Aug 1656
276r-277v	To Portner at [Ratisbon] $1\frac{1}{4}$pp (15x20) Latin fol 277, blank	25 Aug 1656
278r-279v	From [Portner at Ratisbon] $2\frac{1}{3}$pp + ad (15x20) Latin	V Id Aug 1656 [9 Aug 1656]
280r-281v	From [Portner at Ratisbon] 3pp + ad (15x20) Latin	X Kal Sep 1656 [23 Aug 1656]
282r-283v	To Portner at [Ratisbon] 2pp (15x20) Latin fol 283, blank	8 Sep 1656
284r-285v	From [Portner] at Nuremberg 3pp + ad (15x20) Latin	Nonis Sep 1656 [5 Sep 1656]
286r-287v	To Portner at [Ratisbon] 3pp (15x20) Latin *[probably 22 Sep 1656]	undated*
288r-289v	From [Portner at Ratisbon] $2\frac{1}{3}$pp (15x20) Latin	XII Kal Oct 1656 [20 Sep 1656]
290r-291v	To Portner at [Ratisbon] 1p (15x20) Latin fol 291, blank	6 Oct 1656
292r-293v	From [Portner at Ratisbon] $2\frac{1}{2}$pp + ad (15x20) Latin	IV Non Oct 1656 [4 Oct 1656]
294r-295v	To Portner at [Ratisbon] 2pp & 4 lines (15x20) Latin	undated

296r-297v	From [Portner at Ratisbon] 1⅓pp + ad (15x20) Latin	11 Oct 1656
298r-299v	From [Portner at Ratisbon] 2½pp (15x20) Latin	XIV Kal Nov 1656 [19 Oct 1656]
300r-301v	From [Portner at Ratisbon] 3pp + ad (15x20) Latin	VII Kal Nov 1656 [26 Oct 1656]
302r-303v	From [Portner at Ratisbon] To Dupuy at [Paris] 3pp (15x20)	8 Nov 1656
304r-305v	From [Portner at Ratisbon] To Dupuy at Paris 2⅔pp + ad (15x20)	22 Nov 1656
306r-307v	From [Portner at Ratisbon] 4pp (15x20) Latin	XVIII Kal Dec 1656 [14 Nov 1656]
308r-309v	From Portner at [Ratisbon] 2pp + ad (15x20) Latin	III Kal Dec 1656 [29 Nov 1656]
310r-311v	To Portner at [Ratisbon] 1p & 7 lines (18x25) Latin fol 311, blank	24 Nov 1656
312r-313v	From [Portner at Ratisbon] 4pp (15x20) Latin	Idib Dec 1656 [10 Dec 1656]
314r-315v	To Portner at [Ratisbon] 1⅓pp (15x20) Latin fol 315, blank	12 Jan 1657
316r-317v	To Portner at [Ratisbon] 1⅓pp (18x25) Latin fol 317, blank	22 Dec 1656
318r-319v	From [Portner at Ratisbon] 2pp + ad (15x20) Latin	Vigiliâ St Thomae 1656 [10 Dec 1656]
320r-321v	To Portner at [Ratisbon] 2⅔pp (15x20) Latin *[B mistakenly writes 1656]	26 Jan [1657]*
322r-323v	From [Portner at Ratisbon] 3½pp (15x20) Latin	IV Id Jan 1657 [10 Jan 1657]

Folios	Description	Date
324r-325v	From [Portner at Ratisbon] $2\frac{2}{3}$pp + ad (15x20) Latin	17 Jan 1657
326r-327v	From [Portner at Ratisbon] 3pp & 5 lines (15x20) Latin	IV Id Feb 1657 [10 Feb 1657]
328r-329v	From [Portner at Ratisbon] 3pp + ad (15x20) Latin	XII Kal Mar 1657 [18 Feb 1657]
330r-333v	To Portner at [Ratisbon] $6\frac{1}{2}$pp (15x20) Latin Publ extract, URI, pp 221-223	17 Mar 1657
334r-335v	To Portner at [Ratisbon] B at The Hague $3\frac{1}{2}$pp (18x25) Latin	22 May 1657
336r-337v	From [Portner at Ratisbon] 4pp (17x22) Latin	X Kal Jul 1657 [22 Jun 1657]
338r-339v	To Portner at [Ratisbon] B at The Hague $2\frac{2}{3}$pp (18x25) Latin	10 Jul 1657
340r-341v	To Portner at [Ratisbon] B at The Hague $1\frac{2}{3}$pp (18x25) Latin fol 341, blank	3 Aug 1657
342r-343v	From [Portner at Ratisbon] 4pp (15x20) Latin	VII Kal Aug 1657 [26 Jul 1657]
344r-345v	From [Portner at Ratisbon] $1\frac{2}{3}$pp + ad (15x20) Latin	IV Kal Sep 1657 [29 Aug 1657]
346r-347v	To Portner at [Ratisbon] $3\frac{1}{3}$pp (18x25) Latin	12 Oct 1657
348r-349v	From Portner at [Ratisbon] B at The Hague 3pp + ad (15x20) Latin	Idib Nov 1657 [13 Nov 1657]
350r-351v	To Portner at [Ratisbon] 4pp (15x20) Latin	27 Dec 1657

▼ ▼ ▼

COLLECTION BOULLIAU, VOLUME XXIV, FONDS FRANÇAIS 13042

1r-2v	From Luillier at Paris B at Bonneval 1⅔pp + ad (15x20) Publ, LL, pp 489-491	10 Aug 1642
3r-4v	From [Luillier] at Toul 1p + ad (18x25) Publ, LL, pp 495-496	29 [?] 1643
5r-6v	From [Lullier] at Toul 2⅔pp + ad (18x25) Publ, LL, pp 491-495	19 Sep 1643
7r-8v	From [Lullier] at Toul 1p + ad (20x31) Publ, LL, pp 497-498	3 Oct 1643
9r-10v	From [Lullier at Toul] 3pp + ad (18x25) Publ, LL, pp 499-503	20 Oct 1643
11rv	From Luillier at Toul 1p + ad (17x22)	8 Apr 1644
12rv	From Luillier at Paris 1p (17x22) Publ, LL, pp 503-504	18 May 1646
13r-14v	From Luillier at Paris 3pp (17x22) Publ, LL, pp 504-507	16 Jul 1646
15rv	From [Luillier] at Paris 2pp (17x22) Publ, LL, pp 507-508	16 Jul 1646
16r-17v	From [Luillier] at Paris 4pp (18x25) Publ, LL, pp 509-512	11 Sep 1646
18r-19v	From Luillier at Toulon 3pp + ad (20x31) Publ, LL, pp 512-516	26 Dec 1650
20r-21v	From de Valois at Grenoble B at Lyon 3pp + ad (17x22)	21 May 1645

22r-23v	From de Valois at Grenoble B at Venice 3pp + ad (17x22)	5 Nov 1645
24r-25v	From de Valois at Grenoble B at Venice 2pp + ad (20x31) Publ, MC, XIII, p 475, No 1386	7 Sep 1645
26r-27v	From de Valois at Grenoble B at Venice 3pp + ad (20x31)	31 Dec 1645
28r-29v	From de Valois at Grenoble B at [Venice] 2⅔pp (20x31)	28 Jan 1646
30r-31v	From de Valois at Grenoble B at Venice 1⅔pp + ad (17x22)	28 Feb 1646
32r-33v	From de Valois at Grenoble B at Venice 3pp + ad (17x22)	6 May 1646
34r-35v	From de Valois at Grenoble B at Venice 3pp + ad (17x22) Publ, MC, XIV, pp 317-318, No 1483	1 Jul 1646
36r-37v	From Chauvet at Chartreuse 1p + ad (17x22)	1 Nov 1641
38r-39v	From Chauvet at Chartreuse 2pp + ad (18x27)	2 Apr 1642
40r-42v	From Chauvet at Chartreuse 1⅓pp & ½p (20x31) Latin/French [fol 41, book list, Latin] fol 42, blank	24 Oct 1649
43r-44v	From Chauvet at Chartreuse 3pp (17x22)	12 Dec 1649
45rv	From Chauvet at Chartreuse 1⅓pp (20x28)	[26?] Jan 1650
46r-47v	From Chauvet at Chartreuse 2pp + ad (20x28)	18 May 1652

48r-49v	From Bourdelot at Stockholm 2¼pp + ad (17x22)	24 May 1652
50rv	Librarians' notice concerning letters of Bourdelot [13 Mar 1849] (11x22)	
51r-52v	From D. d'Aubray at Paris 3pp + ad (20x31)	11 Apr 1651
53r-54v	From D. d'Aubray at [Paris?] 2⅔pp + ad (20x31)	[29 Aug 1651?]
55r-56v	From D. d'Aubray at [?] 2⅔pp + ad (20x31)	undated
57r-58v	From D. d'Aubray at Paris 3pp + ad (20x31)	22 Sep [1651?]
59rv	From D. d'Aubray at Paris 1p (20x31)	6 Oct [1651?]
60r-61v	From D. d'Aubray at [Paris?] 1½pp + ad (20x31)	20 [Oct 1651?]
62r-63v	From A. d'Aubray at Rome 2⅔pp + ad (20x26)	11 Jan 1655
64r-65v	From A. d'Aubray at Rome 2pp & 5 lines + ad (20x26)	25 Jan 1655
66r-67v	From A. d'Aubray at Rome 1p + ad (20x26)	22 Feb [1655?]
68r-69v	From A. d'Aubray at Rome 1½pp + ad (20x26)	1 Mar 1655
70r-71v	From [A. d'Aubray] at Rome 2pp & 7 lines (20x26)	22 Mar 1655
72r-73v	From A. d'Aubray at Rome 2pp (17x22) fol 73, blank	25 Apr 1655
74r-75v	From A. d'Aubray at Rome 2pp + ad (17x22)	17 May 1655
76r-77v	From A. d'Aubray at Rome 2pp + ad (17x22)	31 May 1655

78r-79v	From A. d'Aubray at Rome $2\frac{1}{2}$pp + ad (17x22)	14 Jun 1655
80r-81v	From A. d'Aubray at Rome 3pp (17x22)	5 Jul 1655
82r-85v	From A. d'Aubray at Rome $5\frac{2}{3}$pp + ad (17x22)	12 Jul 1655
86r-87v	From A. d'Aubray at Rome $3\frac{2}{3}$pp (20x26)	26 Jul 1655
88r-89v	From A. d'Aubray at Rome $2\frac{2}{3}$pp (17x22)	23 Aug 1655
90r-91v	From A. d'Aubray at Rome $1\frac{3}{4}$pp (17x22) fol 91, blank	6 Sep 1655
92r-93v	From A. d'Aubray at Rome 2pp + ad (17x22)	13 Sep 1655
94r-95v	From A. d'Aubray at Vienna 4pp (20x31)	15 Dec 1655
96r-97v	From A. d'Aubray at Vienna 2pp + ad (17x22)	19 Jan 1656
98r-99v	From A. d'Aubray at Vienna 3pp + ad (18x25)	2 Feb 1656
100r-101v	From A. d'Aubray at Vienna $1\frac{2}{3}$pp + ad (18x25)	22 Mar 1656
102r-103v	From A. d'Aubray at Vienna $1\frac{1}{2}$pp (18x25) fol 103, blank	29 Mar 1656
104r-105v	From A. d'Aubray at Danzig 3pp (17x22)	19 [Jul?] 1656
106r-107v	From [Wicquefort] at The Hague To Dupuy at Paris 2pp & 6 lines + ad (17x22)	9 Nov 1656
108r-109v	From Wicquefort at The Hague 1p & 2 lines + ad (17x22)	30 Nov 1656
110r-111v	From Wicquefort at The Hague 2pp + ad (17x22)	14 Dec 1656

112r-113v	From [Wicquefort] at The Hague $1\frac{2}{3}$pp + ad (17x22)	28 Dec 1656
114r-115v	From [Wicquefort] at The Hague 2pp + ad (17x22)	4 Jan 1657
116r-117v	From [Wicquefort] at The Hague $1\frac{2}{3}$pp + ad (17x22)	11 Jan 1657
118r-119v	From [Wicquefort] at The Hague 2pp + ad (17x22)	18 Jan 1657
120r-121v	From [Wicquefort] at The Hague 2pp + ad (17x22)	25 Jan 1657
122r-123v	From [Wicquefort] at The Hague $1\frac{1}{3}$pp + ad (17x22)	1 Feb 1657
124r-125v	From [Wicquefort] at The Hague $2\frac{1}{3}$pp + ad (17x22)	8 Feb 1657
126r	To Ferdinand delle Howe 1p (18x30) Latin	6 Apr 1651
126v	Copy of a letter To Ferdinand delle Howe $\frac{3}{4}$p (20x31) Latin	6 Apr 1651
127r-128v	From d'Abreu at Lisbon 2pp (17x22) Latin fol 128, blank	Kal Sep 1652 [1 Sep 1652]
129r-130v	To d'Abreu at [Lisbon] $2\frac{1}{3}$pp + ad (18x25) Latin	10 Nov 1652
131r-132v	To d'Abreu at [Lisbon] 1p (18x25) Latin fol 132, blank	11 Apr 1652
133r-134v	From Moltken at Gottorp 3pp + ad (18x25)	24 Nov 1654
135r-136v	From Moltken at Geneva 1p & 3 lines + ad (18x25) fol 135r, B writes: "M. Moltken gouverneur du prince de Holstein"	10 Aug 165[8?]
137r-138v	From Gersdorff at [Saumur?] 3pp (17x22)	7 May 1654

139rv	From Gersdorff at [Saumur?] 1p + ad (17x22)	undated
140r-141v	From Gersdorff at [Saumur?] 2½pp + ad (17x22)	15 Jul 1654
142r-143v	From Gersdorff at [Saumur?] 1p + ad (17x22)	12 Sep 1654
144r-145v	From Gersdorff at Dresden 3pp + ad (17x22)	12 Jan 1657
146r-147v	From Moreau at Turin 2pp + ad (20x31)	5 Sep 1658
148rv	From De Laurière at Turin 2pp (15x20)	7 Aug 1658
149r-151v	From Morinière at Paris 3⅓pp (17x22) fol 151, blank (11x19)	29 Aug 1645
152r-153v	From Des Roches at [Paris] 3pp + ad (17x22) Publ, MC, XIII, pp 3-5, No 1242	7 Jan 1644
154r-158v	From Barducci at Rome 9½pp + ad (20x27)	26 Aug 1656
159r-160v	From Barducci at San-Miniato 4pp (15x20)	16 Nov 1657
161r-162v	From Barducci at San-Miniato 2pp + ad (20x31)	19 Apr 1658
163rv&166rv	From De Groot at The Hague 1p + ad (15x20)	7[Feb?] 1658
164r-165v	From De Groot at The Hague 1⅓pp + ad (17x22)	7 Mar 1658
167r-168v	From Vedon at Venice 1p + ad (20x31)	2 Nov 1647
169r-170v	From La Boissière at Venice 4pp (20x31) Latin	6 Oct 1646
171r-172v	From [La Boissière] at Venice 3pp (20x31)	20 [Dec?] 1646

173r-174v	From La Boissière at Venice $1\frac{1}{2}$pp (20x31) fol 174, blank	29 May 1649
175r-176v	From [La Boissière] at Venice 1p (20x31) fol 176, blank	17 Jul 1649
177r-178v	From [La Boissière] at Venice $2\frac{2}{3}$pp (20x31)	30 Oct 1649
179r-180v	From [La Boissière] at Venice $1\frac{1}{2}$pp (20x31) fol 180, blank	[25 Sep?] 1649
181r-182v	From [La Boissière] at Venice 2pp + ad (20x31)	9 Jan 1650
183r-184v	From La Boissière at Venice 3pp (20x31)	6 Mar 1650
185r-186v	From La Boissière at Hamburg $1\frac{1}{2}$pp + ad (17x22)	24 May 1650
187r-188v	From La Boissière at Hamburg $2\frac{1}{2}$pp (20x31)	3 Jun 1650
189r-190v	From La Boissière at London 3pp (20x31)	20 Apr 1651
191r-192v	From D'Anjou at Smyrna 2pp & 2 lines + ad (15x20)	24 Feb 1648
193r-194v	From Lucas at Smyrna 3pp + ad (15x20)	2 Jun 1648
195r-196v	From Lucas at Smyrna $1\frac{1}{2}$pp + ad (15x20)	8 Nov 1648
197r-199v	From Mazerat and Faure at Smyrna 5pp + ad (20x31)	12 Jun 1648
200r-203v	From Faure at Smyrna 7pp + ad (22x35)	26 Jun 1648
204r-205v	From Mazerat and Faure at Smyrna 3pp + ad (22x35)	14 Nov 1648
206r-207v	From Mazerat at Smyrna 1p & 3 lines + ad (15x20)	28 Jan 1654

208r-209v	From [De Fouzibeé?] at Pera To B at Smyrna 1p + ad (15x20)	31 May 1647
210r-211v	From l'Empereur at Constantinople To de Thou at Paris 2pp + ad (20x31)	22 Aug 1639
212r-213v	To l'Empereur at [Constantinople] 3pp (20x31)	25 Jun 1641
214r-215v	From l'Empereur at Constantinople 1p + ad (20x31)	18 Jan 1642
216r-217v	From l'Empereur at Constantinople To B at Venice 3pp + ad (15x20)	[20 Jul?] 1645
218r-219v	From l'Empereur at Constantinople To B at Venice 3pp + ad (15x20)	23 [Sep?] 1645
220r-221v	From l'Empereur at Constantinople To B at Venice 3pp + ad (15x20)	12 Jul 1646
222r-223v	From l'Empereur at Constantinople 3pp + ad (15x20)	2 Nov 1648
224r-225v	From l'Empereur at Constantinople 3pp + ad (15x20)	8 Mar 1650
226r-227v	From l'Empereur at Constantinople 3pp + ad (15x20)	10 Aug 1650
228r-229v	From l'Empereur at Constantinople 3pp + ad (15x20)	26 Sep 1650
230r-231v	From l'Empereur at Constantinople 3pp + ad (15x20)	4 Oct 1650
232r-233v	From l'Empereur at Constantinople 3½pp + ad (15x20)	22 Aug 1651
234r-235v	From l'Empereur at Constantinople 3pp (15x20)	16 Nov 1652
236r-237v	From l'Empereur at Constantinople 3pp + ad (17x22)	4 Jan 1655

238r-239v	From l'Empereur at Constantinople 3pp + ad (17x22)	[5?] Aug 1655
240r-241v	From l'Empereur at Constantinople 3pp (17x22)	13 [Sep?] 1655
242rv	From La Haye-Vantelet 1p & 8 lines (20x35)	8 May 1647
243r-244v	Newsletter Constantinople (Pera) 1½pp (20x31) fol 244, blank	10 Aug 1648
245r-246v	Extract of a letter 3pp (20x31) "Extract d'une lettre de Constantinopli[u] du 12 Aoust 1648."	
247rv	From Anonymous at Constantinople 1p & 6 lines (20x31) Italian	28 May 1649
248r-249v	From Anonymous at Pera 2¾pp (20x31)	1 Jul 1649
250r-251v	From [De la Navel?] at Constantinople 1¾pp + ad (20x31)	2 Jun 1650
252r-253v	From L-G de Thou at Bonneval ⅔p + ad (18x25)	29 Nov NY
254r-255v	From L-G de Thou at Bonneval 2½pp + ad (18x25)	[9?] Dec NY
256r-257v	From Pontac at Bordeaux ½p + ad (18x25)	8 [Oct?] NY
258r-259v	From Pontac at La Riole 1p + ad (18x25)	15 Aug NY
260r-261v	From Valavès at Aix 1p + ad (20x31)	1 Sep 1637
262r-263v	From Valavès at Boisgency 1p + ad (20x31)	3 Oct 1639
264r-265v	From ? at [Rome?] 2pp + ad (13x20)	18 Oct 1655

266r-267v	From Nicéron at [Livorno?] 1p & 6 lines (20x29) fol 267, blank	21 Mar 1646
268r-269v	From De Neuré 1¾pp + ad (17x22)	[1654 or 1655]
270r-271v	From Boullonesi* at Paris B at Venice 2⅔pp + ad (17x22) Italian *[Giorgio Bologneti?]	16 Apr 1646
272rv	From Anonymous at Thouars 1p + ad (20x31)	25 Feb 1650

▼ ▼ ▼

COLLECTION BOULLIAU, VOLUME XXV, FONDS FRANÇAIS 13043

1r-2v	To Hevelius at Danzig 2pp & 4 lines (17x22) Latin	22 Oct 1683
3r-4v	From Hevelius at Danzig To Des Noyers at Warsaw 1p + ad (17x22) Latin	6 Aug 1683
5r-6v	To Hevelius at Danzig 2pp (17x22) Latin fol 6, blank	4 Jun 1682
7r-9v	From Hevelius at Danzig To Mersenne at Paris 4½pp (23x36) Latin In B's hand: "Escrite au Pere Marin Mersenne/& apportee apres sa mort. arrivee/au commencement de Septemb./1648." fol 9r, list of books in Latin	31 Aug 1648
10r-15v	To Hevelius at Danzig 9pp (20x31) Latin/Italian	III Id Dec 1648 [11 Dec 1648]
16r-17v	To Hevelius at Danzig 3pp (20x31) Latin	7 Jan 1650
18r-19v	From Hevelius at Danzig 2pp & 3 lines (20x31) Latin	14 Sep 1651
20r-21v	From Hevelius at Danzig 3¾pp (20x31) Latin	18 Feb 1650
22rv	Copy of a letter (in Hevelius' hand) From Maria Cunitz at Bicini To Hevelius at Danzig fol 22r, in B's hand: "Epistola Mariae Cunitiae ad Joh. Hevelium/scripta Januarii 24. anno 1650. cujus exemplar/..."	24 Jan 1650
23r-24v	To Hevelius at Danzig 2½pp (20x31) Latin	18 Apr 1650
25r-26v	From Hevelius at Danzig 1½pp + ad (20x31) Latin	13 Jan 1652

27r-28v	To Hevelius at Danzig 2½pp (20x31) Latin	25 Apr 1652
29r-30v	To Maria Cunitz at [Bicini] 3½pp (20x31) Latin	25 Apr 1652
31r-32v	Copy of a Letter From Hevelius at Danzig To Albert Linemann at Königsberg 2pp & 5 lines (20x31) Latin	29 Jul 1652
33r-34v	Copy of a Letter From Albert Linemann at Königsberg To Hevelius at Danzig 3pp (20x31) Latin	28 Aug 1652
35r-36v	To Hevelius at Danzig 3⅔pp (20x31) Latin	3 Oct 1653
37r-39v	From Hevelius at Danzig 3pp (20x31) Latin fol 38, PS, 5 lines (5x11)	24 Jan 1653
40r-42v	From Hevelius at Danzig 3pp & 2 lines (20x31) Latin fol 43, PS, 1p (17x21)	19 Jul 1652
43r-44v	From Linemann at Königsberg To Hevelius at Danzig 2¾pp (20x31) Latin	24 Apr 1652
45r-48v	From [Eliae a Leonibus]* at Bicini To Hevelius at Danzig 8pp (20x31) Latin fols 47 & 48, data for eclipse of 8 Apr 1652 (16x22) *Elie Lewen, husband of M. Cunitz	2 Dec 1651
49r-53v&54rv	To Hevelius at Danzig 8¾pp (18x25) Latin fol 54, blank	25 Oct 1652
53rv bis	From Hevelius at Danzig 1½pp (16x22) Latin	1 Apr 1672
55r-56v	To Hevelius at Danzig 1⅔pp (20x31) Latin fol 56, blank	25 Sep 1654

57r-58v	Copy of a letter in B's hand From G.B. Riccioli at Bologna To Hevelius at Danzig 2⅔pp (20x31) Latin	24 Feb 1655
59r-60v	From Hevelius at Danzig 2pp & 2 lines + ad (20x31) Latin	31 Mar 1655
61r-64v	To Hevelius at Danzig 6⅓pp (20x31) Latin	23 Jul 1655
65rv	To Hevelius at Danzig 1p (18x23) Latin	17 Sep 1655
66r-67v	Copy of a letter From Gaston d'Orléans at Blois To Hevelius at Danzig 1p & 7 lines (18x25) fol 67, blank	29 Aug 1655
68r-69v	To Hevelius at Danzig 1p (17x22) Latin fol 69, blank	4 Feb 1656
70r-71v	To Hevelius at Danzig 4pp (17x22) Latin	20 Oct 1656
72r-73v	From Hevelius at Danzig 2½pp + ad (17x22) Latin	21 Jun 1656
74rv	From de Bury at Paris 2pp (17x22)	27 Apr 1657
75r-77v	From de Bury at Paris To Hevelius at Danzig 1⅓pp (17x22) fols 76&77, blank	13 Apr 1657
78r-79v	From Hevelius at Danzig 2pp & 6 lines + ad (17x22) Latin	21 Mar 1657
80r-81v	To Hevelius at Danzig B at The Hague 3½pp (17x22) Latin	4 May 1657
82r-83v	From Hevelius at Danzig 2pp & 3 lines (17x22) Latin	1 Nov 1657

84r-85v	To Hevelius at Danzig 2pp & 3 lines & 2 lines (17x22) Latin	6 Jun 1659
86rv&93rv	From Hevelius at Danzig To Des Noyers at Warsaw 1½pp + ad (20x31) Latin	12 Jul 1659
87r-92v	From Hevelius at Danzig 12pp (20x31) Latin	9 Dec 1659
94r-95v	From Storanus at Rome 3pp + ad (18x25) Latin	XV Kal Nov 1649 [18 Oct 1649]
96r-103v	To Storanus at [Rome] 11pp (20x31) Latin fol 98 (18x25) fols 99&103, blank	Kal Dec 1649 [1 Dec 1649]
104r-105v	From Storanus at Rome 3pp (20x31) Latin	Kal Jan 1650 [1 Jan 1650]
106r-107v	From Storanus at Rome 3pp & 1 line (20x31) Latin	Idib Jan 1650 [13 Jan 1650?]
108r-109v	From Storanus at Rome 3pp (20x31) Latin	X Kal Mar 1650 [20 Feb 1650]
110r-115v	To Storanus at [Rome] 8⅔pp (20x31) Latin fol 115, blank	18 Mar 1650
116r-119v	From Storanus at Rome 8pp (20x31) Latin	X Kal May 1650 [22 Apr 1650]
120r-121v	From Storanus at Rome 3pp (20x31) Latin	X Kal May 1650 [22 Apr 1650]
122r-123v	From Storanus at Ancona 2pp (15x20) Latin fol 123, blank	Idus Sep 1652 [13 Sep 1652]
124r-125v	From Storanus at Ancona To Anonymous 2pp (20x31) Italian fol 125, blank	23 May 1652
126r-127v	From Storanus at Ancona 2pp (20x31) Latin fol 127, blank	VII Kal Jun 1652 [26 May 1652]

128r-129v	From Stelluti at di Casa To Anonymous 1p + ad (20x31) Italian	3 Jun 1652
130r-131v	From Storanus at Ancona To Anonymous 1p (20x31) Italian fol 131, blank	22 Sep 1652
132r-133v	From Storanus at Ancona To Anonymous 1½pp & 3 lines (20x31) Italian	12 Jan 1653
134r-135v	From Storanus at Ancona To Anonymous 1½pp (20x31) Italian fol 135, blank	2 Mar 1653
136r-137v	To Storanus at [Ancona] ¾p (17x22) Latin fol 137, blank	7 Mar 1653
138r-140v	From Storanus at Ancona 4pp & 8 lines (20x28) Latin fol 139, 8 lines (6x20)	Kal Mar 1653 [1 Mar 1653]
141rv	From Storanus at Ancona To Anonymous 1p (18x25) Italian	24 Apr 1653
142r-143v	From Storanus at Ancona To Anonymous 1p (21x28) Italian fol 143, blank	1 Jun 1653
144r-145v	From Storanus at Ancona 3pp (21x28) Latin	Kal Jun 1653 [1 Jun 1653]
146rv	From Storanus at Ancona To Anonymous 1p (19x28) Italian	17 Aug 1653
147r-148v	To Storanus at [Ancona] 3pp (20x33) Latin	11 Jul 1653
149rv	From Storanus at Ancona To Anonymous ⅔p (18x25) Italian	23 Nov 1653

150r-151v	To Storanus at [Ancona] 1⅓pp (15x20) Latin	20 Feb 1654
152rv	From Storanus at Ancona To Anonymous 1p (18x27) Italian	26 Mar 1654
153r-154v	From Storanus at Ancona 1⅔pp (15x20) Latin fol 154, blank	VII Kal Apr 1654 [26 Mar 1654]
155r-156v	From Storanus at Ancona 1p (15x20) Latin fol 156, blank	IX Kal Dec 1653 [23 Nov 1653]
157r-158v	From Storanus at Ancona 1p (15x20) Latin fol 158, blank	X Kal Nov 1655 [23 Oct 1655]
159r-160v	To Storanus at [Ancona] 1½pp (15x20) Latin fol 160, blank	23 Jun 1656
161r-162v	From Storanus at Ancona 2pp & 5 lines + ad (15x20) Latin	XIII Kal Sep 1656 [20 Aug 1656]

▼ ▼ ▼

COLLECTION BOULLIAU, VOLUME XXVI, FONDS FRANÇAIS 13044

Folio	Description	Date
1r-2v	From Hevelius at Danzig B at The Hague 3pp + ad (17x22) Latin	27 Nov 1660
3r-4v	From Hevelius at Danzig 3pp & 5 lines (15x20) Latin	3 Dec 1661
5r-6v	From Hevelius at Danzig B at Warsaw $2\frac{1}{3}$pp + ad (17x22) Latin	5 May 1661
7r-8v	From Hevelius at Danzig B at Warsaw $2\frac{3}{4}$pp + ad (17x22) Latin fol 8, PS	18 Jul 1661
9rv	Book list in Hevelius' hand/B writes: "1661 Sept. 12 die quo Dantisco discessi" $1\frac{1}{4}$pp (20x31) Latin	
10r-11v	"Transitus Saturni/Gedani/Anno 1661, die ♀, 13 may St.n. observatus/à/ Johanne Hevelio" 2pp (20x31) Latin fol 11, drawing "Transitus Saturni"	undated
12rv	From Hevelius at Danzig $1\frac{1}{2}$pp (15x20) Latin	13 Jun 1661
13r-15v	From Hevelius at Danzig $3\frac{3}{4}$pp (17x22) Latin fol 15, blank	21 Jan 1662
16r-17v	From Hevelius at Danzig $3\frac{3}{4}$pp (20x31) Latin	26 May 1662
18r-19v	From Hevelius at Danzig $2\frac{1}{2}$pp (18x25) Latin *Corrected by B from 1663	27 Oct 1662*
20r-21v	From Hevelius at Danzig $2\frac{2}{3}$pp (20x31) Latin	23 Oct 1663
22rv&25rv	From Hevelius at Danzig $1\frac{1}{2}$pp + ad (15x20) Latin	14 Dec 1663

23r-24v	From Hevelius at Danzig 4pp (20x31) Latin	21 Sep 1663
26r-27v	From Hevelius at Danzig 4pp (20x31) Latin	10 Jul 1663 PS 13 Jul 1663
28r-29v	From Hevelius at Danzig 4pp (18x25) Latin	3 Feb 1663
30r-32v	From J.A. Huswedel at Hamburg To ? 4pp (20x31) Latin fol 31v, "Ejusdem Judicium de Origine Cometarum/perscriptum à Filio ejus D. Residente Brandenb./Hamburgi d. 16/26 Dec^{bis} 1664." fol 32, drawing of Comet in Cygnus, 6 Apr 1664	$\frac{13}{23}$ Dec 1664
33r-35v	Extracts of letters [J.M./Johann Müller] at Hamburg 4pp (20x31) Latin "Judicium Dni Ottonis Gericken Cons. Magdeb. de Origine Commetarum/ perscriptum a Filio eius D. Resid. Brandenb. Hamburgi..." fol 35, blank	$\frac{16}{26}$ Dec 1664 $\frac{\text{21 Dec 1665}}{\text{1 Jan 1665}}$
36r-37r	Extracts of letters From J. Müller at Hamburg To Hevelius at [Danzig] 3pp (20x31) Latin	23 Dec 1664 24 Dec 1664 26 Dec 1664
37v-38v	Copy of a letter From Hevelius at Danzig To J. Müller at [Hamburg] 2pp (20x31) Latin fol 38v, drawing titled: "Via cometae per fixas mense Decembri 1664 et Januarii sequenti. Observationibus Varsavien. Bayeri Tabula expressis"	28 Dec 1664
39r-41v	From Hevelius at Danzig $5\frac{2}{3}$pp (20x31) Latin fol 40v, "Ao. 1664 die 6 Augusti/ Observatio ♄, ♃ & ☾ vesi" fol 41v, "Eclips. ☽ ...observ. Rostockii Ao. 1664/die 22 Julii st.v. a D. Casp Marchius"	9 Sep 1664 PS 12 Sep 1664

Folio	Description	Date
42rv	From Hevelius at Danzig ⅔p + ad (20x31) Latin	30 Apr 1664
43r-44v	From Hevelius at Danzig 3½pp (20x31) Latin; data	28 Mar 1664
45rv	From Hevelius at Danzig 1½pp (17x22) Latin *in B's hand	[20 Dec]* 1664
46rv&51rv	From Hevelius at Danzig 2pp (17x22) Latin fol 51, blank	29 May 1666
47rv	From Hevelius at Danzig 1p (17x22) Latin	28 Jun 1666
48rv	To Hevelius at [Danzig] 1½pp (17x22) Latin	17 Apr 1665
49r-50v	From Hevelius at Danzig 4pp (20x31) Latin	19 Feb 1665
52rv	From Hevelius at Danzig 1p + ad (17x22) Latin	7 Apr 1665
53r-54v	From Hevelius at Danzig 4pp (20x31) Latin	9 May 1665
55r-57v	From Hevelius at Danzig 3½pp & 7 lines (17x22) Latin fol 56, 7 line PS (6x14)	27 Jun 1665
58rv&60rv	From Hevelius at Danzig 3⅔pp (17x22) Latin	29 Jun 1686
59rv	From Hevelius at Danzig 1p (11x16) Latin	17 Oct 1665
61r-62v	From Hevelius at Danzig 4pp (20x31) Latin	11 Jul 1665
63rv	From Hevelius at Danzig 1½pp (17x22) Latin/French	5 Dec 1665
64r-65v	From Hevelius at Danzig 3⅔pp (20x35) Latin	28 Nov 1665

66r-67v	From Hevelius at Danzig [B] at Paris 3$\frac{2}{3}$pp (20x35) Latin fol 67r, "Eclipsis Solaris/Gedani/ Obsevata $\overline{\text{ao}}$ 1666, die 2 Julii St.n. mane/à Johanne Hevelio" fol 67v, "Eclipsis Lunae/Ao. 1666, die ♀ 16 Junii observata/Joh. Hevelio"	29 Oct 1666
68rv&72rv	From Hevelius at Danzig 2pp & 3 lines (17x22) Latin	13 Feb 1666
69r-71v	From Hevelius at Danzig 4pp (17x22) Latin fol 71, PS 14 lines (11x16) Latin	23 Jun 1666
73rv&76rv	From Hevelius at Danzig To [Chapelain] at Paris 2pp + ad (17x22) Latin	29 Oct 1666
74r-75v	From Hevelius at Danzig 4pp (17x22) Latin	10 Oct 1665
77r-78v	From Hevelius at Danzig 4pp (17x22) Latin	22 Oct 1667
79r-80v	From Hevelius at Danzig 4pp (22x35) Latin	15 Mar 1667
81r-82v	From Hevelius at Danzig 3pp + ad (17x22) Latin	18 May 1668
83rv	Astronomical data $\frac{3}{4}$p & $\frac{3}{4}$p (20x31) Latin fol 83r, "Eclipsis Solaris/Gedani/ Anno 1668, die 4 Novemb. St. n. observata/à Johanne Hevelio" fol 83v, "Eclipsis Solis/Anno 1668, die 4 Novemb. st. n. observ."	
84rv	Drawing 1p (17x19) "Eclipsis Luna/Ao. 1668, die 18 Novembr. st.n. Gedani observata, à J. Hevelio"	

85rv
Drawing
1p (16x20)
"Anno 1668, die 4 Novembris st.n./
Deliquium Solare/Gedani/observatum/
à/Johanne Hevelio"

86rv
Text and data
2pp (20x28) Latin
"Observatio Eclipsis Lunaris/Gedani/
Ao. 1668, die 18 Novemb. st.n....J.
Hevelio"

87r-88v
From Hevelius at Danzig — 15 Sep 1668
$2\frac{1}{3}$pp + ad (17x22) Latin

89r-90v
From Hevelius at Danzig — 8 Sep 1668
3pp & 4 lines (17x22) Latin

91r-92v
From Hevelius at Danzig — 8 Sep 1668
$3\frac{1}{2}$pp (17x22) Latin

93r-94v
From Hevelius at Danzig — 7 Dec 1668
$3\frac{1}{2}$pp (17x22) Latin

95rv&98rv
Copy of a letter — 28 May 1666
From Hevelius at Danzig
To [T. L. Burattini at Jaszdowa?]
2pp & 2 lines (17x22) Latin

96r-97v
From Hevelius at Danzig — 9 Mar 1669
$2\frac{3}{4}$pp (11x16) Latin

99r-100v
From Hevelius at Danzig — 8 Nov 1669
$3\frac{3}{4}$pp (17x22) Latin

101r-102v
From Hevelius at Danzig — 5 Jul 1670
$2\frac{1}{2}$pp + ad (17x22) Latin

103rv&106rv
From Hevelius at Danzig — 27 Aug 1670
4pp (17x22) Latin

104rv
Drawing
1p (13x14)
"Occultatio media * in caudâ T/Ao.
1671 die 14 Martii st.n. Gedani à
J. Hevelio obs."

105rv	Drawing 1p (17x19) "Eclipsis ☽ du 19/29 Septemb. Ao. 1670 Gedani observata/a Johanne Hevelio."	
107rv	From Hevelius at Danzig ⅔p (17x22) Latin	25 Apr 1671
108rv	From Hevelius at Danzig 1¾pp (17x22) Latin	1 May 1671
109r-110v	Pencil drawing 1p + ad (17x21) In B's hand: "Machine trouvee par M[r] Hevelius/po[r] porter une Lunette de 70 pieds"	undated
111rv&113rv	From Hevelius at Danzig 3pp + ad (17x22) Latin fol 113r, data of ☾-♄ occultation	20 Jun 1671
112rv	Drawing (13x13) Latin "Occultatio Saturni/Ao 1671 die 1 Junii manc observata/Gedani/a/Joh. Hevelio."	
114rv	Post script to fol 115 6 lines (5x16) Latin	undated
115r-117v	From Hevelius at Danzig 8pp (22x35) Latin fol 116, bis fol 117v, "Eclipsis ☽ die 19/29 Septemb. Anno 1670 observata/Gedani/a/Johanne Hevelio"	3 Apr 1671
118rv	From Hevelius at Danzig To [Des Noyers? at Warsaw] 2pp (17x22) Latin	27 Feb 1672
119rv&122rv	From Hevelius at Danzig 2pp & 6 lines (11x16)	23 Apr 1672
120r-121v	From Hevelius at Danzig 2½pp (17x22) Latin	5 Dec 1672

123r-124v	From Hevelius at Danzig 2½pp (17x22) Latin	29 Oct 1672
125r-126v	From Hevelius at Danzig 4pp (20x31) Latin	26 Aug 1672
127rv	Printed illustration 1p (21x34) "Cursus Cometae/Anno 1672 Mens. Mart. et April/Gedani/Observatus, et Delineatus,/à J. Hevelio"	
128r-129v	From Hevelius at Danzig 2½pp (17x22) Latin	2 Apr 1674
130r-131v	From Hevelius at Danzig 3½pp (17x22) Latin	24 Aug 1674
132rv	From Hevelius at Danzig 1½pp (11x16) Latin	30 Oct 1674
133r-136v	Scientific tract 5⅔pp (20x31) Italian "Decchiaratione dello Strumento" [via Burattini; see below] fol 136, blank Publ, BAF, pp 115-118	undated
137r-140v	From Burattini at Jaszdowa To Des Noyers at [Warsaw] 4pp & 7 lines (17x22) Italian fol 138, undated letter Publ, BAF, pp 115-118	24 Jul 1665
141rv&143rv	Copy of a letter From Hevelius at Danzig To Oldenburg at London 3⅔pp (20x31) Latin Publ, OC, XI, pp 458-475, No 2727	[21 Aug 1675]
142rv	Drawing 1p (21x28) "Eclipsis Lunae totalis, cum Occultationibus quarundam Fixarum,/ observata Gedani. à J. Hevelio./Anno 1675, Die ♀ 11 Januarii st. nov. vesp."	
144r-145v	From Hevelius at Danzig 3½pp (20x31) Latin *in B's hand	Vernal Equinox 1675 [20 Mar]*

146r-149v	Copy of a letter From Hevelius at Danzig To Flamsteed at [London?] 8pp (17x22) Latin [see original, Fonds Latin, 10348, T. XII]	24 Jun 1676
150r-151v	From Hevelius at Danzig 3½pp (17x22) Latin	2 Jul 1676
152r-153v	Copies of letters in Hevelius' hand 4pp (20x34) Latin ¾p From Hevelius at Danzig To Des Noyers at Warsaw 3pp From Hevelius at Danzig To Oldenburg at [London] fol 153, astronomical data for Mira Ceti and Cygni	 30 Apr 1677 2 Jan 1677
154r-155v	From Hevelius at Danzig To [Des Noyers* at Warsaw] 3½pp (17x22) Latin *in B's hand	Vernal Equinox 1679 [20 Mar]*
156rv	From Hevelius at Danzig To [Des Noyers* at Warsaw] 1p (16x18) Latin *identified by B, fol 156r	18 Mar 1678
157rv&160rv	From Hevelius at Danzig To [Des Noyers at Warsaw] 1½pp (17x22) Latin fol 160, 1 line (in Des Noyers' hand)	15 Oct 1683
158r-159v& 161r-162v	Latin text 6⅔pp (17x22) Latin "Joh. Hevelii/Historiola/Cometae, Anni 1683/..."	Autumnal Equinox 1683 [23 Sep 1683]
163r-164v	Astronomical tracts & data in Hevelius' hand 3½pp (17x22) Latin "Transitus/Lunae et Stellae in medio corpore Leonis infiniae/Tychoni alias in Ventri Australioris/Anno 1683 die ♄, 13 Novemb. st. n. mane/observ./Gedani/ à/Joh. Hevelio" etc. fol 164, dated 9 Nov 1683	[9 Nov 1683]

165rv	From Hevelius at Danzig To [Des Noyers* at Warsaw] 1½pp (20x31) Latin *identified by B, fol 165r	25 Jun 1683
166r-169v	Astronomical tract in Hevelius' hand at Danzig 7⅔pp (17x22) Latin "Succincta/Historiola/De Tribus Conjunctionibus Magnis,/Saturni scilicet Jovis nec non Martis,/ Gedani/Exeunte Anno 1682, et initio Anni 1683, ad 28 May.../â/Johanni Hevelio/observatis."	21 Jun 1683
170r-171v	From Hevelius at Danzig 3½pp (17x22) Latin	17 Jun 1683
172r-173v	From Hevelius at Danzig 2½pp (17x22) Latin	27 Jun 1681
174r-175v	From Hevelius at Danzig To [Des Noyers* at Warsaw] 4pp (17x22) Latin *identified by B, fol 174r	28 Apr 1679
176r-177v	From Hevelius at Danzig 4pp (17x22) Latin	26 Sep 1675
178r-180v	From Hevelius at Danzig 5⅔pp (18x23) Latin	24 Apr 1679
181r-182v	From Hevelius at Danzig 2½pp (17x22) Latin	31 May 1680
183rv&185rv	From Hevelius at Danzig 3pp & 1 line (17x22) Latin	4 Dec 1680
184rv	Anonymous note ⅔p (12x18)	Lundi matin
186r-187v	From Hevelius at Danzig To [Des Noyers? at Warsaw] 2½pp & 1 line (17x22) Latin	6 Dec 1680
188r-189v	Copies of letters From [Hevelius] at Danzig To [Des Noyers? at Warsaw] 3½pp (17x22) Latin	27 Dec 1680

	fol 188v, "Joannis Hevelii/EPISTOLA/ Ad Amicum/De Vespertina apparitone/ cometae Anno 1680"	
190rv	Printed document & drawing 1p (15x28) German "Abbildung des/Comet = Sternen/Welcher sich in dem Monat Novembr. 1680. hat sehen lassen" [Strasbourg...Friderich Wilhelm Schmuct]	undated
191rv	Printed drawing 1p (12x15) fol 191, same as fol 190	undated
192rv	Drawing 1p (14x28) Crude drawing of Comet and tail; annotations by [Des Noyers?]	undated
193rv	Data from Comet of 1680-1681 [Nov 1680-Jan 1681] 1p (17x22) Latin	undated
194rv	From Hevelius at Danzig To [Des Noyers at Warsaw?] 2pp (11x16) Latin	15 Feb 1680
195r-196v	From Hevelius at Danzig 3pp (17x22) Latin	3 Jan 1682
197r-198v	From Hevelius at Danzig To Des Noyers at [Warsaw] 2½pp (17x22) Latin	10 Dec 1682
199rv	From Hevelius at Danzig To [Des Noyers at Warsaw?] 2pp (17x22) Latin	5 Mar 1682
200r-201v	Astronomical data and text 3½pp (20x31) Latin fol 200, "Eclipsis Lunae Totalis cum mora/Anno 1682, die 21 et 22 Febr... /observata/Gedani/a/Joh. Hevelio" [Data] fol 201, "Animadvertenda/In Eclipsis Anno 1682 die 21 et 22 Febr./Gedani/observatam" [Text]	

202r-204v	From Hevelius at Danzig 4½pp (17x22) Latin	10 Jan 1681
205rv&208rv	Astronomical data 3½pp (20x31) Latin "J. Hevelii/Continuatio Observationum/ Cometae Ao. 1680 et 1681"	undated
206r-207v	Drawings (2) 2pp (14x14) Drawings of stages of ☾ Eclipse of 21 & 22 Feb 1682	
209r-210v	Copy of a letter From Hevelius at Danzig To Des Noyers at [Warsaw] 3pp (17x22) Latin	14 Feb 1681
211r-212v	From Hevelius at Danzig To Des Noyers at Warsaw 4pp (17x22) Latin	28 Jun 1686
213r-214v	Extracts of letters from ? 4pp (17x22) Latin fols 213 & 214r, "Excerptum ex literis ad Cl. Jacobum Golium" fol 214v, "Excerptum ex literis ad Cl. Abrah. de Gran Prof. Mathes. Franekerān" [?]	20 Feb 1665
215r-216v	From Hevelius at Danzig To Des Noyers at Warsaw 2½pp + ad (17x22) Latin	20 Mar 1681
217r-218v	From Hevelius at Danzig 2⅔pp + ad (20x31) Latin	21 Apr 1685
219r-226v	From Burattini at Jaszdowa 12½pp (15x20) Italian fol 226, blank Publ, BAF, pp 106-109	12 Nov 1665
227r-235v	From Burattini at Jaszdowa 12½pp (15x20) Italian fol 234r, drawing fol 235, blank Publ, BAF, pp 103-106	24 Sep 1665

236r-237v	From Burattini at Jaszdowa To [Des Noyers] 4pp (15x20) Italian Publ, BAF, pp 100-101	7 Aug 1665
238r-239v	From Burattini at Jaszdowa 3½pp (15x20) Italian Publ, BAF, p 114	20 Jan 1668
240r-241v	Fragments, goes w/fols 244-245 From Burattini at Jaszdowa 2½pp (15x20)	2 Dec 1667
242rv	From Burattini at [Jaszdowa] ⅔p (15x20) Publ, BAF, p 113	undated
243rv	From Burattini at Jaszdowa To [Des Noyers at Warsaw] 1p + ad (15x20) Italian Publ, BAF, pp 101-103	4 Sep 1665
244r-247v	From Burattini at Jaszdowa 4pp + ad (15x20) w/fols 240-241 fol 246, blank Publ, BAF, pp 111-113	2 Dec 1667
248rv	Large foldout drawing (2 figures) 1p (17x22)&(41x59) from "Decchiaratione dello strumento," See CB, XXVI, fols 133-135 See also BAF, pp 115&117	undated
249r-252v	From Burattini at Warsaw 8pp (17x22) Italian Publ, BAF, pp 126-129 Publ, HSM, Vol I, pp 214-230	7 Oct 1672
253r-254v	From Burattini at Warsaw 3½pp (17x22) Italian Publ, BAF, pp 125-126	8 Jul 1672
255rv	Foldout drawing & text (30x44) Italian [Telescope riggings] Publ, BAF, p 102	undated

256r-257v	From Burattini at Warsaw 2½pp (17x22) Italian Publ, BAF, pp 130-131	4 Oct 1675
258r-259v	From Burattini at Warsaw ½p (17x22) Italian fol 259, blank Publ, BAF, p 131	10 Jul 1676
260r-263v	From Burattini at Warsaw To "li Principi Sovrani" 4½pp (17x22) Italian fol 263, blank Publ, BAF, pp 132-133	8 Jul 1676
264r-265v	From Burattini at Di casa To Des Noyers at [Warsaw] 2½pp + ad (18x23) Italian Publ, BAF, pp 136-137	20 Dec 1680
266rv	Newsletter, anonymous Warsaw 2pp (20x31) Italian	8 May 1662
267r-268v	Newsletter, anonymous Leopoli 2¾pp (20x31) Italian	6 Dec 1662
269rv&276rv &296r-297v	From Viviani at Florence 7½pp (20x28) Italian	21 Apr 1683 [*rec'd* 14 Jun 1683]
270rv&274rv	Greek text 2pp + ad (20x31) Greek fol 270r, from Bk. 8 of Pappus' *Coll. Mathematica* fol 270v, Extract, From Edward Bernard, Oxford, 1683	undated
271rv	Copy of a Greek Text (in B's hand) ½p (17x22) Greek "Ex vinis Codicibus MS A&B/Bibliothecae Savilianae apud Oxoniam./...Pappi Alexandrini ad libr./VII Collectionis Mathematicae"	undated
272rv	Copy of a letter (in B's hand) To Viviani at [Florence?] ¼p (17x22)	20 Sep 1683

273rv&275rv	Notes 6 lines (17x22) Latin 2 lines (13x19) French	undated
277r-278v	From Viviani at Florence 1⅔pp (20x31) Italian fol 278, blank	8 Feb 1660 Ab Inc [8 Feb 1661]
279rv	Fold out drawing 1p (30x41) Italian BR ink & water color of telescope rigging with Italian descriptions	undated
280r-281v	From Anonymous at Vienna 3¼pp (20x31) Italian	12 Nov 1664
282rv	From Anonymous at Vienna 2pp (20x31) Italian	1 Oct 1664
283r-284v	From Anonymous at Vienna 2⅔pp (20x31) Italian	19 Nov 1664
285r-286v	Newsletter 2pp & 2 lines & 1 line (20x31) Italian Vienna Cracow	 10 Jul 1666 4 Jul 1666
287r-288v	Treatise, Anonymous "Dichiaratione del dicontio Pavelio apparso in Danzica alli 20 febr° 1661. a 2. hore A.M." 1½pp + ink drawing (20x31) Italian	20 Feb 1661
289rv	Newsletter Warsaw 1p (20x21) Italian	 23 Jan 1662
290rv	Newsletter Candia 2pp (20x31) Italian	 24 Oct 1667
291rv	Newsletter Candia 1½pp (20x31) Italian	 24 Nov 1667
292rv	Newsletter Candia 1¾pp (20x31) Italian	 4 Nov 1667

292rv bis	From Viviani at Florence 2pp (19x27) Italian	29 Jan 1665 Ab Inc [29 Jan 1666]
293r-294v	From Viviani at Florence 3pp (20x31) Italian	13 Feb 1682 Ab Inc [13 Feb 1683]
295rv&298rv	From Viviani at [Florence?] 1⅔pp + ad (17x22) Italian	undated
299r-303v	Newsletter 8½pp (20x31) Italian Milan Venice Paris	 15,18 Mar 1662 17 Mar 1662 31 Jan 1662
304r-306v	Newsletter 6pp (20x31) Italian Milan Venice	 22,28 Mar 1662 22,24 Mar 1662
307r-308v	Newsletter Venice 4pp (20x31) Italian	 17 Mar 1662

▼ ▼ ▼

COLLECTION BOULLIAU, VOLUME XXVII, FONDS FRANÇAIS 13045

1rv	From Dulaurens at The Hague 1p (20x31)	27 Sep 1657
2rv	From [Dulaurens] at The Hague 1p + ad (20x31) "Aduis de la Haÿe 22 Sep 1657"	
3r-4v	From Janot at The Hague 2¼pp + ad (20x31)	18 Oct 1657
5rv	From Bernart at The Hague ⅔p (20x31)	27 Oct 1657
6rv&8rv	From Janot at The Hague 1p (15x20) fol 8, blank	22 Nov 1657
7rv	Extract of a Letter 1p (21x24)	undated
9r-10v	From Bernart at The Hague 3pp + ad (17x22)	1 Nov 1657
11r-12v	From Bernart at The Hague 3pp + ad (17x22)	6 Dec 1657
13r-14v	From Bernart at The Hague 3pp + ad (11x16)	20 Dec 1657
15r-16v	From Bernart at The Hague 3pp + ad (17x22)	27 Dec 1657
17r-18v	From Bernart at The Hague 2½pp + ad (17x22)	17 Jan 1658
19r-20v	From Bernart at The Hague 3pp + ad (17x22)	21 Feb 1658
21r-22v	From Bernart at The Hague 2½pp + ad (17x22)	14 Mar 1658
23r-25v	From Bernart at The Hague 4pp + ad (17x22) fol 24r, Latin enclosure	28 Mar 1658

26r-27v	From Bernart at The Hague 3pp + ad (17x22)	11 Apr 1658
28r-29v	From Bernart at The Hague $1\frac{2}{3}$pp + ad (17x22)	25 Apr 1658
30r-33v	From Bernart at The Hague 7pp (17x22)	17 May 1658
34r-35v	From Bernart at The Hague 3pp (20x31)	30 May 1658
36r-37v	From Bernart at The Hague 2pp + ad (17x22)	6 Jun 1658
38r-39v	From Bernart at The Hague 2pp & 4 lines + ad (17x22)	13 Jun 1658
40r-41v	From Bernart at The Hague 3pp (17x22)	27 Jun 1658
42rv	From Bernart at The Hague $1\frac{2}{3}$pp (17x22)	4 Jul 1658
43r-45v	From Bernart at The Hague 3pp + ad (17x22)	18 Jul 1658
46r-48v	From Bernart at The Hague 5pp (20x31)	25 Jul 1658
49r-50v	From Bernart at The Hague 3pp (17x22)	1 Aug 1658
51r-52v	From Bernart at The Hague $2\frac{1}{2}$pp (17x22)	15 Aug 1658
53r-54v	From Bernart at The Hague 3pp (17x22)	22 Aug [1658]
55r-56v	From Bernart at The Hague 3pp (17x22)	29 Aug 1658
57rv	PS to fols 58-59 11 lines (10x18)	5 Sep 1658
58r-59v	From Bernart at The Hague 4pp (20x31)	5 Sep 1658

60r-61v	From Bernart at The Hague 3pp (20x31)	12 Sep 1658
62rv	From Bernart at The Hague 2pp (20x31)	19 Sep 1658
63rv	From Bernart at The Hague 1½pp (17x22)	26 Sep 1658
64rv&67rv	From Bernart at The Hague 2pp & 6 lines + ad (17x22)	3 Oct 1658
65r-66v	From Bernart at The Hague 3pp (17x22)	10 Oct 1658
68r-69v	From Bernart at The Hague 2½pp (17x22)	17 Oct 1658
70r-72v	From Bernart at The Hague 4½pp (17x22)	24 Oct 1658
73r-76v	From Bernart at The Hague 6⅓pp (17x22)	31 Oct 1658
77r-78v	From Bernart at The Hague 2½pp (17x22)	7 Nov 1658
79r-81v	From Bernart at The Hague 4pp (17x22)	14 Nov 1658
82r-83v	From Bernart at The Hague 3pp (17x22)	21 Nov 1658
84r-85v	From Bernart at The Hague 2⅓pp + ad (17x22)	28 Nov 1658
86r-87v	From Bernart at The Hague 3pp + ad (17x22)	5 Dec 1658
88r-89v	From Bernart at The Hague 3pp (17x22)	12 Dec 1658
90r-91v	From Bernart at The Hague 2⅔pp (17x22)	19 Dec 1658
92r-93v	From Bernart at The Hague 2pp & 6 lines (17x22)	26 Dec 1658

94r-95v	From Bernart at The Hague 3pp + ad (11x15)	3 Jan 1658
96r-97v	Newsletter in Bernart's hand, The Hague 3pp & 6 lines (20x31)	30 Jan 1659
98rv	From Bernart at The Hague 2pp (17x22)	2 Jan 1659
99r-101v	From Bernart at The Hague 2½pp + ad (17x22)	9 Jan 1659
102r-103v	From Bernart at The Hague 3pp (17x22)	16 Jan 1659
104r-105v	From Bernart at The Hague 3pp (17x22)	23 Jan 1659
106rv	From Bernart at The Hague 1p (17x22)	30 Jan 1659
107r-108v	From Bernart at The Hague 2½pp (17x22)	6 Feb 1659
109rv	From Bernart at The Hague 2pp (17x22)	13 Feb 1659
110r-114v	From Bernart at The Hague 7pp + ad (17x22)	20 Feb 1659
115r-116v	From Bernart at The Hague 3pp (17x22)	27 Feb 1659
117rv	From Bernart at The Hague 2pp (17x22)	6 Mar 1659
118r-120v	From Bernart at The Hague 3pp + ad (17x22)	13 Mar 1659
121r-122v	From Bernart at The Hague 3pp (17x22)	20 Mar 1659
123r-124v	From Bernart at The Hague 3pp (17x22)	27 Mar 1659
125r-126v	From Bernart at The Hague 3pp (17x22)	3 Apr 1659

127rv	From Bernart at The Hague 2pp (17x22)	10 Apr 1659
128r-130v	From Bernart at The Hague 2½pp + ad (17x22) fol 130, blank	17 Apr 1659
131r-132v	From Bernart at The Hague 3pp (17x22)	24 Apr 1659
133r-135v	From Bernart at The Hague 4pp & 4 lines (17x22) fol 134, enclosed newsletter	1 May 1659
136rv&139rv	From Bernart at The Hague 2pp & 5 lines (17x22)	7 May 1659
137r-138v	From Bernart at The Hague 2½pp (17x22)	15 May 1659
140r-141v	From Bernart at The Hague 2½pp (17x22)	22 May 1659
142rv	From Bernart at The Hague 2pp (17x22)	29 May 1659
143rv	From Bernart at The Hague 2pp (17x22)	5 Jun 1659
144r-145v	From Bernart at The Hague 2½pp + ad (17x22)	19 Jun 1659
146rv	From Bernart at The Hague 1p (17x22)	26 Jun 1659
147r-148v	From Bernart at The Hague 2pp & 6 lines (17x22)	3 Jul 1659
149r-150v	From Bernart at The Hague 2½pp (17x22)	10 Jul 1659
151r-152v	From Bernart at The Hague 2½pp (17x22)	17 Jul 1659
153r-154v	From Bernart at The Hague 3pp (17x22)	24 Jul 1659
155rv&158rv	From Wicquefort at The Hague 1p + ad (17x22)	17 Jul 1659

156r-157v	From Bernart at The Hague 2½pp (17x22)	31 Jul 1659
159r-160v	From Bernart at The Hague 3pp (17x22)	7 Aug 1659
161r-162v	From Bernart at The Hague 2pp & 4 lines + ad (17x22)	14 Aug 1659
163r-164v	From Bernart at The Hague 2pp & 4 lines (17x22)	21 Aug 1659
165rv	From Bernart at The Hague 2pp (17x22)	28 Aug 1659
166r-167v	From Bernart at The Hague 2½pp (17x22)	4 Sep 1659
168rv	From Bernart at The Hague 2pp (17x22)	11 Sep 1659
169rv	From Bernart at The Hague 1½pp (17x22)	18 Sep 1659
170rv	From Bernart at The Hague 2pp (17x22)	25 Sep 1659
171r-172v	From Bernart at The Hague 2½pp + ad (17x22)	2 Oct 1659
173r-174v	From Bernart at The Hague 2½pp + ad (17x22)	16 Oct 1659
175rv	From Bernart at The Hague 2pp (17x22)	23 Oct 1659
176rv	From Bernart at The Hague 1½pp (17x22)	30 Oct 1659
177rv	From Bernart at The Hague 1p (17x22)	6 Nov 1659
178rv	From Bernart at The Hague 2pp (17x22)	13 Nov 1659
179rv	From Bernart at The Hague 1⅔pp (17x22)	20 Nov 1659

180rv	From Bernart at The Hague 1⅔pp (17x22)	27 Nov 1659
181r-182v	From Bernart at The Hague 2pp & 6 lines + ad (17x22)	4 Dec 1659
183r-184v	From Bernart at The Hague 1⅔pp + ad (17x22)	11 Dec 1659
185r-186v	From Bernart at The Hague 2⅓pp + ad (17x22)	18 Dec 1659
187rv	From Bernart at The Hague 1⅔pp (17x22)	25 Dec 1659
188r-189v	From Bernart at Amsterdam To [J-A de Thou] 2½pp (20x31)	26 Nov 1660
190rv	From Bernart at The Hague 2pp (17x22)	1 Jan 1660
191rv	From Bernart at The Hague 1p & 7 lines (17x22)	8 Jan 1660
192r-193v	From Bernart at The Hague 2½pp + ad (17x22)	22 Jan 1660
194rv	From Bernart at The Hague 2pp (17x22)	29 Jan 1660
195r-197v	From Bernart at The Hague 2pp & 6 lines + ad (17x22)	5 Feb 1660
198rv	From Bernart at The Hague 1½pp (17x22)	12 Feb 1660
199rv	From Bernart at The Hague 1½pp (17x22)	19 Feb 1660
200rv	From Bernart at The Hague 1⅔pp (17x22)	26 Feb 1660
201rv	From Bernart at The Hague 1⅔pp (17x22)	4 Mar 1660
202rv	From Bernart at The Hague 2pp (17x22)	11 Mar 1660

Folio	Description	Date
203r-204v	From Bernart at The Hague 2pp & 4 lines + ad (17x22)	18 Mar 1660
205rv&208rv	From Bernart at The Hague 2pp + ad (17x22)	25 Mar 1660
206r-207v	From Bernart at The Hague 2pp & 4 lines & PS + ad (17x22)	1 Apr 1660
209r-210v	From Bernart at The Hague 2⅓pp (17x22)	8 Apr 1660
211rv	From Bernart at The Hague 2pp (17x22)	15 Apr 1660
212r-213v	From Bernart at The Hague 2pp & 3 lines + ad (17x22)	22 Apr 1660
214r-215v	From Bernart at The Hague 3pp + ad (17x22) fol 215r, 11 line note from De la Richardiere	29 Apr 1660
216rv	From Bernart at The Hague 1⅔pp (17x22)	6 May 1660
217rv	From Bernart at The Hague 1½pp (17x22)	13 May 1660
218rv	From Bernart at The Hague 2pp (17x22)	20 May 1660
219rv	From Bernart at The Hague 1½pp (17x22)	27 May 1660
220rv	From Bernart at The Hague 2pp (17x22)	3 Jun 1660
221rv	From Bernart at The Hague 1½pp (17x22)	10 Jun 1660
222rv	From Bernart at The Hague 2pp (17x22)	17 Jun 1660
223rv	From Bernart at The Hague 1⅔pp (17x22)	24 Jun 1660
224r-225v	From Bernart at London 2½pp (17x22)	9 Jul 1660

226rv	From Bernart at London 2pp (20x31)	5 Aug 1660
227rv	Newsletter The Hague 2pp (20x31)	30 Jul 1660
228rv	From Bernart at The Hague 1p & 5 lines (17x22)	19 Aug 1660
229rv	From Bernart at The Hague 2pp (17x22)	26 Aug 1660
230r-231v	From Bernart at The Hague 2⅔pp (20x31)	9 Sep 1660
232rv	From Bernart at The Hague 1⅔pp (20x31)	16 Sep 1660
233r-235v	From Bernart at The Hague 2¼pp & 7 lines + ad (20x31) fol 235, 7 lines (5x15) attached to fol 234	23 Sep 1660
236r-237v	From Mme Bernart at The Hague B at Danzig 2pp + ad (17x22)	1 Apr 1661
238rv	Newsletter Danzig 1p (20x21)	19 Nov 1661
239r-240v	From Bernart at The Hague 2⅔pp (20x31)	15 Apr 1661
241r-242v	From Bernart at The Hague 2½pp (20x31)	9 May 1661
243r-244v	From Bernart at The Hague 2⅔pp (20x31)	25 Jul 1661
245r-246v	From Bernart at The Hague 2pp & 4 lines (20x31)	12 Aug 1661
247rv	From Bernart at The Hague 1p (17x22)	13 Jul 1662
248rv	From F. Burrius at The Hague 1p (20x31) Latin	28 Apr 1662

249r-250v	From Bernart at The Hague 3pp (20x31)	27 Jul 1662
251rv&253rv	From Bernart at The Hague 2½pp (20x31)	10 Aug 1662
252rv	Printed News [Smient; Amsterdam] 2pp (14x23)	7 Aug 1662
254r-255v	From Bernart at The Hague 2½pp (20x31)	24 Aug 1662
256rv&258rv	From Bernart at The Hague 2pp & 5 lines + ad (17x22)	7 Sep 1662
257rv	From Bernart at The Hague 2pp (17x22)	21 Sep 1662
259rv	From Bernart at Amsterdam 1p (20x31)	5 Oct 1662
260r-261v	From Bernart at The Hague 2pp & 5 lines (17x22)	12 Oct 1662
262r-263v	From Bernart at The Hague 3pp (20x31)	19 Oct 1662
264rv&266rv	From Bernart at The Hague 2pp & 4 lines (20x31)	26 Oct 1662
265rv	From Bernart at The Hague 2pp (20x31)	2 Nov 1662
267rv	From Bernart at The Hague 2⅔pp (20x31)	9 Nov 1662
268r-269v	From Bernart at The Hague 2⅓pp (20x31)	16 Nov 1662
270r-271v	From Bernart at The Hague 2½pp (20x31)	23 Nov 1662
272rv	From Bernart at The Hague 2pp (20x31)	30 Nov 1662
273rv	From Bernart at The Hague 2pp (20x31)	7 Dec 1662

274rv&278rv	From Bernart at The Hague 2pp & 6 lines (20x31)	14 Dec 1662
275rv	From Bernart at The Hague $1\frac{1}{2}$pp (17x22)	21 Dec 1662
276rv	From Bernart at The Hague 2pp (20x31)	28 Dec 1662
277rv	From Romer at Hamburg To Bernart at The Hague 1p (20x31)	31 Jan 1662

▼ ▼ ▼

COLLECTION BOULLIAU, VOLUME XXVIII, FONDS FRANÇAIS 13046

1rv	From Bernart at The Hague $1\frac{1}{2}$pp (17x22)	4 Jan 1663
2r-3v	From [Bernart] at The Hague 3pp (17x22)	18 Jan 1663
4r-5v	From Bernart at The Hague $2\frac{1}{2}$pp (17x22)	11 Jan 1663
6rv	From Bernart at The Hague $1\frac{2}{3}$pp (17x22)	25 Jan 1663
7rv&9rv	From [Bernart] at The Hague 2pp & 2 lines (17x22)	1 Feb 1663
8rv	From [Bernart] at The Hague $\frac{2}{3}$p (17x22)	8 Feb 1663
10rv&11rv	From Bernart at The Hague $1\frac{1}{2}$pp & 6 lines (17x22)	14 Feb 1663
12r-13v	From [Bernart] at The Hague $2\frac{3}{4}$pp (20x31)	15 Feb 1663
14rv	From [Bernart] at The Hague 1p (20x31)	16 Feb 1663
15rv	From [Bernart] at The Hague $1\frac{1}{2}$pp (20x31)	1 Mar 1663
16rv&20rv?	From Bernart at The Hague 2pp & 5 lines (20x31)	8 Mar 1663
17rv&19rv	From Bernart at The Hague $2\frac{1}{2}$pp (20x31)	15 Mar 1663
18rv	From [Bernart] at The Hague $1\frac{1}{2}$pp (20x31)	22 Mar 1663
21rv	Fragment 9 lines (20x31)	undated
22rv	From [Bernart] at The Hague 2pp (20x31)	29 Mar 1663

23rv&25rv?	From [Bernart] at The Hague 1⅔pp & 6 lines + ad (20x31)	5 Apr 1663
24rv	From [Bernart] at The Hague 1½pp (20x31)	12 Apr 1663
26r-27v	From Bernart at The Hague 2½pp (20x31)	8 Aug 1663
28rv&32rv?	From [Bernart] at The Hague 2pp & 7 lines (20x31)	23 Aug 1663
29rv	From [Bernart] at Amsterdam 2pp (20x31)	6 Sep 1663
30rv	From [Bernart] at Amsterdam 1½pp (20x31)	27 Sep 1663
31rv	From [Bernart] at The Hague 1⅔pp (20x31)	4 Oct 1663
33rv	From [Bernart] at The Hague 2pp (20x31)	10 Oct 1663
34rv	From [Bernart] at The Hague 1p (17x22)	11 Oct 1663
35rv	Note 7 lines (14x15)	undated
36rv	From [Bernart] at The Hague 1⅔pp (20x31)	18 Oct 1663
37r-38v	From [Bernart] at The Hague 2pp & 6 lines (20x31)	25 Oct 1663
39rv&44rv	From [Bernart] at The Hague 2pp & 6 lines + ad (20x31)	1 Nov 1663
40rv&43rv	From [Bernart] at The Hague 2pp & 3 lines + ad (20x31)	8 Nov 1663
41rv	From [Bernart] at The Hague 2pp (20x31)	15 Nov 1663
42rv	Printed document: "Haenen-gekraey" 1p (18x29)	[dated 1663]

45rv	From [Bernart] at The Hague 1½pp (20x31)	22 Nov 1663
46rv	From [Bernart] at The Hague 1½pp (20x31)	29 Nov 1663
47rv	From [Bernart at The Hague] 1½pp (20x31) *dated by B	6 Dec 1663*
48rv	From [Bernart at The Hague] 1½pp (20x31) *dated by B	13 Dec 1663*
49rv&52rv	From [Bernart] at The Hague 1½pp (20x31)	19 Dec 1663
50r-51v	From [Bernart] at The Hague 2pp & 6 lines + ad (20x31)	27 Dec 1663
53rv	From [Bernart] at The Hague 2pp (20x31)	3 Jan 1664
54rv	From [Bernart] at The Hague 1½pp (20x31)	10 Jan 1664
55rv	From [Bernart] at The Hague 1⅔pp (20x31)	17 Jan 1664
56r-57v	From [Bernart] at The Hague 2½pp + ad (20x31)	24 Jan 1664
58rv	From [Bernart] at The Hague 2pp (20x31)	31 Jan 1664
59rv	From [Bernart] at The Hague 1⅓pp (20x31)	7 Feb 1664
60rv	From [Bernart] at The Hague 2pp (20x31)	21 Feb 1664
61rv	From [Bernart] at The Hague 1⅓pp (20x31)	28 Feb 1664
62rv	From [Bernart] at The Hague 1⅔pp (20x31)	6 Mar 1665

63rv	From [Bernart] at The Hague 1½pp (20x31)	13 Mar 1664
64rv	From [Bernart] at The Hague 1⅔pp (20x31)	20 Mar 1664
65rv	From [Bernart] at The Hague 1⅔pp (17x22)	27 Mar 1664
66rv	Fragment 9 lines + ad (20x31)	undated
67rv	From [Bernart] at The Hague 2pp (20x31)	3 Apr 1664
68r-69v	From [Bernart] at The Hague 2½pp + ad (20x31)	10 Apr 1664
70rv	From [Bernart] at The Hague 1½pp (20x31)	17 Apr 1664
71r-72v	From [Bernart] at The Hague 2½pp + ad (20x31)	24 Apr 1664
73rv	From [Bernart] at The Hague 2pp (20x31)	1 May 1664
74rv	From [Bernart] at The Hague 1½pp (20x31)	29 May 1664
75rv&77rv	From [Bernart] at The Hague 2pp & 6 lines + ad (20x31)	5 Jun 1664
76rv	From [Bernart] at The Hague 2pp (20x31)	12 Jun 1664
78rv	From [Bernart] at The Hague 1p (17x22)	19 Jun 1664
79r-80v	From [Bernart] at The Hague 2pp (17x22)	18 Jun 1664
81rv	From [Bernart] at The Hague 1⅔pp (20x31)	26 Jun 1664
82rv	From [Bernart] at The Hague 1⅔pp (20x31)	10 Jul 1664

83rv	From [Bernart] at The Hague 2pp (20x31)	17 Jul 1664
84rv	From [Bernart] at The Hague 2pp (20x31)	24 Jul 1664
85rv	From [Bernart] at The Hague 1½pp (20x31)	31 Jul 1664
86rv	From [Bernart] at The Hague 1½pp (20x31)	7 Aug 1664
87rv	From [Bernart] at The Hague 1⅔pp (20x31)	14 Aug 1664
88r-89v	From [Bernart] at The Hague 2pp + ad (20x31)	21 Aug 1664
90rv&93rv	From [Bernart] at The Hague 2pp + ad (20x31)	4 Sep 1664
91r-92v	From [Bernart] at The Hague 2⅔pp + ad (20x31)	11 Sep 1664
94rv&97rv	From [Bernart] at The Hague 2⅓pp + ad (20x31)	18 Sep 1664
95r-96v	From [Bernart] at The Hague 2½pp + ad (20x31)	25 Sep 1664
98rv	From [Bernart] at The Hague 2pp (20x31)	2 Oct 1664
99rv	From [Bernart] at The Hague 1⅔pp (20x31)	9 Oct 1664
100rv	From Bernart at The Hague ⅔p (17x22)	16 Oct 1664
101rv	From [Bernart] at The Hague 2pp (17x22)	27 Nov 1664
102r-103v	From [Bernart] at The Hague 3pp (20x31)	11 Dec 1664
104r-105v	From [Bernart] at The Hague To [J-A de Thou?] 3pp (20x31)	11 Dec 1664

106rv	From [Bernart] at The Hague 2pp (20x31)	8 Jan 1665
107rv	From [Bernart] at The Hague 1p (17x22)	15 Jan 1665
108r-109v	From [Bernart] at The Hague 2½pp + ad (17x22)	22 Jan 1665
110rv	From [Bernart] at The Hague 1½pp (20x31)	29 Jan 1665
111rv	From [Bernart] at The Hague 2pp (20x31)	5 Feb 1665
112rv	From [Bernart] at The Hague 1½pp (20x31)	12 Feb 1665
113rv	From [Bernart] at The Hague 1p (17x22)	5 Mar 1665
114rv	From [Bernart] at The Hague 2pp (20x31)	12 Mar 1665
115rv	From [Bernart] at The Hague 2pp (20x31)	19 Mar 1665
116rv&119rv	From [Bernart] at The Hague 1⅔pp + ad (20x31)	26 Mar 1665
117r-118v	From [Bernart] at The Hague 2½pp + ad (17x22)	2 Apr 1665
120r-121v	From [Bernart?] at The Hague To [J-A de Thou?] 3pp (20x31)	6 Apr 1665
122rv	From [Bernart] at The Hague 2pp (17x22)	9 Apr 1665
123rv	From [Bernart] at The Hague 1½pp (20x31)	16 Apr 1665
124rv	From [Bernart] at The Hague 2pp (17x22)	30 Apr 1665
125r-126v	From [Bernart] at The Hague 1⅔pp + ad (17x22)	7 May 1665

127rv	From [Bernart] at The Hague 2pp (17x22)	14 May 1665
128rv	From [Bernart] at The Hague 1½pp (17x22)	13 Apr 1665
129rv	From [Bernart] at The Hague 1½pp (17x22)	3 Sep 1665
130rv	From [Bernart] at The Hague 1p (17x22)	17 Sep 1665
131r-132v	From [Bernart] at The Hague 2⅓pp (17x22)	1 Oct 1665
133rv	From [Bernart] at The Hague 2pp (17x22)	8 Oct 1665
134r-136v	From [Bernart] at The Hague 2½pp (17x22) fol 136, blank	22 Oct 1665
137r-138v	From [Bernart] at The Hague 2½pp (17x22)	29 Oct 1665
139rv	From [Bernart] at The Hague 2pp (17x22)	5 Nov 1665
140rv	From [Bernart] at The Hague 2pp (17x22)	12 Nov 1665
141r-142v	From [Bernart] at The Hague 3pp (17x22)	18 Nov 1665
143r-144v	From [Bernart] at The Hague 2pp (17x22)	10 Dec 1665
145rv	From [Bernart] at The Hague 1½pp (17x22)	17 Dec 1665
146rv	From [Bernart] at The Hague 1½pp (17x22)	24 Dec 1665
147rv	From [Bernart] at The Hague 2pp (17x22)	31 Dec 1665
148rv&152rv	From [Bernart] at The Hague 2pp & 4 lines + ad (17x22)	7 Jan 1666

149rv	From [Bernart] at The Hague 1⅔pp (17x22)	14 Jan 1666
150rv	From [Bernart] at The Hague 1½pp (17x22)	21 Jan 1666
151rv	From [Bernart] at The Hague 1⅔pp (17x22)	28 Jan 1666
153r-154v	From [Bernart] at The Hague 2⅔pp + ad (17x22)	4 Feb 1666
155rv&161rv	From [Bernart] at The Hague 2pp & 6 lines + ad (17x22)	11 Feb 1666
156rv	From [Bernart] at Amsterdam 1½pp (17x22)	18 Feb 1666
157rv&160rv	From [Bernart] at The Hague 2pp & 7 lines + ad (17x22)	4 Mar 1666
158r-159v	From [Bernart] at The Hague 1½pp + ad (17x22)	11 Mar 1666
162rv	From [Bernart] at The Hague 1⅔pp (17x22)	18 Mar 1666
163rv	From [Bernart] at The Hague 1½pp (17x22)	25 Mar 1666
164rv	From [Bernart] at The Hague 1½pp (17x22)	8 Apr 1666
165rv	From [Bernart] at The Hague 1½pp (17x22)	15 Apr 1666
166rv&171rv	From [Bernart] at The Hague 2pp + ad (17x22)	22 Apr 1666
167rv&172rv	From [Bernart] at The Hague 2pp & 6 lines + ad (17x22)	6 May 1666
168rv&173rv	From [Bernart] at The Hague 1½pp + ad (17x22)	13 May 1666
169r-170v	From [Bernart] at The Hague 2pp & 6 lines + ad + ad (17x22)	20 May 1666

174r-176v	From [Bernart] at The Hague 3pp & 3 lines + ad (20x31) fol 176, note 3 lines (4x16)	10 Jun 1666
177r-178v	From [Bernart] at The Hague 2⅔pp + ad (17x22)	17 Jun 1666
179rv&183rv	From [Bernart] at The Hague 2½pp + ad (17x22)	24 Jun 1666
180r-182v	From [Bernart] at The Hague 2pp & 11 lines + ad (17x22)	1 Jul 1666
184rv&188rv	From [Bernart] at The Hague 2pp & 4 lines + ad (17x22)	8 Jul 1666
185rv&187rv	From [Bernart] at The Hague 2pp & 6 lines (17x22)	15 Jul 1666
186rv	Copy of a letter From Vanderbeecke To Anonymous 1p (18x27)	7 Jul 1666
189rv	Extract From the register of the Estates Générale des Provinces Unis 1½pp (20x31)	[9 Jul 1666]
190r-191v	From Bernart at The Hague 2½pp (20x31)	15 Jul 1666
192rv	From Bernart at The Hague 2pp (17x22)	22 Jul 1666
193rv&198r-199v	From [Bernart] at The Hague 2pp & 5 lines + ad (17x22)	22 Jul 1666
194rv&197rv	From [Bernart] at The Hague 2½pp + ad (17x22)	29 Jul 1666
195r-196v	From [Bernart] at The Hague 2pp & 6 lines (17x22)	29 Jul 1666
200r-201v	From [Bernart] at The Hague 2½pp (17x22)	5 Aug 1666

202r-204v	From Bernart at The Hague 4½pp (17x22)	5 Aug 1666
205r-206v	From [Bernart] at The Hague 1p & 5 lines & ⅔p (17x22)	12 Aug 1666
207rv	Newsletter The Hague 2pp (20x31)	12 Aug 1666
208r-209v	From [Bernart] at The Hague 3½pp (20x31)	19 Aug 1666
210r-211v	Newsletter The Hague 2⅔pp (17x22)	26 Aug 1666
212rv	From [Bernart] at The Hague 1p (17x22)	26 Aug 1666
213rv	Newsletter The Hague 2pp (17x22) fragment	2 Sep 1666
214rv	From [Bernart] at The Hague 1½pp (17x22)	2 Sep 1666
215rv	Newsletter The Hague 1¾pp (17x22)	9 Sep 1666
216rv	From [Bernart] at The Hague ½p (17x22)	9 Sep 1666
217r-218v	From [Bernart] at The Hague 2⅔pp (20x31)	16 Sep 1666
219r-220v	Newsletter The Hague 2pp & 5 lines (17x22)	16 Sep 1666
221r-222v	From [Bernart] at The Hague 3pp (20x31)	23 Sep 1666
223r-224v	From [Bernart] at The Hague 2⅔pp (20x31)	23 Sep 1666

225r-226v	From [Bernart] at The Hague 2½pp + ad (20x31)	21 Oct 1666
227r-228v	From [Bernart] at The Hague 2pp & 5 lines (17x22)	14 Oct 1666
229r-230v	From [Bernart] at The Hague 2½pp (17x22)	7 Oct 1666
231r-232v	From [Bernart] at The Hague 2⅔pp (17x22)	7 Oct 1666
233rv	From [Bernart] at The Hague 2pp (17x22)	23 Sep 1666
234rv&237rv	From [Bernart] at The Hague 2pp & 6 lines (17x22)	30 Sep 1666
235r-236v&238rv	From [Bernart] at The Hague 3pp (17x22) fol 238, blank	30 Sep 1666
239r-240v	From [Bernart] at The Hague To [J-A de Thou?] 2⅔pp (20x31)	28 Oct 1666
241r-242v	From [Bernart] at The Hague 2pp & 3 lines (17x22)	2 Dec 1666
243r-244v	From [Bernart] at The Hague 2⅔pp (17x22)	25 Nov 1666
245r-246v	From [Bernart] at The Hague 2⅔pp (17x22)	28 Oct 1666
247r-248v	Newsletter The Hague 2⅔pp (17x22)	28 Oct 1666
249rv&251rv	From [Bernart] at The Hague 2pp & 5 lines (17x22)	4 Nov 1666
250rv	Extract of a letter* From Amb. Boreel at Paris To the Estates Général at The Hague ½p (17x22) *[encl w/fol 249]	[22 Oct 1666]

252r-253v	From [Bernart] at The Hague 2⅓pp + ad (17x22)	11 Nov 1666
254r-255v	From [Bernart] at The Hague 2⅔pp + ad (17x22)	18 Nov 1666
256r-257v	From [Bernart] at The Hague 3pp (17x22)	8 Dec 1666
258r-259v	From [Bernart] at The Hague 3⅓pp (20x31)	16 Dec 1666
260r-261v	From [Bernart] at The Hague 1⅔pp & 9 lines (17x22)	23 Dec 1666
262r-263v	From [Bernart] at The Hague 3⅓pp (20x31)	30 Dec 1666

▼ ▼ ▼

COLLECTION BOULLIAU, VOLUME XXIX, FONDS FRANÇAIS 13047

1rv	From [Bernart] at The Hague 2pp (20x31)	6 Jan 1667
2r-3v	From [Bernart] at The Hague 2⅔pp (20x31)	13 Jan 1667
4r-5v	From [Bernart] at The Hague 3pp (20x31)	20 Jan 1667
6rv	From [Bernart] at The Hague 1⅔pp (20x31)	27 Jan 1667
7rv&10rv	From [Bernart] at The Hague 2⅔pp (17x22)	3 Feb 1667
8r-9v	From [Bernart] at The Hague 2½pp + ad (20x31)	10 Feb 1667
11rv	From [Bernart] at The Hague 1¼pp + ad (20x31)	17 Feb 1667
12rv&15rv	From [Bernart] at The Hague 2pp (17x22) fol 15, blank	24 Feb 1667
13r-14v	Newsletter The Hague 2½pp (20x31)	24 Feb 1667
16r-17v	From [Bernart] at The Hague 3pp (20x31)	3 Mar 1667
18r-19v	From [Bernart] at The Hague 2½pp (20x31)	10 Mar 1667
20rv	From [Bernart] at The Hague 2pp (20x31)	17 Mar 1667
21r-22v	From [Bernart] at The Hague 1⅔pp (20x31)	24 Mar 1667
23r-24v	From [Bernart] at The Hague 2½pp (20x31)	31 Mar 1667

25r-26v	From [Bernart] at The Hague 2⅓pp + ad (17x22)	7 Apr 1667
27rv&30rv	From [Bernart] at The Hague 2⅓pp (20x31)	14 Apr 1667
28r-29v	From [Bernart] at The Hague 2⅔pp (20x31)	21 Apr 1667
31rv&34rv	From [Bernart] at The Hague 2½pp (20x31)	28 Apr 1667
32r-33v	From [Bernart] at Paris 3pp (20x31)	5 May 1667
35r-36v	From [Bernart] at The Hague 1⅔pp + ad (20x31)	12 May 1667
37rv&39rv	From [Bernart] at The Hague 2¼pp (20x31)	19 May 1667
38rv	From [Bernart] at The Hague 1p & 4 lines (20x31)	26 May 1667
40rv&45rv	From [Bernart] at The Hague 2¼pp + ad (20x31)	2 Jun 1667
41rv&44rv	From [Bernart] at The Hague 2pp & 3 lines + ad (20x31)	9 Jun 1667
42r-43v	From [Bernart] at The Hague 2pp & 3 lines + ad (20x31)	16 Jun 1667
46r-47v	From [Bernart] at The Hague To [J-A de Thou?] 2⅓pp (20x31)	23 Jun 1667
48r-49v	From [Bernart] at The Hague 3pp (20x31)	23 Jun 1667
50rv	From [Bernart] at Harlem 1½pp (20x31)	31 Jun 1667
51r-52v	From [Bernart] at The Hague 2½pp (20x31)	7 Jul 1667
53rv&58rv	From [Bernart] at The Hague To [J-A de Thou?] 2pp & ½ line (20x31)	14 Jul 1667

54rv&57rv	From [Bernart] at The Hague 2pp & 2 lines + ad (20x31)	14 Jul 1667
55r-56v	From [Bernart] at The Hague To [J-A de Thou?] 2⅔pp (20x31)	21 Jul 1667
59rv&62rv	From [Bernart] at The Hague 2½pp (20x31)	21 Jul 1667
60r-61v	From [Bernart] at The Hague 2½pp (20x31)	28 Jul 1667
63rv	From [Bernart] at The Hague To [J-A de Thou?] 2pp (20x31)	4 Aug 1667
64r-65v	From [Bernart] at The Hague 2⅔pp + ad (20x31)	4 Aug 1667
66rv	From [Bernart] at The Hague 2pp (20x31)	18 Aug 1667
67rv&70rv	From [Bernart] at The Hague To [J-A de Thou?] 2½pp (20x31)	25 Aug 1667
68r-69v	From [Bernart] at The Hague 2½pp + ad (20x31)	25 Aug 1667
71rv	From [Bernart] at The Hague 2pp (20x31)	1 Sep 1667
72rv&74rv	From [Bernart] at The Hague To [J-A de Thou?] 2pp & 5 lines (20x31)	1 Sep 1667
73rv	From [Bernart] at The Hague 1¾pp (20x31)	8 Sep 1667
75rv&78rv	From [Bernart] at The Hague 2⅓pp + ad (20x31)	15 Sep 1667
76rv	From [Bernart] at The Hague 1p (20x31)	29 Sep 1667
77rv&79rv	From [Bernart] at The Hague 1½pp (17x22)	6 Oct 1667

Folio	Description	Date
80rv	From [Bernart] at The Hague 1p (18x31)	20 Oct 1667
81rv	From Bernart at The Hague 1p (17x22)	13 Oct 1667
82r-83v	News account The Hague 3pp (20x31)	13 Oct 1667
84r-85v	News account The Hague 2½pp (20x31)	20 Oct 1667
86rv	From [Bernart] at The Hague 1¼pp (17x22)	27 Oct 1667
87r-88v	News account The Hague 3pp (20x31)	27 Oct 1667
89r-90v	From [Bernart] at The Hague 2pp & 6 lines (17x22)	3 Nov 1667
91r-92v	News account The Hague 3pp (20x31)	3 Nov 1667
93r-95v	From [Bernart] at The Hague 2½pp (17x22) fol 95, newsletter, undated ⅔p (20x31)	10 Nov 1667
96r-97v	News account The Hague 3pp (20x31)	10 Nov 1667
98rv	From J. Meonman? ¾p (18x30)	26 Oct 1667
99r-100v	From [Bernart] at The Hague 2pp (17x22) fol 100, blank	17 Nov 1667
101rv	From [Bernart] at The Hague ⅔p (17x22)	24 Nov 1667

102r-103v	News account The Hague $2\frac{2}{3}$pp (20x31)	17 Nov 1667
104r-105v	News account The Hague $2\frac{2}{3}$pp (20x31)	24 Nov 1667
106r-107v	News account, annotated The Hague 3pp (20x31)	1 Dec 1667
108rv&115rv	From [Bernart] at The Hague 2pp & 8 lines (17x22)	8 Dec 1667
109rv	From [Bernart] at The Hague $\frac{1}{2}$p (17x22)	15 Dec 1667
110rv&114rv	News account The Hague 2pp & 6 lines (20x31)	15 Dec 1667
111r-112v	News account The Hague $2\frac{1}{2}$pp (20x31)	22 Dec 1667
113rv	From [Bernart] at The Hague 1p (17x22)	22 Dec 1667
116r-117v	News account The Hague 3pp + ad (20x31)	29 Dec 1667
118r-119v	News account, annotated The Hague $2\frac{1}{2}$pp + ad (20x31)	5 Jan 1668
120r-121v	News account, annotated The Hague $2\frac{2}{3}$pp + ad (20x31)	12 Jan 1668
122rv&125rv	From [Bernart] at The Hague $1\frac{2}{3}$pp (17x22) fol 125, blank	19 Jan 1668
123r-124v	From [Bernart] at The Hague $2\frac{2}{3}$pp (20x31)	19 Jan 1668

126r-127v	News Account, annotated The Hague 3pp (20x31)	26 Jan 1668
128rv&131rv	From [Bernart] at The Hague 2⅓pp (20x31)	2 Feb 1668
129r-130v	Newsletter The Hague 2½pp (20x31)	9 Feb 1668
132rv&136rv	From [Bernart] at The Hague 2pp & 3 lines + ad (17x22)	16 Feb 1668
133rv	Newsletter The Hague 1½pp (20x31)	23 Feb 1668
134r-135v	News Account, annotated The Hague 2⅔pp (20x31)	1 Mar 1668
137r-138v	News Account, annotated The Hague 3pp (20x31)	8 Mar 1668
139rv&144rv	News Account, annotated The Hague 2⅓pp + ad (20x31)	15 Mar 1668
140rv&143rv	From [Bernart] at The Hague 2pp & 6 lines (20x31)	22 Mar 1668
141r-142v	From [Bernart] at The Hague 2½pp (20x31)	29 Mar 1668
145r-146v&154rv	Fragments fol 145, 1p & 6 lines (20x31) fols 146&154 (17x22) fol 154, blank	undated
147rv&153rv	From [Bernart] at The Hague 2pp & 7 lines (20x31)	5 Apr 1668
148rv&152rv	From [Bernart] at The Hague 2pp & 7 lines (20x31)	19 Apr 1668
149rv&151rv	From [Bernart] at The Hague 2pp & 3 lines (20x31)	12 Apr 1668

150rv	From [Bernart] at The Hague 1p (20x31)	26 Apr 1668
155r-156v	From [Bernart] at The Hague To [J-A de Thou?] 2⅓pp (20x31)	19 Apr 1668
157rv	From [Bernart] at The Hague 1⅔pp (20x31)	3 May 1668
158rv	From [Bernart] at The Hague 1½pp (20x31)	10 May 1668
159rv	From [Bernart] at The Hague 1p (20x31)	17 May 1668
160rv	From [Bernart] at The Hague 1½pp (20x31)	24 May 1668
161rv	From [Bernart] at The Hague ¾p (20x31)	31 May 1668
162rv	From [Bernart] at The Hague 1¼pp (20x31)	7 Jun 1668
163rv	From [Bernart] at The Hague 1½pp + ad (20x31)	14 Jun 1668
164rv	From [Bernart] at The Hague 1½pp + ad (20x31)	21 Jun 1668
165rv	From [Bernart] at The Hague 1½pp (20x31)	28 Jun 1668
166rv	From [Bernart] at The Hague 1½pp (20x31)	5 Jul 1668
167rv	From [Bernart] at The Hague 1¼pp + ad (20x31)	12 Jul 1668
168rv	From [Bernart] at The Hague 1½pp (20x31)	19 Jul 1668
169rv	From [Bernart] at The Hague ⅔p (20x31)	9 Aug 1668
170r-171v	From [Bernart] at The Hague 2⅓pp + ad (20x31)	23 Aug 1668

172rv	From [Bernart] at The Hague $1\frac{1}{3}$pp (20x31)	16 Aug 1668
173rv&176rv	From [Bernart] at The Hague $2\frac{1}{4}$pp + ad (20x31)	30 Aug 1668
174r-175v	From Bernart at The Hague To [J-A de Thou?] $2\frac{1}{2}$pp (20x31)	30 Aug 1668
177r-178v	From [Bernart] at The Hague To [J-A de Thou?] $2\frac{1}{2}$pp (20x31)	6 Sep 1668
179rv	From [Bernart] at The Hague $1\frac{1}{2}$pp (20x31)	6 Sep 1668
180rv&183rv	From [Bernart] at The Hague 2pp & 3 lines + ad (20x31)	6 Sep 1668
181r-182v	From Bernart at The Hague To [J-A de Thou?] 3pp (20x31)	6 Sep 1668
184rv&187rv	From [Bernart] at The Hague 2pp (20x31) fol 187, blank	13 Sep 1668
185r-186v	News account $2\frac{3}{4}$pp (18x25)	13 Sep 1668
188rv	From [Bernart] at The Hague 2pp (20x31)	20 Sep 1668
189r-190v	From [Bernart] at The Hague To [J-A de Thou?] 2pp & 1 line (20x31)	20 Sep 1668
191rv	Missing	
192r-193v	Anonymous newsletter The Hague 3pp (17x22) Italian	26 Sep 1668
194rv	From [Bernart] at The Hague To [J-A de Thou?] 2pp (20x31)	27 Sep 1668

195rv	From [Bernart] at The Hague 1⅔pp (20x31)	27 Sep 1668
196rv	From [Bernart] at The Hague 1p & 6 lines (20x31)	4 Oct 1668
197r-198v	From [Bernart] at The Hague To [J-A de Thou?] 2pp & 2 lines (20x31)	4 Oct 1668
199rv	"Mémoire" ¾p (20x31)	undated
200r-201v	From [Bernart] at The Hague 1½pp + ad (20x31)	11 Oct 1668
202rv	From [Bernart at The Hague] 1⅔p (20x31)	16 Oct 1668
203rv	From [Bernart] at The Hague 1½pp (20x31)	25 Oct 1668
204rv&207rv	From [Bernart] at The Hague 2pp & 4 lines (20x31)	1 Nov 1668
205r-206v	From [Bernart] at The Hague 2pp + ad (20x31)	8 Nov 1668
208rv	From [Bernart] at The Hague 2pp (20x31)	15 Nov 1668
209rv	From [Bernart] at The Hague 2pp (20x31)	29 Nov 1668
210rv	From [Bernart] at The Hague 1½pp (20x31)	22 Nov 1668
211rv	From [Bernart] at The Hague 1¾pp (20x31)	6 Dec 1668
212rv	From [Bernart] at The Hague 1⅔pp (20x31)	20 Dec 1668
213rv	From [Bernart] at The Hague 1p & 5 lines (20x31)	13 Dec 1668
214r-215v	From [Bernart] at The Hague 1½pp + ad (20x31)	27 Dec 1668

216rv	From [Bernart] at The Hague 1½pp (20x31)	3 Jan 1669
217r-218v	From [Bernart] at The Hague 2⅓pp (20x31)	24 Jan 1669
219rv	From [Bernart] at The Hague 1p (20x31)	31 Jan 1669
220r-223v	From [Bernart] at The Hague* 4pp & 5 lines + ad (20x31) *[includes copy of a letter]	7 Feb 1669
224r-225v	From [Bernart] at The Hague 3pp (20x31)	14 Feb 1669
226rv	From [Bernart] at The Hague 1½pp (20x31)	21 Feb 1669
227rv	From [Bernart] at The Hague 1p (20x31)	28 Feb 1669
228rv&232rv	Newsletter The Hague 1½pp + ad (20x31)	28 Feb 1669
229rv	From [Bernart] at The Hague 1p + ad (20x31)	7 Mar 1669
230r-231v	From [Bernart] at The Hague 1⅓pp + ad (20x31)	14 Mar 1669
233rv&236rv	News account, annotated The Hague 1½pp & 6 lines + ad (20x31)	21 Mar 1669
234r-235v	News account, annotated The Hague 2⅔pp & 6 lines + ad (20x31)	28 Mar 1669
237r-238v	News account The Hague 2½pp + ad (20x31)	4 Apr 1669
A239rv*	From [Bernart] at The Hague ⅔p (15x20) *[two fols are numbered 239]	4 Apr 1669

B239rv*&242rv	News account, annotated The Hague 2pp & 6 lines + ad (20x31) *[two fols are numbered 239]	11 Apr 1669
240r-241v	News account, annotated The Hague $2\frac{1}{3}$pp + ad (20x31)	18 Apr 1669
243rv&246rv	News account The Hague $2\frac{1}{4}$pp (20x31)	25 Apr 1669
244r-245v	From [Bernart] at The Hague $2\frac{1}{3}$pp + ad (20x31)	2 May 1669
247rv	From [Bernart] at The Hague $\frac{2}{3}$p (20x31)	16 May 1669
248rv	Newsletter The Hague 1p & 4 lines (20x31)	16 May 1669
249r-250v	News account The Hague $2\frac{2}{3}$pp (20x31) *[dated by Bernart]	23 May 1669*
251r-252v	News account The Hague $2\frac{2}{3}$pp (20x31) *[dated by Bernart]	29 May 1669*
253rv	From [Bernart] at The Hague $\frac{2}{3}$p (17x22)	6 Jun 1669
254r-255v	News account The hague $2\frac{1}{2}$pp (20x31)	6 Jun 1669
256r-257v	News account The Hague $2\frac{2}{3}$pp (20x31)	13 Jun 1669
258r-259v	News account $2\frac{1}{2}$pp (20x31)	20 Jun 1669

260rv	From [Bernart] at The Hague 1p (17x22)	27 Jun 1669
261r-262v	News account The Hague $2\frac{1}{3}$pp (20x31)	27 Jun 1669
263r-264v	News account annotated The Hague $2\frac{2}{3}$pp (20x31)	4 Jul 1669
265r-266v	News account annotated The Hague 3pp (20x31)	11 Jul 1669
267r-268v	News account annotated The Hague 3pp (20x31)	18 Jul 1669
269r-270v	News account The Hague $2\frac{1}{2}$pp (20x31)	25 Jul 1669
271rv&274rv	From [Bernart] at The Hague $2\frac{1}{3}$pp (20x31)	22 Aug 1669
272r-273v	News account The Hague $2\frac{1}{3}$pp (20x31)	26 Aug 1669
275r-276v	Newsletter The Hague 3pp (20x31)	16 Oct 1669
277rv&279rv	From [Bernart] at The Hague 1p (17x22) fol 279, blank	16 Oct 1669
278r-A278v*	News account The Hague $2\frac{1}{2}$pp (20x31) *[fol A278, unnumbered folio]	24 Oct 1669
280rv	From Bernart at The Hague $\frac{2}{3}$p (15x20)	31 Oct 1669
281r-282v	News account $2\frac{1}{2}$pp (20x31)	31 Oct 1669

283r-284v	From [Bernart] at The Hague 2½pp (20x31)	7 Nov 1669
285r-286v	News account, annotated The Hague 2⅓pp (20x31)	14 Nov 1669
287r-288v	News account The Hague 2⅓pp (20x31)	21 Nov 1669
289r-290v	News account The Hague 2⅓pp (20x31)	28 Nov 1669
291r-292v	News account The Hague 3½pp (20x31)	[19?] Nov 1669
293r-294v	News account The Hague 3pp (20x31)	5 Dec 1669
295r-296v	News account The Hague 2⅓pp (20x31)	12 Dec 1669
297r-298v	News account The Hague 2⅓pp (20x31)	19 Dec 1669

▼ ▼ ▼

COLLECTION BOULLIAU, VOLUME XXX, FONDS FRANÇAIS 13048

Folio	Description	Date
1r-2v	From [Bernart] at The Hague 3pp (20x31)	2 Jan 1670
3rv	From [Bernart] at The Hague $1\frac{1}{3}$pp (20x31)	9 Jan 1670
4rv	From [Bernart] at The Hague $1\frac{1}{2}$pp (20x31)	16 Jan 1670
5rv	From [Bernart] at The Hague $1\frac{1}{2}$pp (20x31)	30 Jan 1670
6rv	From [Bernart] at The Hague $1\frac{2}{3}$pp (20x31)	6 Feb 1670
7r-8v	From [Bernart] at The Hague $2\frac{1}{2}$pp (17x22)	13 Feb 1670
9rv	From [Bernart] at The Hague 2pp (17x22)	20 Feb 1670
10r-11v	News account, annotated The Hague $2\frac{2}{3}$pp (20x31)	27 Feb 1670
12r-13v	News account, annotated The Hague $2\frac{2}{3}$pp (20x31)	6 Mar 1670
14rv	From [Bernart] at The Hague $\frac{3}{4}$p (17x22)	20 Mar 1670
15r-16v	News account The Hague $2\frac{1}{2}$pp (20x31)	undated
17r-18v	News account, annotated The Hague $2\frac{2}{3}$pp (20x31)	13 Mar 1670
19rv	News account The Hague $1\frac{2}{3}$pp (17x22)	17 Apr 1670

20r-21v	News account, annotated The Hague 2½pp (20x31)	10 Apr 1670
22r-23v	News account, annotated The Hague 2⅔pp (20x31)	3 Apr 1670
24r-25v	News account, annotated The Hague 2⅔pp (20x31)	27 Mar 1670
26rv	From [Bernart] at The Hague 1½pp (17x22)	24 Apr 1670
27r-28v	From [Bernart] at The Hague 2⅔pp (20x31)	1 May 1670
29rv	From [Bernart] at The Hague 1p & 3 lines (20x31)	8 May 1670
30rv	From [Bernart] at The Hague 1½pp (17x22)	16 May 1670
31rv	From [Bernart] at The Hague 1⅓pp (17x22)	22 May 1670
32rv	From [Bernart] at The Hague 2pp (20x31)	29 May 1670
33rv	From [Bernart] at The Hague 1⅔pp (20x31)	5 Jun 1670
34rv&40rv	From [Bernart] at The Hague 2pp & 7 lines + ad (20x31)	12 Jun 1670
35rv	From [Bernart] at The Hague 1⅔pp (20x31)	19 Jun 1670
36rv&41rv	From [Bernart] at The Hague 2pp & 2 lines + ad (20x31)	26 Jun 1670
37rv	From [Bernart] at The Hague 2pp (20x31)	3 Jul 1670
38r-39v	From [Bernart] at The Hague 2⅓pp + ad (20x31)	10 Jul 1670

42rv	From [Bernart] at The Hague 1p & 4 lines (17x22)	17 Jul 1670
43r-44v	From [Bernart] at The Hague 2½pp (20x31)	24 Jul 1670
45rv	From [Bernart] at The Hague 2pp (20x31)	31 Jul 1670
46rv	From [Bernart] at The Hague 1⅔pp (20x31)	7 Aug 1670
47r-48v	From [Bernart] at The Hague 3pp (20x31)	12 Aug 1670
49rv	News account The Hague 1½pp (20x31)	4 Sep 1670
50rv	From [Bernart] at The Hague ½p + ad (20x31)	4 Sep 1670
51rv	News account The Hague 2pp (20x31)	11 Sep 1670
52rv&55rv	News account The Hague 2pp & 8 lines + ad (20x31)	18 Sep 1670
53r-54v	News account, annotated The Hague 2⅓pp + ad (20x31)	25 Sep 1670
56r-57v	From [Bernart] at The Hague 2½pp (20x31)	2 Oct 1670
58r-59v	From [Bernart] at The Hague 3pp (20x31)	9 Oct 1670
60r-61v	News account, annotated The Hague 2½pp (20x31)	16 Oct 1670
62r-63v	News account, annotated The Hague 2⅓pp + ad (20x31)	23 Oct 1670

64r-65v	From [Bernart] at The Hague 3pp (20x31)	30 Oct 1670
66rv	From [Bernart] at The Hague 2pp (20x31)	6 Nov 1670
67r-68v	From [Bernart] at The Hague 3pp (20x31)	13 Nov 1670
69r-70v	From [Bernart] at The Hague 3pp (20x31)	20 Nov 1670
71r-72v	From [Bernart] at The Hague 2½pp (20x31)	27 Nov 1670
73r-74v	News account The Hague 3pp (20x31)	4 Dec 1670
75rv&80rv	From [Bernart] at The Hague 2pp & 6 lines + ad (20x31)	11 Dec 1670
76rv	From [Bernart] at The Hague 2pp (20x31)	18 Dec 1670
77rv	From [Bernart] at The Hague 2pp (20x31)	24 Dec 1670
78r-79v	From [Bernart] at The Hague 2⅓pp + ad (20x31)	1 Jan 1671
81r-82v	From [Bernart] at The Hague 2½pp + ad (20x31)	8 Jan 1671
83r-84v	From [Bernart at The Hague] 2⅔pp + ad (20x31)	15 Jan 1671
85rv	From [Bernart] at The Hague 2pp (20x31)	22 Jan 1671
86rv	From [Bernart] at The Hague 1⅔pp (20x31)	29 Jan 1671
87r-88v	From [Bernart] at The Hague 2⅔pp + ad (20x31)	5 Feb 1671
89r-90v	From [Bernart] at The Hague 3pp (20x31)	12 Feb 1671

91r-92v	News account, annotated The Hague 2¾pp (20x31) *[corrected from 1670 by B]	19 Feb 1671*
93r-94v	News account, annotated The Hague 2¾pp (20x31)	26 Feb 1671
95r-96v	From [Bernart] at The Hague 2½pp + ad (20x31)	5 Mar 1677
97rv	News account The Hague 2pp (20x31)	12 Mar 1671
98r-99v	News account The Hague 2⅔pp + ad (20x31)	2 Apr 1671
100rv	From [Bernart] at The Hague 1⅔pp (17x22)	9 Apr 1671
101r-102v	News account, annotated The Hague 2½pp + ad (20x31)	16 Apr 1671
103r-104v	News account The Hague 2⅔pp (20x31)	23 Apr 1671
105r-106v	News account, annotated The Hague 3pp (20x31)	30 Apr 1671
107r-108v	News account, annotated The Hague 2½pp + ad (20x31)	7 May 1671
109r-110v	News account, annotated The Hague 2¾pp (20x31)	14 May 1671
111r-112v	News account, annotated The Hague 2⅔pp + ad (20x31)	21 May 1671

Folios	Description	Date
113r-114v	News account, annotated The Hague 2½pp (20x31)	28 May 1671
115r-116v	News account, annotated The Hague 2¾pp (20x31)	4 Jun 1671
117r-118v	News account, annotated The Hague 2¾pp (20x31)	11 Jun 1671
119r-120v	News account, annotated The Hague 2½pp + ad (20x31)	18 Jun 1671
121r-122v	News account, annotated The Hague 2½pp (20x31)	25 Jun 1671
123r-124v	News account, annotated The Hague 2½pp (20x31)	2 Jul 1671
125r-126v	News account, annotated The Hague 2⅓pp (20x31)	9 Jul 1671
127r-128v	News account, annotated The Hague 2½pp (20x31)	16 Jul 1671
129r-130v	News account, annotated The Hague 2⅔pp + ad (17x22)	23 Jul 1671
131r-132v	News account, annotated The Hague 2½pp (20x31)	6 Aug 1671
133rv&137rv	News account, annotated The Hague 2pp & 7 lines (20x31) fol 137, 7 lines by Bernart	13 Aug 1671
134rv	News account The Hague 2pp (17x22)	20 Aug 1671

135r-136v	News account, annotated The Hague 4pp (17x22)	27 Aug 1671
138rv&141rv	News account, annotated The Hague 2½pp + ad (17x22)	3 Sep 1671
139r-140v	News account, annotated The Hague 4pp (17x22)	10 Sep 1671
142r-143v	News account The Hague 2⅔pp (17x22)	17 Sep 1671
144r-145v	News account, annotated The Hague 3⅔pp (17x22)	24 Sep 1671
146r-147v	News account, annotated The Hague 2⅔pp (17x22)	1 Oct 1671
148r-149v	News account, annotated The Hague 3pp (17x22)	8 Oct 1671
150r-151v	News account, annotated The Hague 3½pp (17x22)	15 Oct 1671
152r-153v	News account, annotated The Hague 2½pp (17x22)	22 Oct 1671
154r-155v	News account, annotated The Hague 3pp (17x22)	29 Oct 1671
156r-157v	News account, annotated The Hague 3pp (17x22)	5 Nov 1671
158r-159v	News account, annotated The Hague 3pp (17x22)	12 Nov 1671

Folio	Description	Date
160rv	News account, annotated The Hague 2pp (17x22)	26 Nov 1671
161rv	News account, annotated The Hague 2pp (17x22)	3 Dec 1671
162rv&165rv	News account, annotated The Hague 2pp & 9 lines (17x22)	10 Dec 1671
163r-164v	News account, annotated The Hague 3pp (17x22)	17 Dec 1671
166r-167v	News account, annotated The Hague 3pp (17x22)	24 Dec 1671
168r-169v	News account, annotated The Hague 3pp (17x22)	31 Dec 1671
170r-171v	News account, annotated The Hague 3pp (17x22)	7 Jan 1672
172r-173v	News account, annotated The Hague 3pp & 6 lines (17x22)	14 Jan 1672
174r-175v	News account, annotated The Hague $3\frac{2}{3}$pp (17x22)	21 Jan 1672
176r-177v	News account, annotated The Hague $3\frac{3}{4}$pp (17x22)	28 Jan 1672
178r-179v	News account, annotated The Hague 3pp & 5 lines (17x22)	4 Feb 1672
180r-181v	News account The Hague $3\frac{1}{2}$pp (17x22)	11 Feb 1672

182r-183v	News account, annotated The Hague 3½pp (17x22)	18 Feb 1672
184r-185v	News account, annotated The Hague 3pp (17x22)	25 Feb 1672
186r-187v	News account, annotated The Hague 2⅔pp (17x22)	3 Mar 1672
188r-189v	News account The Hague 2½pp (17x22)	10 Mar 1672
190r-191v	News account, annotated The Hague 2½pp (17x22)	17 Mar 1672
192r-193v	News account, annotated The Hague 3pp (17x22)	24 Mar 1672
194r-195v	News account, annotated The Hague 2½pp (17x22)	31 Mar 1672
196r-197v	News account, annotated The Hague 3pp (17x22)	7 Apr 1672
198r-199v	News account, annotated Utrecht 3pp (17x22)	9 Aug 1672
200r-201v	News account, annotated Utrecht 3pp (17x22)	16 Aug 1672
202r-203v	News account, annotated Utrecht 2⅔pp (17x22)	23 Aug 1672
204rv	From [Bernart] at Utrecht 1½pp (17x22)	26 Aug 1672

205rv&207rv	News account Utrecht 3pp (17x22)	30 Aug 1672
206rv	From [Pieter Pels] at The Hague 1p + ad (16x20)	28 Aug 1672
208r-209v	News account, annotated Utrecht 3pp (17x22)	6 Sep 1672
210r-211v	News account, annotated Utrecht 3pp (17x22)	13 Sep 1672
212r-213v	News account, annotated Utrecht 3pp (17x22)	20 Sep 1672
214rv&218rv	News account, annotated Utrecht 2pp & 4 lines (17x22)	27 Sep 1672
215rv	From [Bernart] at Utrecht 2pp (17x22)	7 Oct 1672
216r-217v	From [Bernart] at Utrecht 2½pp (17x22)	14 Oct 1672
219rv	From [Bernart] at Utrecht 1½pp (17x22)	25 Oct 1672
220rv	From [Bernart] at Utrecht 1½pp (17x22)	1 Nov 1672
221rv	From [Bernart] at Utrecht 2pp (17x22)	15 Nov 1672
222r-223v	From [Bernart] at Utrecht 2½pp (17x22)	22 Nov 1672
224r-225v	From [Bernart] at Utrecht 2¾pp + ad (17x22)	29 Nov 1672
226rv&231rv	From [Bernart] at Utrecht 2pp & 5 lines + ad (17x22)	6 Dec 1672

227rv	From [Bernart] at Utrecht 2pp (17x22)	13 Dec 1672
228rv&230rv	From [Bernart] at Utrecht 2pp & 5 lines + ad (17x22)	20 Dec 1672
229rv	Newsletter Cologne ½p (17x22)	13 Dec 1672
232r-233v	From [Bernart] at Utrecht 2⅓pp + ad (17x22)	27 Dec 1672
234rv	From [Bernart] at Utrecht 1⅔pp (17x22)	3 Jan 1673
235rv&239rv	From [Bernart] at Utrecht 2pp & 8 lines + ad (17x22)	10 Jan 1673
236rv&238rv	From [Bernart] at Utrecht 2pp & 3 lines + ad (17x22)	24 Jan 1673
237rv	From [Bernart] at Utrecht 1½pp (17x22)	31 Jan 1673
240rv	From [Bernart] at Utrecht 1½pp (17x22)	7 Feb 1673
241rv	From [Bernart] at Utrecht 1⅔pp (17x22)	14 Feb 1673
242rv	From [Bernart] at Utrecht 1⅔pp (17x22)	21 Feb 1673
243rv	From [Bernart] at Utrecht 2pp (17x22)	28 Feb 1673
244rv	From [Bernart] at Utrecht 2pp (17x22)	7 Mar 1673
245rv	From [Bernart] at Utrecht 1½pp (17x22)	14 Mar 1673
246rv	From [Bernart] at Utrecht 2pp (17x22)	21 Mar 1673
247rv	From [Bernart] at Utrecht 1⅓pp (17x22)	28 Mar 1673

248rv	From [Bernart] at Utrecht 1½pp (17x22)	4 Apr 1673
249rv	From [Bernart] at Utrecht 2pp (17x22)	11 Apr 1673
250rv	From [Bernart] at Utrecht 1½pp (17x22)	17 Apr 1673
251rv	From [Bernart] at Utrecht 2pp (17x22)	25 Apr 1673
252r-253v	From [Bernart] at Utrecht 3pp (20x31)	23 Apr 1673
254rv	From [Bernart] at Cologne 1⅔pp (17x22)	13 Jun 1673
255rv	From [Bernart] at Cologne 2pp (20x31)	17 Jun 1673
256rv	From [Bernart] at Cologne 2pp (17x22)	24 Jun 1673
257rv	From [Bernart] at Cologne 1¾pp (17x22)	8 Jul 1673
258rv	From [Bernart] at Cologne 1⅔pp (17x22)	22 Jul 1673
259rv&263rv	From [Bernart] at Cologne 2pp & 7 lines + ad (17x22)	28 Jul 1673
260rv	From [Bernart] at Cologne 1⅓pp (17x22)	4 Aug 1673
261rv	From [Bernart] at Cologne 1½pp (17x22)	11 Aug 1673
262rv	From [Bernart] at Cologne 1p & 5 lines (17x22)	18 Aug 1673
264rv	From [Bernart] at Cologne 1½pp (17x22)	2 Sep 1673
265rv	From [Bernart] at Cologne 1½pp (17x22)	8 Sep 1673

266rv	From [Bernart] at Cologne $1\frac{3}{4}$pp (17x22)	16 Sep 1673
267rv	From [Bernart] at Cologne $1\frac{2}{3}$pp (17x22)	23 Sep 1673
268rv	From [Bernart] at Cologne $1\frac{1}{2}$pp (17x22)	30 Sep 1673
269rv	From [Bernart] at Cologne $1\frac{1}{2}$pp (17x22)	7 Oct 1673
270rv	From [Bernart] at Cologne $1\frac{2}{3}$pp (17x22)	14 Oct 1673
271rv	From [Bernart] at Cologne $1\frac{1}{2}$pp (17x22)	21 Oct 1673
272rv	From [Bernart] at Cologne $1\frac{1}{2}$pp (17x22)	28 Oct 1673
273rv	From [Bernart] at Cologne $1\frac{2}{3}$pp (17x22)	11 Nov 1673
274r-275v	From [Bernart] at Cologne $2\frac{1}{2}$pp + ad (17x22)	24 Nov 1673
276rv	From [Bernart] at Cologne 2pp (17x22)	2 Dec 1673
277rv	From [Bernart] at Cologne 2pp (17x22)	12 Dec 1673
278rv	From [Bernart] at Cologne $1\frac{1}{2}$pp (17x22)	19 Dec 1673
279rv	From [Bernart] at Cologne $1\frac{3}{4}$pp (17x22)	26 Dec 1673
280rv	From [Bernart] at Cologne 2pp (17x22)	6 Jan 1674
281rv	From [Bernart] at Cologne $1\frac{3}{4}$pp (17x22)	13 Jan 1674
282rv&289rv?	From [Bernart] at Cologne 2pp & 7 lines + ad (17x22)	20 Jan 1674

283rv	From [Bernart] at Cologne 1½pp (17x22)	26 Jan 1674
284rv	From [Bernart] at Cologne 2pp (17x22)	2 Feb 1674
285rv&288rv	From [Bernart] at Cologne 1½pp + ad (17x22)	9 Feb 1674
286r-287v	From Bernart at Cologne 2¾pp + ad (17x22)	16 Feb 1674
290rv	From [Bernart] at Cologne 1⅔pp (17x22)	20 Feb 1674
291rv	From [Bernart] at Cologne 1¾pp (17x22)	27 Feb 1674
292r-293v	From [Bernart] at Cologne 2½pp + ad (17x22)	9 Mar 1674
294r-295v	From [Bernart] at Cologne 2½pp + ad (17x22)	16 Mar 1674
296r-297v	From [Bernart] at Cologne 2⅓pp + ad (17x22)	23 Mar 1674
298rv&302rv	From [Bernart] at Cologne 2pp & 6 lines + ad (17x22)	30 Mar 1674
299rv	From [Bernart] at Cologne 1⅔pp (17x22)	6 Apr 1674
300rv	From [Bernart] at Cologne 2pp (17x22) B notes: "Mr Bernartz arriva a Paris au mois de May 1674."	13 Apr 1674
301rv	From Bernart at The Hague 2pp (17x22)	19 Oct 1679
303r-305v	From [Bernart] at The Hague 4½pp + ad (17x22)	2 Nov 1679
306rv&309rv	From [Bernart at The Hague] 1p + ad (17x22)	9 Nov 1679

307r-308v	From [Bernart at The Hague] 2½pp + ad (17x22)	18 Jan 1680
310rv	From [Bernart at The Hague] 1⅓pp (17x22)	8 Feb 1680
311rv	From Bernart at The Hague ¾p (17x22)	19 Feb 1680
312r-313v	From [Bernart at The Hague] 3pp + ad (17x22)	2 May 1680
314rv	From [Bernart at The Hague] 1¾pp (17x22)	9 May 1680
315rv	From [Bernart at The Hague] 2pp (17x22]	16 May 1680
316r-317v	From [Bernart at The Hague] 3pp + ad (17x22)	23 May 1680
318r-319v	From [Bernart] at The Hague 2⅔pp + ad (17x22)	30 May 1680
320r-321v	From [Bernart at The Hague] 3pp + ad (17x22)	6 Jun 1680
322rv	From [Bernart at The Hague] 2pp (17x22)	13 Jun 1680
323rv	From [Bernart at The Hague] 1½pp (17x22)	20 Jun 1680
324r-325v	From [Bernart at The Hague] 3pp + ad (17x22)	27 Jun 1680
326r-327v	From [Bernart at The Hague] 3pp + ad (17x22)	4 Jul 1680
328r-329v	From [Bernart at The Hague] 2⅔pp + ad (17x22)	18 Jul 1680
330r-331v	From [Bernart at The Hague] 2pp & 5 lines + ad (17x22)	23 Jul 1680
332rv&334rv	From [Bernart at The Hague] 2pp & 5 lines + ad (17x22)	1 Aug 1680

333rv	Newsletter 1⅔pp (17x22)	23 Jul 1680
335r-336v	From [Bernart at The Hague] 2½pp + ad (17x22)	8 Aug 1680
337r-338v	Newsletter, Anonymous London 2½pp (17x22)	6 Aug 1680
339r-340v	From [Bernart at The Hague] 2⅔pp + ad (17x22)	12 Aug 1680
341r-342v	From [Bernart at The Hague] 3pp + ad (17x22)	15 Aug 1680
343rv	From [Bernart at The Hague] 1¾pp (17x22)	22 Aug 1680
344rv&346rv	From [Bernart at The Hague] 2⅓pp (20x31)	29 Aug 1680
345rv	From [Bernart at The Hague] 1⅓pp (17x22)	2 Sep 1680
347rv&350rv	From [Bernart at The Hague] 2½pp + ad (20x31)	5 Sep 1680
348r-349v	From [Bernart at The Hague] 2⅔pp + ad (17x22)	12 Sep 1680
351r-352v	From [Bernart at The Hague] 2½pp + ad (20x31)	19 Sep 1680
353rv&356rv	From [Bernart at The Hague] 2½pp + ad (17x22)	23 Sep 1680
354rv	Printed document 1p (7x12) "Charge Generale...de la Compagnie des Indes Orientales"	15 Mar 1680
355rv	Printed document 1p (15x22) "L'Année 1680/La Compagnie des Indes Orientales..."	

357r-358v	From [Bernart at The Hague] $2\frac{2}{3}$pp + ad (15x22) [358r, B notes: "Mr Bernards mourut l' 11e Octobre cette annee 1680"]	26 Sep 1680

▼ ▼ ▼

COLLECTION BOULLIAU, VOLUME XXXI, FONDS FRANÇAIS 13049

Folio	Description	Date
1r-2v	Newsletter Paris 3¼pp (18x25)	1 Oct 1660
3rv	From Valeran at Rome 1p (20x27)	18 Oct 1655
4rv	From Valeran at Rome 1p (20x27)	19 Dec 1656
5rv	From Valeran at Rome ⅔p (20x27)	2 Jan 1657
6r-7v	From Valeran at Rome 1⅓pp + ad (20x27)	30 Jan 1657
8rv	From Valeran at Rome 1p (20x27)	13 Feb 1657
9rv	From Valeran at Rome 1⅔pp (20x27)	20 Feb 1657
10rv	From [Valeran] at Rome 1p (20x27)	5 Nov 1657
11rv	From Valeran at Rome 1½pp (20x27)	12 Nov 1657
12rv	From [Valeran] at Rome 1p (20x27)	19 Nov 1657
13rv	From Valeran at Rome 1⅓pp (20x27)	26 Nov 1657
14rv	From [Valeran] at Rome 1p (20x27)	3 Dec 1657
15rv	From Valeran at Rome 1p (20x27)	10 Dec 1657
16rv	From [Valeran] at Rome 1p (20x27)	7 Jan 1658
17rv	From Valeran at Rome 1p (20x27)	22 Jan 1658

18rv	From [Valeran] at Rome 1½pp (20x27)	12 Feb 1658
19rv	From [Valeran] at Rome 1p & 6 lines (20x27)	19 Feb 1658
20rv	From [Valeran] at Rome 1p (20x27)	25 Feb 1658
21rv	From [Valeran] at Rome 1p (20x27)	5 Mar 1658
22rv	From [Valeran] at Rome 1p (20x27)	12 Mar 1658
23rv	From [Valeran] at Rome ⅔p (20x27)	19 Mar 1658
24rv	From [Valeran] at Rome 1p (20x27)	26 Mar 1658
25rv	From [Valeran] at Rome 1p (20x27)	2 Apr 1658
26rv	From [Valeran] at Rome 1p & 6 lines (20x27)	9 Apr 1658
27rv	From [Valeran] at Rome 1p (20x27)	16 Apr 1658
28rv	From [Valeran] at Rome 1p & 6 lines (20x27)	23 Apr 1658
29rv	From Valeran at Rome 1p (20x27)	30 Apr 1658
30rv	From Valeran at Rome 1p (20x27)	7 May 1658
31rv	From [Valeran] at Rome 1p (20x27)	14 May 1658
32rv	From [Valeran] at Rome 1p (20x27)	21 May 1658
33rv	From [Valeran] at Rome 1p (20x27)	28 May 1658

34rv	From [Valeran] at Rome 1p (20x27)	11 Jun 1658
35rv	From [Valeran] at Rome 1p & 5 lines (20x27)	2 Jul 1658
36rv	From [Valeran] at Rome 1p & 7 lines (20x27)	9 Jul 1658
37rv	From [Valeran] at Rome 1p (20x27)	15 Oct 1658
38rv	From [Valeran] at Rome 1p (20x27)	3 Dec 1658
39rv	From [Valeran] at Rome 1p & 7 lines (20x27)	10 Dec 1658
40rv	From [Valeran] at Rome 1⅔pp (20x27)	17 Dec 1658
41rv	From [Valeran] at Rome 1p (20x27)	24 Dec 1658
42rv	From Valeran at Rome 1p (20x27)	15 Jan 1659
43rv	From [Valeran] at Rome 1p (20x27)	29 Jan 1659
44rv	From [Valeran] at Rome 1p (20x27)	5 Feb 1659
45rv	From [Valeran] at Rome 1⅔pp (20x27)	26 Feb 1659
46rv	From [Valeran] at Rome 1p (20x27)	5 Mar 1659
47rv	From [Valeran] at Rome 1p (20x27)	1 Apr 1659
48rv	From [Valeran] at Rome 1p (20x27)	8 Apr 1659
49rv	From [Valeran] at Rome 1p (20x27)	15 Apr 1659

50rv	From [Valeran] at Rome 1p (20x27)	22 Apr 1659
51rv	From [Valeran] at Rome 1p (20x27)	29 Apr 1659
52rv	From [Valeran] at Rome 1p (20x27)	24 Jun 1659
53rv	From [Valeran] at Rome 1p (20x27)	1 Jul 1659
54rv	From [Valeran] at Rome 1p (20x27)	8 Jul 1659
55rv	From [Valeran] at Rome 1p (20x27)	15 Jul 1659
56rv	From [Valeran] at Rome 1p (20x27)	26 Aug 1659
57rv	From [Valeran] at Rome 1p & 8 lines (20x27)	2 Sep 1659
58rv	From Valeran at Rome 1p (20x27)	16 Sep 1659
59rv	From [Valeran] at Rome 1p (20x27)	8 Sep 1659
60rv	From [Valeran] at Rome 1⅔pp (20x27)	21 Oct 1659
61rv	From [Valeran] at Rome 1p (20x27)	28 Oct 1659
62rv	From [Valeran] at Rome 1p (20x27)	4 Nov 1659
63rv	From [Valeran] at Rome 1p & 3 lines (20x27)	11 Nov 1659
64rv	From [Valeran] at Rome 1p & 8 lines (20x27)	18 Nov 1659
65rv	From Valeran at Rome 1p (20x27)	25 Nov 1659

66rv	From [Valeran] at Rome 1p & 4 lines (20x27)	1 Dec 1659
67rv	From Valeran at Rome 1p (20x27)	9 Dec 1659
68rv	From [Valeran] at Rome 1p (20x27)	16 Dec 1659
69rv	From [Valeran] at Rome 1p (20x27)	23 Dec 1659
70rv	From Valeran at Rome 1p (20x27)	30 Dec 1659
71rv	From Valeran at Rome 1p (20x27)	6 Jan 1660
72rv	From [Valeran] at Rome 1p & 2 lines (20x27)	13 Jan 1660
73rv	From [Valeran] at Rome 1p (20x27)	20 Jan 1660
74rv	From [Valeran] at Rome 1p (20x27)	27 Jan 1660
75rv	From [Valeran] at Rome 1p (20x27)	3 Feb 1660
76rv	From [Valeran] at Rome 1p (20x27)	10 Feb 1660
77rv	From Valeran at Rome 1p (20x27)	17 Feb 1660
78rv	From Valeran at Rome 1p (20x27)	24 Feb 1660
79rv	From Valeran at Rome 1p (20x27)	2 Mar 1660
80rv	From [Valeran] at Rome 1p (20x27)	9 Mar 1660
81rv	From [Valeran] at Rome 1p (20x27)	16 Mar 1660

82rv	From [Valeran] at Rome 1p (20x27)	23 Mar 1660
83rv	From [Valeran] at Rome 1p (20x27)	30 Mar 1660
84rv	From Valeran at Rome 1p (20x27)	6 Apr 1660
85rv	From Valeran at Rome 1p (20x27)	13 Apr 1660
86rv	Unidentified note ⅔p (15x22)	undated
87rv	From Valeran at Rome 1p (20x27)	20 Apr 1660
88rv	From [Valeran] at Rome 1p (20x27)	27 Apr 1660
89rv	From [Valeran] at Rome 1p & 8 lines (20x27)	4 May 1660
90rv	From [Valeran] at Rome 1¾pp (20x27)	11 May 1660
91rv	From Valeran at Rome 1p & 4 lines (20x27)	18 May 1660
92rv	From Valeran at Rome 1p (20x27)	25 May 1660
93rv	From [Valeran] at Rome 1p (20x27)	1 Jun 1660
94rv	From [Valeran] at Rome 1p & 4 lines (20x27)	8 Jun 1660
95rv	From [Valeran] at Rome 1p (20x27)	15 Jun 1660
96rv	From [Valeran] at Rome 1p (20x27)	22 Jun 1660
97rv	From [Valeran] at Rome 1p (20x27)	29 Jun 1660

98rv	From Valeran at Rome 1p (20x27)	6 Jul 1660
99rv	From Valeran at Rome 1p (20x27)	13 Jul 1660
100rv	From [Valeran] at Rome 1⅓pp (20x27)	27 Jul 1660
101rv	From [Valeran] at Rome 1p (20x27)	20 Jul 1660
102rv	From Valeran at Rome 1p & 8 lines (20x27)	3 Aug 1660
103rv&108rv	From de Bacher To Valeran at [Rome] 1½pp + ad (14x20)	undated
104rv	From [Valeran] at Rome 1p (20x27)	31 Aug 1660
105rv	From [Valeran] at Rome 1p (20x27)	24 Aug 1660
106r-107v	From [Valeran] at Rome 1p (20x27) fol 107, blank	10 Aug 1660
109rv	From Leopoldo* at Florence 1p & 2 lines (20x27) Italian *[Cardinal de' Medici]	1 Mar 1667 Ab Inc [1 Mar 1668]
110rv	From Leopoldo* at Florence ½p (20x27) Italian *[Cardinal de' Medici]	15 Feb 1668 Ab Inc [15 Feb 1669]
111rv	From Leopoldo* at Florence 1p & 2 lines (20x27) Italian *[Cardinal de' Medici]	16 Sep 1672
112rv	Brief tract 1p & 6 lines (20x27) Italian	undated
113rv	From Leopoldo* at Florence ⅔p (20x27) Italian *[Cardinal de' Medici]	11 Nov 1672

114rv	Extract of a letter From Leopoldo at [Florence] To Anonymous 1p (20x27) Italian	undated
115rv	From Leopoldo at Florence 1p (20x27) Italian	18 Jan 1660 Ab Inc [18 Jan 1661]
116rv	Missing	
117rv	From Leopoldo at Florence 1p (20x27) Italian	1 Oct 1660
118r-119v	From [Leopoldo at Florence] 3⅔pp (20x27) Italian [Mémoire describing four physics instruments]	undated
120rv	From Leopoldo at Florence 1p & 3 lines (20x27) Italian	15 Sep 1662
121rv	From Leopoldo at Florence 2pp (20x27) Italian	3 Mar 1661 Ab Inc [3 Mar 1662]
122rv	From Leopoldo at Florence 7 lines (20x27) Italian	3 Nov 1662
123rv	From Leopoldo at Florence ¾p (20x27) Italian	15 Jun 1663
124r-125v	From Leopoldo at Florence 1p & 4 lines (20x27) Italian fol 125, blank	7 Sep 1663
126r-127v	Drawing: Path of Comet of 1664 2pp (29x39) Italian	[18-27] Dec 1664
128rv	Text of Observations of Comet of 1664 Obs at Bologna 1p (15x26) Italian	
129r-130v	Foldout: data list, location, time, degrees, minutes, sec, etc. Obs at Pisa 2pp (26x28) Italian	
131rv	Observations of Comet of 1664 Obs at Bologna & Rome 1½pp (20x37) Italian	

132rv&134rv	Folded drawing: Path of Comet of 1664 Obs at Rome 2pp (29x39) Italian	
133rv	Drawing: Close-up drawing of Comet of 1664 for Dec 27, 28, & 29 Obs at Bologna 1p (13x19) Italian	
135r-136v	Tract & data concerning the Comet of 1664 $2\frac{1}{2}$pp (20x27) Italian	
137rv&139rv	From [?] at Pisa $\frac{1}{2}$p + ad (20x27) Italian	25 Feb 1664 Ab Inc [25 Feb 1665]
138rv	From Leopoldo at Florence 1p & 5 lines (20x27) Italian	20 Feb 1664 Ab Inc [20 Feb 1665]
140rv	Sketch of Comet [by Des Noyers] 6 lines + drawing (17x22)	8 Apr 1665
141rv	From Leopoldo at Florence 7 lines (20x27) Italian	2 Jan 1664 Ab Inc [2 Jan 1665]
142rv	From Leopoldo at Florence 8 lines (20x27) Italian	8 Aug 1664
143rv	Observations of ☾ Eclipse by Grand Duke [Ferdinand] Obs at Florence 1p (20x27) Italian	8 Aug 1664
144rv	From Leopoldo at Florence $\frac{3}{4}$p (20x27) Italian	14 Nov 1664
145rv	From Leopoldo at Florence $1\frac{2}{3}$pp (20x27) Italian	27 Jun 1664
146rv	From Leopoldo at Florence $\frac{2}{3}$p (20x27) Italian	6 Mar 1664 Ab Inc [6 Mar 1665]
147rv	From Leopoldo at Florence $\frac{2}{3}$p (20x27) Italian	22 May 1665
148rv	From Leopoldo at Florence $1\frac{2}{3}$pp (20x27) Italian	31 Jul 1665
149rv	From Leopoldo at Florence $1\frac{1}{2}$pp (20x27) Italian	2 Oct 1665

150rv	From Leopoldo at Florence ⅔p (20x27) Italian	7 May 1666
151rv	From Leopoldo at Florence 1p & 2 lines (20x27) Italian	11 Feb 1666 Ab Inc [11 Feb 1667]
152rv	From Leopoldo at Florence ⅔p (20x27) Italian	12 Nov 1666
153rv	From Leopoldo at Florence ½p (20x27) Italian	20 May 1667
154rv	From Leopoldo at Florence ½p (20x27) Italian	20 Aug 1667
155rv	From Leopoldo* at Florence ½p (20x27) Italian *signed Cardinal de' Medici	27 Jan 1667 Ab Inc [27 Jan 1668]
156rv	Copy of a letter in B's hand 1p (20x31) "Lettre de la Princesse d'Oxholder a/Mad. la Princesse Louyse."	undated
157rv	Copy of a letter in B's hand 1⅔pp (20x31) "Lettre de la Princesse d'Oxholder" [A Sa Majesté la Reyne de Bohesme... De Bergues...Elisabeth, Princesse de Solern]	28 Dec 1657
158r-160v	Copy of a letter 5½pp (17x22) "Copia de la lettre de la Princesse Louise a S.A.E. Palatine"	undated
161rv	Printed letter fol 161r, ½p (14x18) "Lettre de Madame la Princesse LOUISE PA/LATINE, a la Reyne de Boheme./du 19. Decemb, 1657" fol 161v, ½p (14x18) "Lettre de Monseigneur le Prince Edouart Palatin à Madame la/Princesse Louise Palatine."	
162r-163v	From Louise at Olavie To Mme Dönhof ["Mon Astre"] 1½pp + ad (14x20)	15 Apr NY

Folio	Description	Date
164rv	From Mirame [Princess of Brieg] To Mme Dönhof ["Divine Astreè"] 1p + ad (14x20)	29 [?] NY
165rv	From Louise To Mlle de Blois at Glogan 7 lines + ad (14x20)	20 Dec 1655
166rv&170rv	From Louise To Mme Dönhof ["Mon Astre"] 8 lines & 3 lines + ad (14x20)	17 Apr NY
167rv	From Paolo [Giambotto? at Padua?] To Anonymous 1p (20x31) Italian	6 Nov 1658
168rv	Copy of a letter in B's hand 1p (20x31) Latin "Lettre de la Reine de Suede au Parlement de Paris" [Christina, Stockholm, 10 Apr 1652]	
169rv	Copy of a note in B's hand ½p (20x31) concerning "l'assemblée de l'armée de Bourdeaux qui a faict cet acte en 1652. le 14e May"	undated
171rv	Copy of a letter ½p (20x31) Italian "Copia di Lettera scritta dalla Regina/di Suetia in Roma sotto li 31 Maggio 1659/al Marchese Santinelli di Vienna"	
172rv	Copy of a letter in B's hand 1p & 5 lines (18x25) "Lettre de la Reine de Suede a Mr Gassendi/De Stockholme le 27 Septembre 1652"	
173r-174v	Copy of a letter in B's hand 2½pp (15x20) "Lettre de la Reine de Suede a Madame/Saumaise sur la mort de son mari"	undated
175r-176v	From Jean Blaeu [a fils] at Amsterdam 1⅔pp + ad (17x22)	11 Jul 1669
178rv	From Robert Scott at London 1p (20x34)	3 Oct 1677

179rv&177rv	From E. Sherburne at London 1p + ad (20x31) Latin	Prid Non Oct [1677] [16 Oct 1677]
180rv	From Blaeu at Amsterdam B at The Hague ½p + ad (21x36)	19 May 1662
181r-183v	News accounts: Rome, Genoa, Madrid, Paris, Naples, Venice, Milan 5¾pp (20x27) Italian	30 Jul 1667
183v-185v	News accounts: Cologne, Vienna, Poland, Venice, Milan 3½pp & 3 lines (20x27) Italian	11 Jul 1667
185v-186v	News accounts: Cologne, Vienna, Cracow 2pp (20x27) Italian	29 Sep 1667
187r-189v	News accounts: Rome, Genoa, Paris, Madrid, Milan, Naples, Venice 5pp & 4 lines (20x27) Italian	24 Sep 1667
189v-190v	News accounts: Cologne, Paris, London, Vienna, Warsaw, Danzig 1½pp & ¾p (20x27) Italian	4 Sep 1667
191r-193v	News accounts: Rome, Genoa, Madrid, Paris, Livorno, Venice, Milan 5¾pp (20x27) Italian	25 Feb 1668
193v-194v &201rv	News accounts: Cologne, Hamburg, Vienna, Warsaw 2¼pp & ⅔p (20x27) Italian	5 Feb 1668
195r-196v	News accounts: Rome, Genoa, Madrid, Paris, Naples 4pp (20x27) Italian	22 Oct 1667
197r-199v	News accounts: Rome, Genoa, Livorno, Paris, Madrid, Naples, Venice, Milan 5½pp (20x27) Italian	17 Sep 1667

199v-200v	News accounts: Cologne, Vienna 1½pp & ⅓p (20x27) Italian	23 Aug 1667

▼ ▼ ▼

COLLECTION BOULLIAU, VOLUME XXXII, FONDS FRANÇAIS 13050

Folio	Description	Date
1r-4v	From F-A de Thou at Venice To Dupuy 7pp (17x22)	8 Dec 1626
5rv	From [J-A de Thou] at Charlons 1p (17x22)	25 [Sep?] 1652
6rv	From [J-A de Thou] at Meslay 1p (17x22)	Lundi
7rv&10rv	From [J-A de Thou] at Meslay 2pp & 2 lines + ad (17x22)	5 Nov 1652
8r-9v	From [J-A de Thou] at [Salle?] 2⅔pp + ad (17x22)	26 Nov 1652
11rv	From [J-A de Thou] at Meslay 1p (17x22)	Vendredi
12rv	From [J-A de Thou] at Meslay ¾p (17x22)	5 Dec 1652 Jeudi
13rv	From [J-A de Thou at Meslay] 1p (17x22)	undated
14r-15v	From [J-A de Thou] at Meslay 3pp + ad (17x22)	Lundi
16r-17v	From [J-A de Thou] at Meslay 3pp + ad (17x22)	15 Dec 1652 Dimanche
18r-19v	From [J-A de Thou] at Meslay 2½pp + ad (17x22)	Mercredi
20rv	From [J-A de Thou] at Meslay 1½pp (17x22) *in B's hand	Mardi [10^{e}]*
21r-22v	From Mme L-G de Thou at [Salle?] 2½pp + ad (17x22)	[30] Jun NY
23r-24v	From Mme L-G de Thou at [Salle?] 3pp + ad (17x22)	25 Jun NY

25r-26v	From Mme L-G de Thou at [Salle?] 3pp + ad (17x22)	24 Jul NY
27rv	From [J-A de Thou at Meslay] 1p (17x22)	Vendredi
28r-29v	From J-A de Thou at [Meslay] 2½pp + ad (17x22)	Jeudi
30rv	From [J-A de Thou at Meslay] 1p (17x22)	8 Feb 1653
31rv&34rv	From [J-A de Thou] at Meslay ½p + ad (17x22)	Dimanche
32r-33v	From [J-A de Thou] at Meslay 2½pp + ad (17x22)	2 Feb NY
35rv	From [J-A de Thou at Meslay] 1p (17x22)	20 Aug 1653 Dimanche
36r-37v	From [J-A de Thou at Meslay] 2½pp + ad (17x22)	Jeudi
38rv	From [J-A de Thou at Meslay] 1p (17x22)	Lundi
39rv	From [J-A de Thou at Meslay] ⅔p + ad (17x22)	Mardi
40rv	From [J-A de Thou] at Meslay 1p (17x22)	4 Jan NY
41rv	From [J-A de Thou] at Meslay 1p (17x22)	21 Jan NY
42rv	From J-A de Thou at Meslay 1p (20x31)	29 Sep 1655 Mercredi
43rv	From [J-A de Thou] at Meslay 1p (20x31)	Samedi
44rv	From [J-A de Thou at Meslay] 1p (20x31)	14 Oct 1655 Jeudi
45rv	From [J-A de Thou] at Meslay 1p + ad (17x22)	12 Jan 1653

46r-47v	From [J-A de Thou] at Meslay 2pp & 5 lines + ad (17x22)	19 Feb 1653 Mercredi
48rv	From [J-A de Thou at Meslay] 1p (17x22)	[16?] Feb 1653 Dimanche
49rv	From Picardet* at [Meslay] To [B?] 1p (17x22) *[Madame de Thou]	8 Feb NY
50r-51v	From [J-A de Thou at Meslay] 2½pp + ad (17x22)	17 Jun 1653
52rv	From [J-A de Thou] at Meslay 2pp (17x22)	10 Jun 1653
53rv&57rv	From [J-A de Thou] at Meslay 1⅔pp + ad (17x22) fol 57, 15 lines in B's hand	2 Apr 1653 Mercredi
54rv&58rv	From [J-A de Thou at Meslay] ⅔p + ad (17x22)	5 Apr 1653
55r-56v	From [J-A de Thou at Meslay] 2⅔pp & 15 lines + ad (17x22)	23 Mar 1653
59rv	From [J-A de Thou at Meslay] 1½pp (17x22)	11 Mar 1653
60r-61v	From [J-A de Thou at Meslay] 2pp + ad (17x22)	18 Mar 1653
62r-63v	From [J-A de Thou at Meslay] 2½pp + ad (17x22)	9 Mar 1653
64rv	From [J-A de Thou at Meslay] 1p (17x22)	26 Feb 1653
65rv&68rv	From [J-A de Thou at Meslay] 2pp & 5 lines + ad (17x22)	23 Feb 1653
66r-67v	From [J-A de Thou at Meslay] 2pp + ad (17x22)	24 Jun 1653
69rv	From [J-A de Thou at Meslay] 1p (17x22)	8 May 1653

70rv	From J-A de Thou at [Meslay] 1p (17x22)	Samedi
71rv	From [J-A de Thou at Meslay] 1p (17x22)	Mardi
72r-73v	From [J-A de Thou] at Meslay 1p + ad (17x22)	4 May 1653
74rv	From [J-A de Thou at Meslay] 1½pp (17x22)	27 Apr 1653
75rv&78rv	From [J-A de Thou at Meslay] 2pp & 4 lines + ad (17x22)	1 May 1653
76r-77v	From [J-A de Thou at Meslay] 2½pp + ad (17x22)	Dimanche
79rv	From [J-A de Thou at Meslay] 1p (17x22)	Jeudi
80r-81v	From [J-A de Thou at Meslay] 2¾pp + ad (17x22)	30 Mar 1653 Dimanche
82rv	From [J-A de Thou at Meslay] 1⅔pp (17x22)	4 Feb 1653 Mardi
83rv	From [J-A de Thou at Meslay] 1½pp (17x22)	26 Mar 1653
84rv	From [J-A de Thou at Meslay] 1p (17x22)	17 Jul 1653
85r-86v	From [J-A de Thou at Meslay] 2½pp + ad (17x22)	3 Aug 1653 Dimanche
87rv	From [J-A de Thou at Meslay] 1⅔pp (17x22)	7 Aug 1653
88rv	From [J-A de Thou] at Meslay 1p (17x22)	26 Jun 1653
89rv	From [J-A de Thou at Meslay] 1p (17x22)	6 Jul 1653
90r-91v	From [J-A de Thou at Meslay] 1p & 1 line + ad (17x22)	Mardi

92r-93v	From [J-A de Thou at Meslay] 3pp + ad (17x22)	Dimanche
94r-95v	From [J-A de Thou at Meslay] 2⅓pp + ad (17x22)	Vendredi
96rv	Copy of a letter [in de Thou's hand] To M. d'Espernon 1p (17x22)	28 Jul 1653
97r-98v	From [J-A de Thou at Meslay] 2pp & 7 lines + ad (17x22)	Jeudi
99rv	From J-A de Thou at [Meslay] 1p (17x22)	Lundi
100r-101v	From [J-A de Thou at Meslay] 1p & 7 lines + ad (17x22) fol 101, 7 lines in B's hand	Mardi
102rv	From J-A de Thou at Meslay 2pp (17x22)	Dimanche
103r-104v	From [J-A de Thou] at Meslay ⅔p + ad (17x22)	28 Oct 1655
105rv	From [J-A de Thou] at Meslay 1p (20x31)	Jeudi
106rv	From [J-A de Thou] at [Orléans?] ⅔p (17x22)	4 [Nov 1655]
107rv	From [J-A de Thou] at Meslay 1p (20x31)	23 Nov 1655
108rv	From [J-A de Thou] at Meslay ¾p (17x22)	11 Nov 1655
109rv	From J-A de Thou at Meslay 1½pp (17x22)	Jeudi
110rv	From J-A de Thou at Charlons 1p (17x22)	Lundi
111r-112v	From J-A de Thou at The Hague To Rivière at Paris 1p & 7 lines + ad (20x31)	1 Aug 1657

113r-114v	From Picardet* at Paris B at [The Hague] 2pp & 10 lines + ad (20x31) *B writes "Mad. de Thou, 11 May 1657"	[11 May 1657]
115rv&119rv	From Courtin at Montreuil B at [The Hague] 1p & 3 lines + ad (12x18)	24 May 1657
116rv	From Soëfve at Meslay B at [The Hague] 2pp (17x22)	27 May 1657
117r-118v	From Picardet* at Meslay B at The Hague 2pp & 5 lines + ad (17x22) *B writes: "28 May 1657 Mad./de Thou"	28 May 1657*
120r-121v	From Picardet at Paris B at The Hague 2pp & 5 lines + ad (17x22)	13 Jul 1657
122r-123v	From J-A de Thou at Maestricht B at The Hague 2⅔pp + ad (20x31)	10 Jun 1662 Samedi
124rv	From [J-A de Thou] at The Hague 1p (20x31)	11 Jan 1661
125rv	From [J-A de Thou] at The Hague 1p (20x31)	24 Jan 1661
126r-127v	From [J-A de Thou] at The Hague 2pp (20x31) fol 127, blank	19 Dec 1660
128rv	From [J-A de Thou] at The Hague 1p (20x31)	6 Jun 1661
129rv	From [J-A de Thou] at The Hague 1p (20x31)	27 Jun 1661
130r-131v	From [J-A de Thou] at The Hague 2¾pp + ad (15x20) *in B's hand	5 [Jan 1660]* Mercredi
132rv	From [J-A de Thou] at The Hague 2pp (20x31)	7 Jul 1661

133rv	From [J-A de Thou] at The Hague 1p (20x31)	14 Feb 1662
134rv	From [J-A de Thou] at The Hague 1¾pp (15x20)	8 Jan NY Samedi
135rv	From [J-A de Thou] at The Hague 1p (15x20) *in B's hand	[9 Jan 1661]* Dimanche
136rv	From [J-A de Thou] at The Hague 2pp (20x31) *in B's hand	[16 Jan 1661]* Dimanche
137rv	From J-A de Thou at The Hague 1p (20x31)	Lundi
138rv	From [J-A de Thou] at The Hague 1p (20x31)	12 Jan 1661 Mercredi
139rv	From J-A de Thou at Paris 1⅔pp (17x22) *in B's hand	[15 Dec 1661]* Jeudi
140r-142v	From [J-A de Thou] at The Hague 2⅔pp + ad (20x31)	4 Apr 1661
143rv	From [J-A de Thou] at The Hague 1p (20x31)	3 Feb 1661
144rv	From [J-A de Thou] at The Hague 1p (20x31)	20 May 1661
145r-146v	From [J-A de Thou] at The Hague B at Amsterdam 3pp + ad (20x31)	14 Jan 1661 Vendredi
147rv	From [J-A de Thou] at The Hague 1p (20x31)	17 Oct 1660 Dimanche
148rv	From [J-A de Thou] at The Hague 1p (20x31)	22 Dec 1660
149r-150v	From L-A de Thou at [Rome]* 3pp + ad (17x22) *in B's hand	9 Aug 1667
151r-152v	From [L-A de Thou] at Rome 3pp + ad (17x22)	19 Jul 1667

153r-154v	From [L-A de Thou] at Rome 3½pp (17x22)	30 Aug 1667
155r-156v	From [L-A de Thou] at Rome 3pp + ad (17x22)	13 Sep 1667
157r-158v	From [L-A de Thou at Rome] 2¾pp (17x22)	20 Sep 1667
159r-160v	From [L-A de Thou at Rome] 3pp (17x22)	27 Sep 1667
161r-162v	From [L-A de Thou] at Rome 3pp (17x22) PS, pasted to fol 162r [unnumbered] (6x15) 7 lines	11 Oct 1667
163r-164v	From [L-A de Thou] at Rome 3pp (17x22)	18 Oct 1667
165r-166v	From [L-A de Thou] at Rome 3pp (17x22)	25 Oct 1667
167r-168v	From [L-A de Thou at Rome] 4pp (17x22)	8 Nov 1667
169rv	From [L-A de Thou] at Rome 1p (17x22)	15 Nov 1667
170r-171v	From [L-A de Thou] at Rome 3pp (17x22)	29 Nov 1667
172r-173v	From [L-A de Thou] at Rome 3pp (17x22)	6 Dec 1667
174r-175v	From [L-A de Thou] at Rome 1½pp (17x22)	3 Jan 1668
176r-177v	From [L-A de Thou] at Rome 2pp (17x22)	6 Feb 1668
178r-179v	From [L-A de Thou] at Venice 2pp (17x22)	24 Mar 1668
180r-181v	From L-A de Thou 1½pp (17x22)	Dimanche

182rv	From Rivière at Meslay 1p (17x22)	14 Jul 1653
183rv	From Rivière at Meslay ⅔p (17x22)	21 Jul 1653
184rv	From Rivière at Paris B at [The Hague] 1p (17x22)	13 Apr 1657
185rv	From Rivière at Paris B at [The Hague] 1⅔pp (17x22)	4 May 1657
186rv	From Rivière at Paris B at [The Hague] 2pp (17x22)	11 May 1657
187r-188v	From [Rivière] at Paris B at [The Hague] 2½pp (17x22)	18 May 1657
189rv	From Rivière at Paris B at [The Hague] 1p (17x22)	25 May 1657
190rv	From [Rivière] at Paris B at [The Hague] 2pp (17x22)	1 Jun 1657
191rv	From [Rivière] at Paris B at [The Hague] 1p (17x22)	8 Jun 1657
192rv	From Rivière at Paris B at [The Hague] 2pp (17x22)	15 Jun 1657
193r-194v	From Rivière at Paris B at [The Hague] 3pp (17x22)	22 Jun 1657
195rv&198rv	From Rivière at Paris B at [The Hague] 2pp + ad (17x22)	29 Jun 1657
196r-197v	From Rivière at Paris B at The Hague 2⅔pp (17x22)	6 Jul 1657

199r-200v	From Rivière at Paris B at The Hague 3pp + ad (17x22)	13 Jul 1657
201r-202v	From Rivière at Paris B at [The Hague] 3pp (17x22)	20 Jul 1657
203r-204v	From Rivière at Paris B at [The Hague] 3pp + ad (17x22)	27 Jul 1657
205rv	From Rivière at Paris 2pp (17x22)	18 Nov 1657
206rv	Copy in B's hand "Exploict de la Reine Christine." 2/3p + ad (17x22)	15 Nov 1657
207rv	From Rivière at Paris B at [The Hague] 1p (17x22)	22 Nov 1657
208rv	From Rivière at Paris 1p (17x22)	25 Nov 1657
209rv	From Rivière at Paris 1p (17x22)	29 Nov 1657
210r-211v	From Soëfve at Paris B at The Hague 2 1/2pp + ad (17x22)	13 Jul 1657
212r-215v	From Soëfve at Paris B at The Hague 6pp + ad (17x22)	6 Jul 1657 Vendredi
216r-219v	From Soëfve at Paris B at The Hague 6pp + ad (17x22)	29 Jun 1657
220r-221v	From Soëfve at Paris B at The Hague 2 1/3pp + ad (17x22)	22 Jun 1657

222r-223v	From Soëfve at Paris B at The Hague 3pp + ad (17x22)	15 Jun 1657
224r-225v	From Soëfve at Paris B at The Hague 3pp + ad (17x22) *corrected by B; Apr was written	8 Jun* 1657 Vendredi
226r-228v&231rv	From Soëfve at Paris B at [The Hague] 5pp & 3 lines + ad (17x22)	15 Apr 1657
229r-230v	From Soëfve at Paris B at The Hague 2pp & 5 lines (17x22)	20 Apr 1657
232r-236v	From Soëfve at Paris B at The Hague 7½pp + ad (17x22)* *fol 235 (8x14)	27 Apr 1657
237r-238v	From Soëfve at Paris B at [The Hague] 3pp & 4 lines (17x22)	4 May 1657
239r-240v	From Soëfve at Paris B at The Hague 2¾pp + ad (17x22)	10 May 1657 Jeudi
241r-242v	From Soëfve at Paris B at The Hague 3⅓pp + ad (17x22)	17 May 1657
243rv	From Soëfve at Paris B at [The Hague] 1p (17x22)	21 May 1657
244rv	Fragment [Soëfve] 1p (16x19)	undated
245rv	From Soëfve at Paris B at The Hague 1p & 8 lines [in B's hand] + ad (20x31)	27 Jul 1657
246r-247v	From Soëfve at Paris B at The Hague 2⅓pp + ad (20x27)	20 Jul 1657

248rv	From Anonymous at [London?] 1p + ad (15x20)	27 Mar 1684
249rv&254rv	From Aston at London 1p & 3 lines + ad (17x22) Latin	10 Mar 1683/1684 [20 Mar 1684]
250rv	From Gronovius at Leiden 1p + ad (20x31)	Prid Non Quintil 1659 [7 May 1659]
251rv	From Smith at London 1p + ad (15x20)	12 May NY
252r-253v	From Ménage at Paris B at The Hague 1p + ad (15x20)	1 Apr 1662
255r-256v	Extracts of letters in B's hand 3½pp (17x22) Italian/French Domenico Guidi & Charles de Brun:	 15 Dec 1675 22 Dec 1675 16 Dec 1675 22 Dec 1675
257rv	Copy of a letter in B's hand From Charles de Brun at Paris To [Guidi] 1p (17x22)	31 Jan 1676
258r-260v	To Gassendi at [Digne] 4pp (17x22) Latin fol 260, blank Publ, GOO, III, pp 410-413 Publ, MC, III, pp 447-452, No 260	21 Jun [1633]
261r-262v	From Schooten at Leiden 2¾pp + ad (15x20)	22 Jul [1658?]
263rv	From S. Ward at Oxford 1p (15x20) Latin	19 Mar 1659 OS [29 Mar 1660]

▼ ▼ ▼

COLLECTION BOULLIAU, VOLUME XXXIII, FONDS FRANÇAIS 13051

1r-2v	"Recueil des pieces trouvées/Dans les manuscrits de M. Bouilliaud, Lequel avoit/este secretaire de M. de Thou dans ses ambassades et est au nombre de homes illustres de ce siècle" 3½pp (24x34) "Pieces Contenues au Vol. A" [48 entries]
2v-6v	"Pieces Contenues au Vollume B" 7½pp & 3 lines (24x34) [123 entries]
6v	"Pieces Contenues au Vollume C" ⅓p (24x34) [7 entries]
6v-8r	"Pieces Contenues au Vollume D" 3pp (24x34) [56 entries]
8r-9v	"Pieces Contenues au Vollume E" 3pp (24x34) [63 entries]
9v	"Pieces Contenues au Vollume F" ⅓p (24x34) [6 entries]
9v	"Pieces Contenues au Vollume H" ¼p (24x34) [3 entries]
10r	"Pieces d'Astronomie" (24x34) [Note:] "Astronomie divisée en quatre paquets--le 1er contient les horoscopes--le 2 contient les problemes le 3me les dissertations mathématiques le quatrième les observations celestes sans comprendre les lettres concernant ses matières qui seront dans l'articles des lettres ci-dessous." "Pieces du pacquet marqué G 1er" [1 entry, horoscopes]

10r "Pieces du pacquet marqué G 2^{d}"
½p (24x34)
[11 entries]

10r-11r "Pieces du paquet marqué G 3me"
1¾pp (24x34)
[30 entries]

11r-14r "Pieces du paquet marqué G 4me"
1½pp (24x34)
[Note:] "Une infinite d'observations parue manuscrites parues imprimes faites sur l'elevation du pol de plusieurs endroits sur la qualite de l'air sur les comets--sur les planetes, sur les eclipses du solleil et de la lune...toutes ses observations ont est faites par M. Bouillau, sont lires des escrits des plus savans astronomes depuis [1520] comme Ticobrahé, Copernic, Gassendi, Kepler, Sheneir, Cujlly, Ritccioli, Bourdin, Hevelius, Longomontanus, Cassiny, et plusiers autres avec quelques lettres de ses messieurs rellies avec les observations"
[12 entries]
fols 12 & 13, blank

14rv "Pieces Contenues au Paquet I"
2pp (24x34)
[38 entries]

15r-16r "Pieces Contenues au Paquet L"
2⅓pp (24x34)
[39 entries]

16r-17r "Pieces Contenues au Paquet M"
2⅓pp (24x34)
[51 entries]

17r-22r "Pieces Contenues Au Paquet N"
9½pp (24x34)
[147 entries]

22rv "Pieces Contenues au paquet O"
1⅓pp (24x34)
[19 entries]

22v	"Pieces Contenues au paquet P" $\frac{1}{3}$p (24x34) [3 entries]
22v-26r	"Pieces du paquet Numero 9" 7$\frac{1}{3}$pp (24x34) [150 entries]
26v-28v	"Pieces Contenues au paquet R" 5pp (24x34) [86 entries]
29r-32v	"Pieces Du Paquet marque S" 7$\frac{1}{4}$pp (24x34) [145 entries]
32v-40r	"Pieces du paquet marque T" 15$\frac{1}{4}$pp (24x34) [221 entries]
40r	"Pieces du paquet marque V" $\frac{1}{2}$p (24x34) [11 entries]
40r-43v	"Pieces du paquet marque X" 7$\frac{1}{4}$pp (24x34) [141 entries]
44r-48r	"Pieces du paquet marque Y" 8$\frac{2}{3}$pp (24x34) [155 entries]
48r-53r	"Pieces du paquet marque Z" 10$\frac{1}{3}$pp (24x34) [160 entries]
53v-57v	"Pieces du paquet marque K" 8$\frac{1}{4}$pp (24x34) [85 entries]
57v-59r	"Pieces adioustés aux paquets ci-dessus" 3$\frac{1}{4}$pp (24x34) [49 entries]

59v "Plusieurs paquets de lettres"
1/2p (24x34)
[Note:] "Lettres à monsieur de Peiresc par diverse personnes illustres depuis 1600 iusquen 1650
Les autres sont de M. Bouillaud qu'il a escrit ou qu'on lui a adressé de Rome de Pologne de Venise de Constantinople sur les sciences sur les affaires d'estat elles sont des plus illustres personnes de ces roiaumes comme on le vera par les abregés [20 paquets].
Il y en [a?] d'autres de diverses persones ramassées par monsieu Bouillaud sur l'astronomie et sur differentes matieres.
20 paquets de lettres a M. de Peresque de plusieurs persones
22 paquets de M. Bouillaud ou qu'il a escrite ou recues ou ramasees
4 grots paquets de lettres meles
3 vollumes de lettres rellues--deux d'astronome par les meilleurs astronome deux autres du sieur Matarel [Matharel] resident du roy a Venise in 1648-50-51; un volume des lettres du pere Dupuis Chartreuse a Rome depuis 1641 iusquen 53--un autre des lettres du Prince Leopold de Toscane avec les responces depuis 1657 iusquen 1660--un autre vollume de celles de ses amis--un autre des lettres de Portener avec les responces depuis 1652 iusquen 1657"

60r-64v "Catalogue/Des Livres trouves dans les papiers de feu/M. Bouillaud"
8 1/4pp (24x34)
"Liste des Livres Imprimes" [90 entries]
"Liste des Livres manuscripts" [26 entries]
"Livres Grecs" [19 entries]

65r-66v "Pieces a Voir Contenues au Vol. A" [and B,C,D,I,L,N,O,No.9,T,V,X,Y,Z,K et Livres trouves dans les papiers de M. Bouillaud]
3pp (24x34)
[fols 65-69, in different hand]

67r-69v "Pieces a avoir Contenues au Vol. B"
[and D,I,L.,No.9,X,Y,Z,K et Catalogue
des livres dans les papiers de M. Bouillaud]
4½pp (24x34)

▼ ▼ ▼

COLLECTION BOULLIAU, VOLUME XXXIV, FONDS FRANÇAIS 13052

1r-2v	Extract in B's hand Lucian's Lucius or the Ass 3¼pp (15x18) Greek/Latin
2v	Extract in B's hand Lucian's Somnium 1p (15x18) Greek/Latin
3r	Extract in B's hand Lucian, "Prometseus es in verbis" 1p (15x18) Greek/Latin
3v-5v	Extract in B's hand Lucian's Nigrinus 4¼pp (15x18) Greek/Latin
5v-9r	Extract in B's hand Lucian's Timon 5¾pp (15x18) Greek/Latin fol 7, blank
9v	Notes in B's hand "Halycon," "Prometseus seu Caucasus" 1p (15x18) Greek/Latin
10r-16v	Notes in B's hand Lucian's Dialogues of the Gods, etc. 14pp (15x18) Latin "Dial. Jovis & Amoris," etc. "Apollonius et Vulcan," "Veneris et Luna," "Apollonius et Mercurii," "Tritonis Amymonis et Neptuni," "De Sacrificiis, Vitarum Auctio," etc.
17r-25r	Notes cont'd, classical authors w/brief sketches & extracts 15pp (15x18) Greek/Latin
25r-26v	Extracts in B's hand Lucian's True History 2pp (15x18) Greek/Latin
27r-30v	Notes in B's hand Lucian's "Phalaris" 8pp (15x18) Greek/Latin

31r-33v	Notes in B's hand Apis Aegyptus 6pp (15x18) Greek/Latin
34r-43r	Extracts in B's hand Lucian's Imagines, Toxaris, etc. 19pp (15x18) Greek/Latin
43v-48r	Notes in B's hand De Luctu, Hippias, Bacchus, etc. 8pp (15x18) Greek/Latin
48v-54v	Notes in B's hand Macrobius' Saturnalia, etc. 13pp (15x18) Greek/Latin
55r	Notes in B's hand "Lapithae seu Symposium" 1p (15x18) Greek/Latin
55v-62r	Notes in B's hand Lucian's On the Syrian Goddess 5pp (15x18) Greek/Latin fols 58-61, blank
62v-66v	Extracts in B's hand, from Ennius [Quintus Ennius of Rudiae] 9pp (15x18) Greek/Latin
67r	Extract in B's hand Euripides' Medea transl. by Ennius 1p (15x18) Latin
67v	Notes in B's hand "Achille" 1p (15x18) Latin
68rv	Notes in B's hand "Andromache" [probably of Euripides] 2pp (15x18) Latin
69rv	Notes in B's hand "Phoenice," [Euripides' Phoenissae?] 2pp (15x18) Latin
70r-77v	Extracts in B's hand From Ausonius [Ausonius, Decimus Magnus] Epigrammatum Liber, Epistolae 14pp (15x18) Latin fol 77, blank

78r-79r	Extracts in B's hand Plautus' Amphitruo 3pp (20x30) Latin
79v-102v	Extracts in B's hand Plautus' Varronian plays, Asinaria, Captivi, Casina, Cistellaria, Epidicus, Bacchides, Mostellaria, Menaechmi, Miles, Mercator, Pseudolus, Paenulus, Persa 47pp (20x30) Latin
103r-107v	Extracts in B's hand Plautus' Rudens, Stichus & Trinummus 10pp (20x30) Latin
108r-109v	Extracts in B's hand Plautus' Truculentus 4pp (20x30) Latin
110r-111v	Notes in B's hand Plautus' Amphitruo 2pp (20x30) Latin fol 111, blank
112r-125v	Extracts & notes in B's hand "Terentii Comaediae," Eunuchus, Heauton-timoroumenos, Adelphi, Hecyra, Phormio 28pp (20x30) Latin
126r-147v	Extracts in B's hand T. Cari Lucretti; De Rerum Natura; Liber I, II, III, IV 40pp (20x30) Latin fols 146 & 147, blank
148r	Extracts in B's hand "Juvenalis...et Lucanus..." 1p (20x30) Latin
148v-152r	Extracts in B's hand Virgil's Georgica, Bks. I, II, III, & IV 8pp (20x30) Latin/Greek
152r-175v	Extracts in B's hand Virgil's Aeneid, Bks I, II, III, IV, V, VI, VII, VIII, IX, X, XI, XII 42pp (20x30) Latin fols 173-175, blank

176r-179v	List of works & notes in B's hand "Cod. Ms Thuanus..." 6pp (20x30) Latin/Greek fol 179, blank
180r-183v	Notes in B's hand Theocritus, Idyls 1-27 (Scholia) 7pp (20x30) Greek/Latin
184r-187v	Notes & extracts in B's hand "Catulli" (Epithalamium, Peleus & Thetis; Coma Berenices, etc.) 8pp (20x30) Latin
188r-190r	Extracts & notes in B's hand Plautus, Lucan, Virgil, Servius, etc. 4pp (20x30) Latin/Greek
190r-191r	Brief extracts in B's hand, cont'd Lucan 2½pp (20x30) Latin
191v-192r	Notes in B's hand "De dextris, et Sinistris partibus juxta Astrologos veteres, et recentiores." 2pp (20x30) Latin
192v-196v	Notes in B's hand "De partibus dextris, et sinistris/ Astronomorum" 8pp (20x30) Latin/Greek

▼ ▼ ▼

COLLECTION BOULLIAU, VOLUME XXXV, FONDS FRANCAIS 13053

1r-22v	News extracts in B's hand	
	44pp (16x20) Italian	
	Rome	5,12,19,27 Jul 1659
	Vienna (Extract, Siri to Des Noyers)	22 Sep 1659
	Vienna	29 Sep 1659
	Rome	20,27 Sep 1659
	Rome	4 Oct 1659
	Vienna	6,13 Oct 1659
	Rome	11,18 Oct 1659
	Vienna	20,27 Oct, 3,10 Nov 1659
	Vienna	10,17,24 Nov 1659
	Rome	25 Oct 1659
	Rome	1,8,15 Nov 1659
	Vienna	1,8,15 Dec 1659
	Rome	22 Dec 1659
	Vienna	8,22 Dec 1659
	Rome	29,6 Dec 1659
	Vienna	29 Dec 1659
	Rome	20,27 Dec 1659
	Rome	17 Jan 1660
	Vienna	31 Jan 1660
	Rome	24 Jan 1660
	Vienna	15 Feb 1660
	Rome	31 Jan 1660
	Rome	21,28 Feb 1660
	Rome	27 Mar 1660
	Vienna	12 Apr 1660
	Rome	13,10,17,24 Apr 1660
	Rome	1 May 1660
	Vienna	17,24,31 May 1660
	Vienna	7 Jun 1660
	Rome	29 May 1660
	Vienna	14 Jun 1660
22v-25r	Copies of letters	undated
	M. Molino, Ambassador of Venice	
	5½pp (16x20) Italian	
25r-28r	News accounts in B's hand	
	7pp (16x20) Italian	
	Vienna	20 Jun 1660
	Rome	5,12 Jun 1660
	Vienna	20 Jun 1660
	Vienna	5 Jul 1660

28rv	Copy of a letter	15 Jun 1660
	To Sig.^{r} A. Maÿer	
	$1\frac{1}{4}$pp (16x20) Italian	
28v-38v	News accounts in B's hand	
	20pp (16x20) Italian	
	Rome	19,26 Jun 1660
	Rome	1,17 Dec 1663
	Udine	28 Nov 1663
	Genoa	2 Jan 1664
	Rome	1,17 Dec 1663
	Venice	22 Mar 1664
	Rome	5 Apr 1664
	Milan	7 May 1664
	Rome	3,10 May 1664
	Naples	3 May 1664
	Rome	7 Jun 1664
	Extract of a letter	9 Jun 1664
	Petrovia	26 May 1664
	Naples	14 Jun 1664
	Venice	27 Jun 1664
	Milan	2 Jun 1664
	Rome	5 Jun 1664
	Petrovia	29 Jun 1664
	Rome	19 Jul 1664
	Venice	18 Jul 1664
	Rome	26 Jul 1664
	Venice	16 Aug 1664
	Milan	13 Aug 1664
	Rome	9,16,16 Aug 1664
	Milan	20 Aug 1664
	Genoa	20 Aug 1664
	Genoa	6 Sep 1664
	fols 30-32, blank	
39rv	Verse & text	[1664]
	2pp (16x20) Italian/French	
40rv	News accounts in B's hand	
	2pp (16x20) Italian	
	Venice	20 Sep 1664
	Milan	1 Oct 1664
	Genoa	15 Oct 1664
	Milan	11 Oct 1664
	Germany	16 Oct [1664]
41r-42v	Newsletter	20 Sep 1664
	From Anonymous at Vienna	
	$3\frac{2}{3}$pp (20x31) Italian	

43rv	Copy [concerns M. Pasquin] ½p (16x20) Italian	30 Jan 1663
44r-54r	Extracts & news accounts in B's hand 21pp (16x20) Italian	
	Venice	7 Nov 1664
	Rome	1 Nov 1664
	Genoa	10 Dec 1664
	Vienna	29 Nov 1664
	Rome	6,13 Dec 1664
	Milan	25 Mar 1665
	Genoa	20 May 1665
	Rome	16,30 May 1665
	Vienna	23,24 May 1665
	Extract re: a Jesuit	undated
	Genoa	1 Jul 1665
	Rome	20 Jun 1665
	Florence	30 Jun 1665
	Genoa	29 Jul 1665
	Rome	25 Jul 1665
	Venice	1 Aug 1665
	Vienna	11 Jul 1665
	Warsaw	1 Jul 1665
	Rome	5,12 Aug 1665
	Genoa	19 Aug 1665
	Rome	15,22 Aug 1665
	Genoa	2 Sep 1665
	Venice	5 Sep 1665
	Venice	13 Feb 1666
	Vienna	13 Feb 1666
	Vienna	6 Mar 1666
	Rome	1 May 1666
	Venice	28 Apr 1666
	Venice	5,8 May 1666
	Rome	21 Jul 1666
	Vienna	7 Aug 1666
	Rome	14,21,28 Aug 1666
	Rome	18 Sep 1666
	Genoa	13,6 Oct 1666
	Rome	16 Oct 1666
	Genoa	24 Nov 1666
	Vienna	4,18 Dec 1666
	Vienna	ND Mar 1667
	Rome	2 Apr 1667
	Rome	19 Mar 1667
	Naples	22 Mar 1667
	Rome	26 Mar 1667

54rv	Account in B's hand Concerning Colonello Pardi Luguese 1p (16x20)	undated
54v-55v	News accounts & extracts in B's hand 3pp (16x20) Italian	
	Rome	2 Apr 1667
	Quotation of Alexander VII	15 Apr 1667
	Rome	30 Apr 1667
	Rome	7,14 May 1667
56r-57r	Description in B's hand 2¾pp (16x20) [concerns Comte de Durazzo, Ambassador Extr. de Genes]	[18 Dec 1666]
57r-59r	Copy of a letter in B's hand From Galata at Constantinople 3¼pp & 8 lines (16x20)	18 Dec 1668
59r-62r	Extracts & news accounts in B's hand 7pp (16x20) Italian	
	Genoa	24,12,15 May 1667
	Genoa	21 Jun 1667
	Vienna	18 Jun 1667
	Rome	25 Jun 1667
	Rome	2 Jul 1667
	Candia	27 May 1667
	Vienna	16,9 Jul 1667
	Naples	12 Jul 1667
	Rome	16 Jul 1667
	Genoa	14,21 Jul 1667
	Venice	29 Jul 1667
62rv	Copy of a letter in B's hand From Anonymous at Candia 1p (16x20) Italian	11 Jun 1667
62v-63r	Extracts & news accounts in B's hand 3pp (16x20) Italian	
	Venice	3,6 Aug 1667
63rv	Copy of a letter in B's hand From Anonymous at Venice ⅔p & 6 lines (16x20)	10 Aug 1667
63v-68v	Extracts & news accounts in B's hand 11pp (16x20) Italian	
	Venice	13 Aug 1667

	Rome	6 Aug 1667
	Vienna	23,30 Jul 1667
	Milan	7 Aug 1667
	Candia	2 Jul 1667
	Vienna	6 Aug 1667
	Madrid	23 Jul 1667
	Rome	13,20,27 Aug 1667
	Naples	23 Aug 1667
	Candia	25 Jul 1667
	Venice	3 Sep 1667
	Vienna	20 Aug 1667
	Madrid	30 Jul 1667
	Turin	24 Jul 1667
	Vienna	19 Aug 1667
68v-69r	Copy of a letter in B's hand From Marquis Guido Villa To M. le Mar[al] de Grancey de Candie 1½pp (16x20)	15 Aug 1667
69r	Extract in B's hand To the Ambassador of Venice 1p (16x20)	rec'd 17 Oct 1667
69rv	Extracts & news accounts in B's hand 2pp (16x20) Italian	
	Venice	8 Oct 1667
	Genoa	1 Oct 1667
70rv	Copy of a letter From Anonymous at Venice 2pp (20x31) Italian	14 Nov 1665
71r	News account in B's hand 1p (16x20) Italian	
	Vienna	24 Nov 1667
71rv	Extract of a letter in B's hand 1¼pp (16x20) Italian "Ecc.[mo] Marchese [Guido] Villa, Generale dell' Armi in Candia della Ser.[ma] Republica Veneta..."	7 Sep 1667
71v-78r	Extracts & news accounts in B's hand 14pp (16x20) Italian	
	Venice	22 Oct 1667
	Vienna	8,8,16 Oct 1667
	Venice	18 Oct 1667
	Venice	16 Nov 1667

	Rome	10 Mar 1668
	Venice	30 Mar 1668
	Rome	30 Mar 1668
	Venice	4 May 1668
	Vienna	13,20 May 1668
	Rome	9 Jun 1668
	Vienna	10 Jun 1668
	Florence	17 Jul 1668
	Rome	21 Jul 1668
	Venice	25 Jul 1668
	Vienna	14 Jul 1668
78r-79r	Extract of a letter in B's hand To Marchese Villa From [Agostino Bianchi Segretaris] 1⅔pp & 6 lines (16x20) Italian	4 Jul 1668
79r-80v	Extract of a letter in B's hand From [M. le Marquis St. André Montbrun de Candie To Justiniani, Amb. of Venice] 3¼pp (16x20)	26 Jun 1668
80v-83v	News accounts in B's hand 7pp (16x20) Italian	
	Venice	1 Aug 1668
	Vienna	21 Jul 1668
	Rome	4 Aug 1668
	Vienna	28 Jul 1668
	Vienna	11 Aug 1668
	Rome	18 Aug 1668
	Vienna	11,18 Aug 1668
	Rome	25 Aug 1668
	Venice	29 Aug 1668
	Rome	1 Sep 1668
	Venice	5,12 Sep 1668
83v-84r	Copy of a letter From [Breuz del Papa] To [Louis XIV, King of France] 2pp (16x20) Latin/Italian	9 Sep 1668
84r-85r	News accounts in B's hand 3pp (16x20) Italian	
	Rome	22 Sep 1668
	Vienna	29 Sep 1668
	Rome	29 Sep 1668
	Rome	6,9 Oct 1668

85rv	Copy of a letter in B's hand From Pope Clement IX To [Louis XIV] 1p (16x20) Italian	26 Aug 1668
85v-86r	News accounts in B's hand 2pp (16x20) Italian	
	Rome	16 Oct 1668
	Rome	3 Nov 1668
86rv	Copy of an account in B's hand Candia 2pp (16x20)	15 Jul 1668
86v-91r	Extracts & news accounts in B's hand 10pp (16x20) Italian	
	Rome	20 Nov 1668
	Venice	24 Nov 1668
	Rome	27 Nov 1668
	Brief history	undated
	Rome	1 Dec 1668
	Madrid	19 Dec 1668
	Rome	22 Dec 1668
	Vienna	15,29 Dec 1668
	Rome	29,18,22,25,29 Dec 1668
	Venice	26 Dec 1668
	Rome	5 Jan 1669
	Vienna	11 Jan 1669
	Milan	4 Jan 1669
	Madrid	2 Jan 1669
	Rome	15 Jan 1669
	Venice	19,16 Jan 1669
	Rome	29 Jan 1669
	Rome	12,16 Feb 1669
	Venice	2 Mar 1669
91v	Copy of an account in B's hand 1p (16x20) Italian/French [concerns French Ambassador in Venice]	undated
92r-95r	Extracts & news accounts in B's hand 7pp (16x20) Italian	
	Rome	9,23 Mar 1669
	Venice	10 Apr 1669
	Genoa	15 May 1669
	Vienna	27 Apr 1669
	Rome	15 Jun 1669
	Genoa	ND Jun 1669
	Rome	8 Jun 1669

Folio	Description	Date
	Genoa	30 Jun 1669
	Naples	4 Jun 1669
	Rome	6,20 Jul 1669
	Rome	3 Aug 1669
95rv	Copy of a letter in B's hand $1\frac{1}{4}$pp (16x20) Italian ["dell' Auditor delle Galere Pontificie scritta a Nostro Sig^{re}"]	1 Aug 1669
95v	News account in B's hand $\frac{1}{2}$p (16x20) Italian	
	Rome	31 Aug 1669
95v-97v	Copy of a letter in B's hand Candia 4pp (16x20)	ND Aug 1669
97v-98r	News accounts in B's hand 2pp (16x20) Italian	
	Rome	14 Sep 1669
	Madrid	28 Aug 1669
	Rome	21,21,28 Sep 1669
98v-99v	Copy of a letter in B's hand Venice $2\frac{2}{3}$pp (16x20) Italian	23 Oct 1669
99v-100v	News accounts in B's hand 3pp (16x20) Italian	
	Venice	2 Nov 1669
	Rome	29 Oct 1669
100v-103r	Copy of a letter in B's hand $5\frac{3}{4}$pp (16x20) "Lettre de M. de St. André Montbrun à Mr le Prince, de Candie..."	7 Sep 1669
103r-104r	Copy of a letter in B's hand $1\frac{3}{4}$pp (16x20) "Lettre de Mr le Marquis de Frontenac à Mr Priandi, de Candie..."	12 Sep 1669
104r-109v	News accounts in B's hand 12pp (16x20) Italian	
	Vienna	19 Oct 1669
	Venice	2 Nov 1669
	Bergamo	4 Nov 1669
	Vienna	2 Nov 1669

	Venice	15 Nov 1669
	Madrid	6 Nov 1669
	Bergamo	12 Nov 1669
	Rome	2,9 Nov 1669
	Venice	16 Nov 1669
	Vienna	16 Nov 1669
	Rome	16 Nov 1669
	Venice	23 Nov 1669
110rv	Copy of a letter in B's hand 1⅓pp (16x20) Italian* *translated from Turkish	undated
111r-112r	Copy in B's hand 2¾pp (16x20) Italian "Oratione...dal Sig^{er} Conte di Frontenac Luogotenente Gñale. dell' Armi della Ser^{ma} Republica di Venetia li 7 Febraio 1670"	7 Feb 1670
112r-116v	Copy in B's hand 9¼pp (16x20) Italian "Memoriale dato dal Sig^{r} Conte di Frontenac..."	undated
117r-153v	News accounts in B's hand 74pp (16x20) Italian	
	Venice	6 Jun 1670
	Venice	31 May 1670
	Oration in French	
	Rome	7 Jun 1670
	Lucerne	13 Jun 1670
	Candia	19 May 1670
	Venice	26 Jul 1670
	Vienna	12 Jul 1670
	Bergamo	21 Jul 1670
	Rome	12 Jul 1670
	Bergamo	1,4 Sep 1670
	Vienna	6,10 Sep 1670
	Bergamo	22 Sep 1670
	Venice	13,20 Sep 1670
	Madrid (w/list of Counts, Barons, Dukes, etc.)	10 Sep 1670
	Vienna	13 Sep 1670
	Bergamo	29 Sep 1670
	Rome	4 Oct 1670
	Vienna	4 Oct 1670
	Venice	4,11,25 Oct 1670
	Venice	1 Nov 1670

Bergamo	3 Nov 1670
Bergamo	20 Oct 1670
Vienna	9,11,18 Oct 1670
Warsaw	8 Oct 1670
Rome	11,18,25 Oct 1670
Venice	8 Nov 1670
Vienna	25 Oct 1670
Bergamo	10 Nov 1670
Rome	18 Nov 1670
Bergamo	24,26 Nov 1670
Venice	26 Nov 1670
Vienna	15,22,29 Nov 1670
Bergamo	9 Dec 1670
Venice	7,15 Dec 1670
Bergamo	22 Dec 1670
Venice	20 Dec 1670
Rome	20,27 Dec 1670
Bergamo	29 Dec 1670
Rome	3,10 Jan 1671
Venice	17 Jan 1671
Bergamo	17,26 Jan 1671
Vienna	15 Feb 1671
Rome	24,21 Feb 1671
Bergamo	3 Mar 1671
Venice	28 Feb 1671
Bergamo	16 Mar 1671
Vienna	8,15 Mar 1671
Bergamo	6 Apr 1671
Venice	4 Apr 1671
Vienna	21 Mar 1671
Rome	4,11 Apr 1671
Vienna	5,12 Apr 1671
Venice	25 Apr 1671
Rome	18 Apr 1671
Genoa	6 May 1671
Venice	2 May 1671
Rome	2 May 1671
Venice	9 May 1671
Vienna	26 Apr 1671
Vienna	6,3 May 1671
Venice	13 May 1671
Rome	9,9,16 May 1671
Venice	20 Apr 1671
Vienna	17 May 1671
Bergamo	1 Jun 1671
Rome	23 May 1671
Venice	30 May 1671
Vienna	24 May 1671
Venice	13 Jun 1671

	Vienna	6 Jun 1671
	Venice	20 Jun 1671
	Rome	20 Jun 1671
	Venice	27 Jun 1671
	Bergamo	29 Jun 1671
	Bergamo	6 Jul 1671
	Rome	27 Jun 1671
	Rome	4,4 Jul 1671
	Vienna	27 Jun 1671
	Bergamo	13 Jul 1671
	Vienna	4 Jul 1671
	Venice	15,18 Jul 1671
	Bergamo	26 Jul 1671
	Venice	25,26 Jul 1671
	Rome	18 Jul 1671
	Bergamo	3 Aug 1671
	Venice	1 Aug 1671
	Venice [marginalia]	29 Jul 1671
	Vienna	18,25 Jul 1671
	Bergamo	13,17 Aug 1671
	Venice	8,12,15 Aug 1671
	Rome	8 Aug 1671
153v-154v	Copy of a letter in B's hand 2⅔pp (16x20) Italian	16 Aug 1671
154v-155v	Copy of a letter in B's hand 2⅓pp (16x20) French/Italian "Relation du mesme faict envoyee par M. le Command[r]. de Grémonville, de par luy composee"	undated
155v-206r	News accounts in B's hand 102pp (16x20) Italian	
	Bergamo	29 Aug 1671
	Bergamo	4 Sep 1671
	Venice	26 Aug 1671
	Venice	4 Sep 1671
	Rome	22,29,29 Aug 1671
	Vienna	29 Aug 1671
	Rome	6 Sep 1671
	Venice	12 Sep 1671
	Rome	12,12 Sep 1671
	Rome	19,19 Sep 1671
	Venice	16,26,12 Sep 1671
	Venice	2 Oct 1671
	Rome	26 Sep 1671
	Rome	3 Oct 1671
	Vienna	27 Sep 1671

Venice	10 Oct 1671
Bergamo	5,12 Oct 1671
Vienna	10 Oct 1671
Bergamo	26 Oct 1671
Venice	24 Oct 1671
Rome	17,17 Oct 1671
Vienna	17 Oct 1671
Bergamo	9 Nov 1671
Rome	31 Oct 1671
Rome	2 Nov 1671
Venice	14 Nov 1671
Rome	7,14 Nov 1671
Zurich	7 Nov 1671
Bergamo	23 Nov 1671
Vienna	7 Nov 1671
Venice	21 Nov 1671
Rome	21 Nov 1671
Vienna	14 Nov 1671
Venice	25 Nov 1671
Rome	28 Nov 1671
Venice	5,12 Dec 1671
Vienna	23,29 Nov 1671
Genoa	29 Nov 1671
Venice	16,19 Dec 1671
Vienna	6 Dec 1671
Rome	5,12 Dec 1671
Vienna	13 Dec 1671
Bergamo	28 Dec 1671
Bergamo	4,11 Jan 1672
Vienna	19,26 Dec 1671
Rome	26 Dec 1671
Venice	18 Jan 1672
Bergamo	25 Jan 1672
Vienna	9 Jan 1672
Bergamo	2,8,15 Feb 1672
Vienna	16,23,30 Jan 1672
Venice	3 Feb 1672
Venice	30 Jan 1672
Venice	6,24,27 Feb 1672
Rome	6,20,27 Feb 1672
Rome	5 Mar 1672
Genoa	14,21 Feb 1672
Bergamo	22,29 Feb 1672
Bergamo	7,14 Mar 1672
Vienna	20 Feb 1672
Venice	2,9 Mar 1672
Genoa	27 Mar 1672
Bergamo	28 Mar 1672
Warsaw	25 Mar 1672

Rome	12,26 Mar 1672
Vienna	13,20,27 Mar 1672
Vienna	2,9 Apr 1672
Bergamo	4,11,18,25 Apr 1672
Venice	13 Apr 1672
Vienna	16,24 Apr 1672
Venice	29 Apr 1672
Venice	4,6 May 1672
Bergamo	2 May 1672
Rome	16,23,30 Apr 1672
Venice	20 May 1672
Vienna	15 May 1672
Venice	25 May 1672
Bergamo	23,31 May 1672
Bergamo	6 Jun 1672
Vienna	22 May 1672
Genoa	26 May 1672
Bergamo	13 Jun 1672
Venice	10 Jun 1672
Vienna	4 Jun 1672
Bergamo	20 Jun 1672
Rome	22 Jun 1672
Vienna	22 Jun 1672
Bergamo	27 Jun 1672
Vienna	18,25 Jun 1672
Vienna	2 Jul 1672
Bergamo	11,18 Jul 1672
Venice	13 Jul 1672
Constantinople	9 May 1672
Rome	2 Jul 1672
Venice	20,23 Jul 1672
Vienna	9,16,23 Jul 1672
Bergamo	28,31 Jul 1672
Bergamo	8 Aug 1672
Rome	23 Jul 1672
Venice	29 Jul 1672
Rome	30 Jul 1672
Vienna	30 Jul 1672
Vienna	6 Aug 1672
Warsaw	27 Jul 1672
Vienna	14 Aug 1672
Warsaw	3 Aug 1672
Bergamo	29 Aug 1672
Leghorn	10 Aug 1672
Venice	24 Aug 1672
Rome	13,20 Aug 1672
Genoa	13 Aug 1672
Vienna	20 Aug 1672
Bergamo	5 Sep 1672

	Venice	31 Aug 1672
	Genoa	28 Aug 1672
	Vienna	28 Aug 1672
	Milan	10 Sep 1672
	Venice	9 Sep 1672
	Rome	3 Sep 1672
	Vienna	4 Sep 1672
	Venice	17 Sep 1672
	Rome	10 Sep 1672
	Vienna	11 Sep 1672
	Venice	21,24 Sep 1672
	Vienna	18,25 Sep 1672
	Milan	18 Oct 1672
	Warsaw	14,21 Sep 1672
	Vienna	2 Oct 1672
	Bergamo	20,27 Oct 1672
	Venice	19 Oct 1672
	Vienna	9 Oct 1672
	Rome	15 Oct 1672
	Bergamo	3 Nov 1672
	Vienna	16 Oct 1672
	Venice	19 Oct 1672
	Rome	22 Oct 1672
206v-207r	Extract of a letter in B's hand From Mr. Ranucci at Warsaw To M^{r}. Priandi 1p (16x20) Italian	14 Oct 1672
207r-235v	News accounts in B's hand 48pp (16x20) Italian/French	
	Vienna	29 Dec 1672
	Vienna	4 Mar 1673
	Vienna	28 Sep 1673
	Graz	23 Oct 1673
	Milan	7 Nov 1673
	Vienna	9 Nov 1673
	London	24 Nov 1673
	Extract of a letter from Madrid	8 Nov 1673
	London	25 Jan 1673
	London	4,8 Feb [1674 NS]
	Rome	3 Apr 1674
	Vienna	11,15,22 Nov 1674
	Vienna	13,20,27 Dec 1674
	Vienna	3 Jan 1675
	Vienna	21 Feb 1675
	Naples	18 Feb 167?
	fols 231-235, blank	

236r-249v	Extracts in B's hand 22pp (17x22) Italian "Relatione de gli ultimi successi di Candia. . .di Bartholemaeo Varisano Grimaldi" fols 247-249, blank	undated
250r-253v	Copy in B's hand 5¾pp (17x22) Italian "Relatione de Candia del Sig.[r] Marchese Villa, Aprile 1668" fol 253, blank	
254rv	Newsletter in B's hand 2pp (17x22) Candia	24 Oct 1667
255rv	Newsletter in B's hand 1½pp (17x22) Candia	4 Nov 1667
256r-257v	Newsletter in B's hand 1½pp (17x22) Candia fol 257, blank	24 Nov 1667
258r-259r	Copy of a letter 2½pp (23x33) "Mr de St. André Montbrun a Mr le Prince. . ."	7 Sep 1669
259r-260v	Copy of a letter 2pp (23x33)* [follows preceding letter] *fol 260 (20x29)	undated
261r-262v	Copy in B's hand 2½pp (19x29) Italian "Testamento della Citta' di Candia/ Coppia tratta dalli atti di Mr Pasquino nottaro/publico"	undated
263r-266r	Copy 6½pp (17x22) [Descriptions, military & political]	undated

266rv	Copy in B's hand 1½pp (17x22) "Ordre tenu a la reception & traictement fait au/Roy de Navarre par Mad. L'Abesse de Fonteurault/sa tante, ledit Roy l'ayant esté visiter"	undated
267r-268v	Copy/account From [Brugis?] To Anonymous 3pp (20x31) Latin	undated
269r-270v	News account in B's hand 2pp (15x20) Maestricht Vienna, etc.	 28 Aug NY 2 Sep NY
271r-272v	Account (fragment) in B's hand 1p (17x23) fol 272, blank	undated
273rv	From Anonymous To Anonymous 1p (19x20)	17 Dec 1627
274r-275v	Fragment, Anonymous 2⅓pp (20x27) Italian	[ca 1625]
276r-277v	From Marcello Zacherly To T. L. Burattini 3pp + ad (17x22) Italian	19 Nov 1655
278r-287v	Copy of an account, Anonymous 19pp & 3 lines (20x27) "Relation de ce qui s'est passé de cinq esclaves/que le Duc de Montalte avait amenes a Rome/qui s'y sont faits Chrestiens" [Rome; 9 Aug 1639]	
288r-289v	Copy of an account in B's hand 2pp (20x31) "Relation du Prince d'Ethiopie, qui est venu a Rome l'an/1633 sur le recit d'un Pere Recollect qui l'a accompagné/de Nazareth jusques a Rome," etc.	Feb 1633
290r-291v	Copy [concerns a Greek text] in B's hand 1p (17x22) Latin/Greek fol 291, blank	undated

Folio	Description	Date
292r-293v	List in B's hand $1\frac{1}{3}$pp (17x22) "Ministres des Princes qui ont eu un absolu pouvoir/dans les affairs de leurs Maistres, que nous appellons/Favorits" fol 293, blank	undated
294r-295v	Account in B's hand 1p (17x28) [concerns execution of five seditionaries at Rouen, 1640] fol 295, blank	7 Jan 1640
296rv	Newsletter $1\frac{1}{2}$pp (20x31) Venice	16 May 1654
297rv	Newsletter 1p (13x16) Latin Vienna	1 Jun 1656
298r-301v	Copy of a letter in B's hand 6pp (20x31) Italian [concerns the plague of Naples, 1656] fol 301, blank	29 Jun 1656
302r-304v	From Bordeaux at London To Anonymous 4pp & 3 lines (17x22)	26 Oct 1657
305rv	Copy & translation of a letter From Hermann Wrangel [Charles Gustaf Vranguel] To the Prince Palatine 1p (20x31)	24 Oct 1657
306r-307v	From Bordeaux at London To Anonymous 3pp & 2 lines (20x31)	9 Nov [1658?]
308r-309v	From Bordeaux at [London] To Anonymous $2\frac{2}{3}$pp (23x34)	3 Jan 1658
310r-311v	From Bordeaux at London To Anonymous 4pp (17x22)	29 Nov 1658

312r-313v	Copy of a letter Anonymous at Brussels 1p & 4 lines (17x22) fol 313, blank	29 Jun 1658
314r-315v	From Anonymous at Poznań To Anonymous 2pp (20x31) fol 315, blank	6 May 1658
316r-317v	From [Gehloër?] at Frankfurt To Anonymous 3⅔pp (17x22)	9 May 1658
318r-319v	Copy From Jehan [?]* To Anonymous ⅔p (17x22) *[Ambassador of Magdeburg] fol 319, blank	 undated
320r-321v	From [Bordeaux?] at London To Anonymous ⅔p (18x31) fol 321, blank	$\frac{16}{26}$ Apr 1658 [26 Apr 1658]
322r-323v	Extracts of letters 4pp (13x16) French/Italian	5,8,21,1 Oct 1658
324r-325v	Copy of a letter Anonymous at Pyrennes 2⅓pp (22x33)	28 Jun 1660
326r-328v	Newsletters 3pp (17x22) Cassovia, Latin [Hungary] Italian fol 327, blank	 5 Jul 1660 8 Aug 1660
329r-330v	Account in B's hand ⅔p (17x18) fol 330, blank	[20 Sep 1660]
331rv	Copy of a letter in B's hand From L. Warner at Pera To Statibus Gñalibus, foed. Belgii 1⅓pp (17x22) Latin	18 Feb 1662

332r-336v	Extracts of letters 6pp	
	Antwerp (20x31) Italian	3 Mar 1662
	Cologne (20x31) Italian	5 Mar 1662
	Rome, Madrid (20x31) Italian	25 Mar 1662
	Rome (17x22) Italian	27 Mar 1662
	[Geneva] (17x22) Italian	5 Apr 1662
	Spirâ [Speyer] (17x22) Latin	1 Apr 1662
	Vienna (17x22) Latin	23 Mar 1662
	Nuremberg (17x22) Latin	$\frac{31}{21}$ Mar 1662
	fol 336, blank	
337rv	Anonymous letter 1⅔pp (17x22)	12 Apr 1662
338rv	News account 1p (17x22)	
	Spain	30 Mar 1663
339rv	From Gamarra at The Hague To Anonymous ⅔p (20x31)	9 Jul 1663
340r-343v	News account 6⅔pp (17x22)	
	Rome	25 Sep 1663
	Genoa	3 Oct 1663
344r-345v	Newsletter 3pp (17x22)	
	Lyon	12 Oct 1663
346r-347v	Copy of a Mémoire* 2pp & 9 lines (20x31) *From Card. Landgrave of Hesse To the Duke of York	[dated 1664]
348r-349v	Copy of a letter in B's hand 2pp (17x22) Latin "Reur.do...Johanni [Grammans] Soc. Jesu... à R.P. Gruebec Româ 8° Martii 1664" fol 349, blank	
350r-353v	Printed text 6pp (14x20) "Relation veritable de ce qui s'est passé en l'Armée du Roy, commandée par M. le/ Duc. de Beaufort, depuis son arivée à/	undated

	Port-Mahon, jusques à la descente qu'el-/ le a fait à Gigiry en Afrique" fol 353, blank	
354r-359v	Printed text 10pp (15x21) "RELATION/DE CE QUI S'EST PASSE/A LA DESCENTE DES TROUPES/du Roy à Gigiry en Afrique...a Aix/chez Charles Nesmoz, rue du/College..." [1664]	
360r-363v	Copy of an account 3pp (17x22) [concerns above expedition] fols 362&363, blank	undated
364r-373v	Copies of letters & documents in B's hand 20pp (17x22) Extract of a letter From Anonymous at The Hague Copy of a letter From M. Van Beuninghen To the Estates Génerale des Prov. Unis, etc. [concerns vessels lost by the Dutch to the English, 13 Jun 1665] Extracts of letters From Anonymous at The Hague	 18 Jun 1665 19 Jun 1665 2,9,16,18 Jul 1665 22,23,30 Jul 1665 6 Aug 1665 17 Sep 1665 29 Oct 1665
374r-375v	From Anonymous at The Hague To Anonymous 1p & 1 line (19x30)	28 Oct 1666
376r-377v	Copy of a letter in B's hand 2pp (17x22) Italian "Relazione dell' horribile Terremoto sequito nella/città di Ragusa..." fol 377, blank	6 Apr 1667
378r-380v	Newsletter 4pp & 5 lines (20x31) The Hague [Camp near Douai]	 19 May 1667 [5?] Jul 1667

381rv	News accounts, military	18 Apr 1667
	2pp (17x22)	
382r-407v	Newsletters	
	31⅔pp (20x31)*	
	The Hague	22 Sep 1667
	The Hague	6,13,27 Oct 1667
	The Hague	11,4 Aug 1667
	The Hague *(17x22)	20 Sep 1668
	The Hague	4 Apr 1669
	The Hague	27 Mar 1670
	Paris	20 Jun 1672
	The Hague *(17x22)	14 Apr 1672
	[Camp of Dieren?] *(17x22)	25 Jun 1672
	[Mayoner?]	14 Jul 1674
408r-409v	Account	ca 9 Jan 1675
	1p (17x22)	
	fol 409, blank	
410r-415v	Newsletters	
	7pp & 9 lines (20x31)*	
	Camp of Cambray *(17x22)	28 Mar 1677
	The Hague	16 Dec 1677
	Metz, etc	22,28 Mar 1677
	Metz	5 Sep 1677
	Other camps	27,30,31 Aug 1677
416r-423v	Copies of letters in B's hand	
	10pp & 4 lines (17x22) Italian	
	Madrid	10,24 Feb 1677
	Madrid, etc	5 May 1677
	fols 422&423, blank	
424r-425v	Newsletter	
	1⅔pp (17x22) Latin	
	Leopoldi, etc	18 Jun 1677
		27 May 1677
	fol 425, blank	
426r-427v	Newsletter, extracts	1 Oct 1677
	2⅓pp (17x22)	
	[concerns armies of Créquy]	
428rv	Account	
	1p (17x22)	
	London	29 Nov 1677

429r-434v	Extracts of letters	
	10¼pp (20x31)* Latin/French	
	[Poland]	22 Apr 1673
	Warsaw *(17x22)	15 Dec 1673
	Heidelberg	25 Oct 1673
	Camp near Philisbourg	29 Oct 1673
	Camp near Neustadt?	2 Nov 1673
435r-439v	Newsletter	
	8pp (17x22)* Latin/French/German	
	Strasbourg	[21?] Oct 1675
	Cologne	17 Oct 1675
	Ratisbon *(20x31)	22 Mar [1665?]
	Venice *(20x31)	24 Oct 1669
440r-441v	Copy of a letter in B's hand	
	From Anonymous at D'Ispaham	1 Mar 1680
	3pp (17x22)	
442r-443v	Newsletter	
	2⅓pp (17x22) Italian	
	Vienna	10 Jul 1683

▼ ▼ ▼

COLLECTION BOULLIAU, VOLUME XXXVI, FONDS FRANÇAIS 13054

Folio	Description	Date
1r-2v	Copies of letters in B's hand $2\frac{1}{2}$pp (17x22) Latin "Innocentius Papa XI ad perpetuam rei memoriam" [J. Slusius, Rome, 23 Feb 1677] "Charissimo in Christo filio nostro Carlo Hispaniarum Regi/Catholico Innocentius Papa XI" [Rome, 6 Feb 1677]	
3rv	Copy of a letter (translated) $1\frac{3}{4}$pp (20x31) "Innocent XIe Pape, au Roy Tres Chrestien" [Rome, 24 Sep 1676]	
4r-5v	Account $2\frac{2}{3}$pp (20x31) [concerns Cardinal Odescalchy]	21 Sep 1676
6r-7v	List of Cardinals $\frac{3}{4}$p (17x22) Italian "Cardinali d'Innocentio XI" fol 7, blank	undated
8r-9v	Copy of a letter From Innocent XI at Rome To Louis XIV at Paris 2pp (17x22) Latin fol 9, blank	13 Nov 1685
10r-11v	Discourse $2\frac{3}{4}$pp (20x31) Latin "Discours fait au Pape dans/Le consistoire, par Mr. Le Cardinal/ D'estreés, sur la promotion des/ Cardinaux"	undated
12rv	List of Cardinals in B's hand $1\frac{1}{4}$pp (17x22) Italian "Nomi de'Cardinali viventi quando Papa Clemente X. morì,/...Julii 20 anno 1676"	
13r-15v	Copy in B's hand $5\frac{1}{2}$pp (17x22) Italian "Rimonstranze fatte al Sig.^{r} di Pompone Ministro/di Stato di S. Mta. Christianissma, sotto il di/2 Settemb. 1676" [B writes: "l'auteur de ce discours est le Sr. Abbé Atto Melani."]	

16r-23v — Copy of a letter in B's hand
2⅓pp (17x22) Italian
"Lettre au Roy Très Chr. du Pape./
Innocent XI. sur son election/au
Pontificat. 24 Sept. 1676..."
fols 18-23, blank

24r-45v — Printed text
43pp (20x31) Latin/French
"Recueil/Des Indults/Accordez au Roy/
par les Papes/Alexandre VII./et/Clement
IX./.../A Paris/.../1670/..."

46r-47v — Copy of a letter — undated
2⅓pp & 2 lines (20x31)
"Copie de la lettre du Roy Tres
Chrestien au Pape Clement IX"
[annotated by B]

48r-49v — Copy in B's hand — undated
⅓p (17x22) Latin
"Oratio habita a Dño. nřo. Clemente IX
in Consistorio ad fratres/purpuratos
circa nuncium pacis inter Christianiss.
et Catholicum."
fol 49, blank

50r-55v — Printed text
9½pp (18x26)
"Lettre/ecrite/au Roy/par Mr. De Lionne,/
sur l'Election de nostre S.P. le Pape/
Alexandre VII/Du 7.Avril 1655/.../a
Paris/.../1655..."

56r-57v — Copy in hand of [J-A de Thou?]
lp & 3 lines (20x31)
"Harengue du M. Duc du Chaulnes/au
Conclave" [Jan 1670]

58r-59v — Printed letter
2¼pp (21x30) Latin/French
"Bref de N.S.P. Le Pape/Alexander VII./
Envoyé à Monsieur/'Abbé Comte de Richelieu,/
au suject de son voyage d'Allemange."
[24 Mar 1664]

60rv	Printed page 1p (21x33) "Insigne, Nomen, Cognomen/S.R.E. Cardinalis creati à Sanctiss. D.N./ Innocentio Papa X./in Consistorio secreto,/sub die 14 Novembris 1644. et publicati in eodem Consi-/storio die 4 Decembris 1645.../Romae...[1645]"	
61r-62v	Copies in B's hand 3¼pp (20x31) Latin "Elogium" etc. [for various Popes]	undated
63r-64v	Copy in B's hand ⅔p (20x31) Latin "Verba summi Pontificis Alexandri VII. in consistorio/die 24. Aprilis 1656." fol 64, blank	
65r-68v	Printed document 6½pp (19x27) Latin "S.mi D.N.D./Innocentii/Divina Providentia/Papae X/Constitutio/.../ Romae.../[1646]"	
69rv	Copy in B's hand ½p (17x22) Latin/French "These proposee par un bacchelier appellé Guillou, &/qui devoit estre disputee le 10. Janvier 1656..."	
70r-71v	Copy of a letter in B's hand 1p & 4 lines (17x22) Latin "Venerabilibus fratribus Archiepiscopis,/ Episcopis, aliisque nomine cleri/ Gallicani Parisiis congregatis/Alexander Papa. VII." [20 Mar 1656] fol 71, blank	
72r-73v	Copy of a letter From Anonymous To Pope Urban VIII 2¾pp & 1 line (20x31) Latin	[20 May 1626]

74r-75v	Copy 4pp (20x31) Latin/French "Bulle du Pape Urbain VIII par/Laquelle il donne pouvoir au Cardinal/Barberini son Nepueu de venir en France/pour disposer Le Roy a la Paix/avec les autres Princes Chrestiens"	undated
76r-77v	Copy of a letter From Louis XIII at Paris To Parlement of France 1p & 5 lines (20x31) fol 77, blank	[12 May 1625]
78r-95v	Extracts of letters 26pp (15x19) French/Latin [Extracts from letters and books concerning diverse Church officials, 5 Aug 1594-Nov 1596.] fols 92-95, blank	
96r-101v	Extract 10pp (20x31) "Pierre Gonçalez de Mendoze, Car-/dinal, et Archevesque de Tolede./Extraict de l'Histoire du dict Cardinal, Mise par/escript par Pierre de Salazar, Chanoine de l'Eglise/de Tolede, Imprimée à Tolede, l'an 1625." fol 101, blank	
102r-104v	Bibliographic notes (fragments) ⅔p (20x31) French/Latin 2pp & 4 lines (14x15) French/Latin	undated
105r-110v	Copy of a discourse by Mgr Nuncio [Altoviti?] 12pp (20x31) Italian	undated
111r-122v	List & extracts of letters in B's hand 23⅓pp (16x22) French/Latin "Assemblee des Evesques & autres Notables a Orleans/l'an 1478, pendant les mois de Septembre et/Octobre." [followed by extracts of letters and mémoires]	

123r-124v	Copy of a letter in B's hand 2pp (16x22) Latin "Lettre du Cardinal Bessarion Legal envoyé par/le Pape au Roy Louiis XI." [15 Aug 1472] fol 124, blank	
125r-127v	Copy of a letter in B's hand 4⅔pp (17x22) Latin/French "Copie d'un escrit tiré de la Chambre des Comptes, qui/concerne les choses, qui furent avisees touchant la conclusion/ de l'assemblee d'Orleans, qui fut tenue soubs Louiis XI l'an."	undated
127r-136v	Copies and extracts, cont'd 15½pp (17x22) "Copie de l'appellation faicte par le Procureur/du Roy a l'Assemblee d'Orleans" and other extracts fols 135 & 136, blank	undated
137rv	Copy of a document 1p (20x31) Latin	[dated 14 Sep 1632]
138r-139v	Copy in B's hand 3½pp (17x22) Latin/French "Erection du Prieuré de la Chaise Dieu en Abbaye par le Roy/Henri I. l'an 1052· 12 kal. Octob. le 20 Septemb." [in B's hand: "20 Sept."]	
140rv	Printed fragment 1p (14x19) Latin "Extraict des Archives de Sainct Remy de Reims"	
141rv	Bibliographic fragment in B's hand ⅓p (17x22)	undated
142rv	Declaration of Henry III 2pp (22x56) French/Latin [folding at center]	29 Aug 1586
143rv	From Anonymous to [J-A de Thou?] 1p (20x31)	undated

144rv&147rv	Mémoire 2pp & 3 lines (20x31) "Memoire de L'affaire de la brande/ Chartreuse au grand Conseil de/La Haye En hollande"	undated
145r-146v	From Le Tellier at Fontainbleau To J-A de Thou 1p fol 146, blank	18 [?] 1659
148r-150v	Fragments 2pp (16x21 & 20x31) [concerns the affair, fol 144 above] fol 150, blank	
151r-152v	Printed letter and extracts (3) 4pp (20x31) "Copie de la Lettre Ecrite au Roy/par l'Assemblée generale de l'Oratoire, en datte du 23./Septembre 1678." Extract, Assembly Act Extract, order of General Assembly	
153r-154v	Copy of a petition 2pp & 6 lines (20x31) Latin "Illustrissimo et Reverendissimo Dn'o./ Domino Nuncio Apostolio"	undated
155r-160v	Copy of an Act of Henry III 8pp & 10 lines (20x31) Latin	[dated 1582]
161r-162v	Copy of an account in B's hand 2½pp (17x28) "Le Cardinal d'Este élu Abbe de Cluny à la place de Mazarin"	[24 Sep 1677]
163r-164v	Extract from archives of Cluny in B's hand 3⅓pp (17x28) Latin [concerns document of Pierre Du Perron]	
165r-166v	Extract from archives of Cluny in B's hand 1¾pp (17x28) fol 166, blank	

167r-170v	Copy in B's hand, Papal Bull at Rome 5⅔pp (17x28) Latin [signed Johannes Slusius] fol 170, blank	[1 Jun]
171r-172v	Copy in B's hand 2¾pp (17x28) [re: Cluny]	16 Apr 1669
173r-174v	Copy of a letter in B's hand by Johannes Gualterus Slusius, Sec of Innocent XI 3⅓pp (17x28) Latin	V Id Jan 1676 [9 Jan 1676]
175r-176v	Extracts in B's hand 3pp (20x31) [re: various Catholic orders]	undated
177r-178v	Copy in B's hand 1⅓pp (17x22) Latin "Charta donationis a Dagoberto Francorum Rege anno 706..." [Dagobert III] fol 178, blank	
179r-180v	Notes in B's hand From *Rev. Germ. Pistorii*, Tome I 3⅓pp (17x22) Latin	undated
181r-182v	Copy* of a document of Dagobert III 3pp (20x31) Latin *[copied 19 Nov 1664]	
183r-184v	Extract in B's hand 3⅔pp (20x31) Latin "Ex verti MS chronicorum excerpta contiente, una cum Pauli/Diaconi Historia miscella"	undated
185r-186v	Printed text 4pp (19x30) "Addition de Factum./Et obversations, concernant trois inscriptions de faux & moyens/fournis·/Par Maistre André de Murard..."	undated

187r-188v	Printed text 3pp (19x30) "Motifs de l'Arrest Rendu/au Parlement de Bourgongne" [concerns M. Vincent de St. Jacques arrested 10 Jul 1640]	undated
189r-192v	Printed text 8pp (21x31) "Remarques sur Quatre Missives..."	[ca 1658]
193r-194v	Printed list 4pp Latin [list of Ecclesiastics assembled 19 Jun 1669]	undated
195r-198v	Printed text 6pp (20x32) "Extrait des Pieces Produites dans l'Affaire/des cinq Abbayes de la Congregation de Chezalbenoist,..." [1488-1671] fol 198, blank	undated
199r-204v	Printed text 10pp (20x30) "Au Roy/Et à Nosseigneurs de son Conseil"/ [concerns 5 abbeys]	[ca 1677]
205r-217v	Notes & extracts in B's hand 22pp (17x22) "Domaine du Roy" "Noms d'Autheurs qui ont parlé de l'Amnistie" "Exemples de la Condamnation a mort de quelques Evesques" "Sur les Articles XV et XVI des Libert· de l'Eglise Gallic de M. P. Pithou" [Pierre Pithou], etc. "Remarques de Mr. Blondel touchant les merites et l'excellence de l'Eglise Gallicane" [followed by extracts from printed sources] fols 216&217, blank	

218r-219v	Copy of a letter From Louis XI, To Anonymous 1p & 4 lines (17x22) [concerning Cardinal Balüe] fol 219, blank	1 Apr 1475
220r-227v	Extracts of letters and documents in B's Hand 11pp French/Latin [Ecclesiastical affairs July 1315- 12 Apr 1456] fols 223&227, blank	
228r-283v	Copy 108½pp (20x31) "Relation de L'ambassade/de Messire Guillaume Cosinot/Chevallier seigneur de Montereuil/.../Pierre Gruel/.../et Guillaume le Franc,/.../envoyes devers le Pape Paul IIe pour obtenir/vicariat pour faire le proces aux Cardinal d'Angers [Jean Balue] et/Evesque de Verdun [G. de Haraucourt], criminelze/ de leze-majesté en l'année/1469." fol 229, blank	
284r-293v	Copy and notes in B's hand 20pp* (17x22) [concerns St. Ambrose, Valentinia II, Theodosius, etc.] *w/marginalia; page sequence inconsistent	
294r-299v	Extracts of letters on Socinianism From Theodore Bèze To Laurent [Durnhoffer?] and others 11pp & 6 lines (15x20) Latin 12 Apr 1579; 7 Feb 1584; 11 Feb 1584; 3 May 1584; 26 Oct 1584; 2 Apr 1572; Document: 14 Feb 1581	
300r-305v	Printed text 10¼pp (16x21) Latin "Libellus/Ad Univer-/sos et Singu-/los Senatores Prima-/rios...Hagae-Comtīs,/ Ex Officinâ Alberti Henrici/cIↄ.IↄC.	

306r-311v	Printed text 9¾pp & 8 lines (19x26) Latin "Joannis/Carmvelis Lubovischi/ad Petrum Gassendum•/Epistola/...Cum Petri Gassendi Responsione..."	
312r-313v	Copy in B's hand "Suppressio Congregationum S.ti Georgii in Alga Venetiarum/ac fratrum Jesuatarum, et Sti Hieronymi de Faesalis" 4pp (17x22) Latin	[ca 18 Dec 1668]
314r-316r	Copy in B's hand 4pp (17x22) Italian "Coppia d'attestatione fatta dalli sottoscritti testimonii"	6 Jan [NY]
316r-319v	Copy in B's hand, cont'd 5½pp (17x22) Italian "Copia del comandamento del Gran. Sig.re Tradotto/di Turcheseo in Italino" fol 319, blank	[28 Mar 1675]
320r-321v	Deposition 2pp & 8 lines (17x22) "De Sluck en la Russe blanche..."	[2 Jul 1667]
322r-329v	Mémoire 16pp (17x22) "Memoire pour Monseig.r Le President de Thou,/...vers Mess.rs les/Estats Gnaux. des Provinces unies du pays bas" [concerns Bishop of Utrecht]	undated
330r-339v	Copy of mémoire in the hand of J-A de Thou II 18¼pp (20x31) [Mémoire sur les catholiques des Provinces-Unies] "Memoire fait par order du Roy a St. Germain en Laye au mois du Juillet 1662..." [dated 14 Jul 1662, St. Germain en Laye, signed Jacques-Auguste de Thou]	
340r-341v	Mémoire Addressed to [J-A de Thou] 2½pp (20x31)	undated

342r-351v	Mémoire, Anonymous 17½pp & 5 lines (20x31) Italian	undated
352r-353v	From Soler at Delft To J-A de Thou at The Hague ⅓p + ad (20x31)	1 Mar 1659
354r-355v	Mémoire by J-A de Thou 3pp (20x31) [concerns the Affair of Alexandre More (Morus); see H. Harrisse, *Le President de Thou et ses Descendants...*, Paris, 1905, p 176, fn 1]	[25 Feb 1660]
356r-357v	Secret Mémoire 2½pp & 2 lines (20x31) [concerns the Conseil des Conferences, 1663]	undated
358rv	Extract in B's hand 1½pp (17x22) Latin "Excerptum ex epistola/...Joh. Christoph Wagenselii/ad D. H. Protestt R.P.R. Paris." [Altdorf, Non Kal Feb 1684]	
359r-360v	Extract in B's hand ¾p & 3 lines (11x17) "Extraict d'une lettre escrite de Strasbourg a Nuremberg par un Aleman qui y passa a la fin de Mars 1684," etc.	
361r-362v	Extract 10 lines (8x11) Latin "Ex Litteris Electoris Saxoniae ad Electorum Bran-/denburgicum" [Dresden, 12 Feb] fol 362, blank	
363rv	Copy in B's hand 1½pp (17x22) "Dans les Articles proposez en l'Assemblee de Nancy pour estre arrester/en la generale de Mars prochaine 1588-ceux·cy sont compris"	
364rv	Copy of a letter in B's hand ¾p (17x22) "Lettre de la Reine Marguerite au Roy Henri IV."	

365r-366v	Extracts 3pp & 5 lines (20x31) Latin "Summa Legationes Guisiana ad Pont. Maximum,/..." "Memoires de l'Advocat David"	undated
367r-378v	Copy in B's hand 21pp & 6 lines (20x31) "La Pragmatique Sanction," etc. [yrs. 1418-1585] [373r, "Des Elections;" fol 374r, "Investitures" relates material ca 682-ca 1159] fol 378, blank	
379r-389v	Printed text 20⅔pp (19x29) Latin "Manifestum/Pro Regno/Lusitaniae/ [1647]"	
390r-391v	Copy From Jean IV, King of Portugal To Pope Innocent X 2⅓pp (20x31) Latin	undated
392r-395v	Printed text, dated: Lisbon, 1649 5⅓pp (16x23) Latin [re: Jean IV & Innocent X] fol 395, blank	
396r-403v	Autograph Copy in B's hand 15pp (20x31) Latin "Pro Ecclesiis Lusitanicis..."	
404r-419v	Printed text 31pp (19x24) Latin "Pro/Ecclesiis/Lusitanicis/Libelli Duo/...Parisiis,/Ex Officina Cramosiana/ M.DC.LV" [section by Ismaël Boulliau]	
420r-427v	Printed text 13pp & 7 lines (14x19) Latin "Responsum Praecipuorum Doctorum/ Parisiensis/Academie Sorbonicae/ Potentissimo Portugalliae Regno." fol 427, blank	
428rv	Copy in B's hand, untitled ¾pp (20x31) Latin	undated

429r-430v
Copy of a letter — Prid. Id. Apr [1652?] [1 Apr (1652?)]
From [Tubeuf?] at Paris
To Pope [Alexander VII]
3pp* (20x31) Latin
fol 430, blank
*[there is an unnumbered folio between 429 & 430]

431r-432v
Copy of an account — [7 Jun 1641]
2⅔pp & 3 lines (20x31)
[concerns Assemblée du Clergé, 1641]

433r-440v
Printed text
12⅓pp (21x27)
fol 440, blank
"Protestation/de Monseigneur l'Illustrissime/et Reverendissme/ Messire Charles Maurice/Le Tellier,/ archeveque du de Reims/.../contre la Pretendue Erection,/de l'Eglise de Cambray en Metropole; avec la signification/de la dite Protestation/ Du 14me Février 1678/a Paris/.../[1678]"

441r-444v
Printed text
7⅓pp (21x31)
"Factum/pour Messire Claude Auvry, Evesque de Coustances/.../14 Decembre 1652..."

445r-455v
Notes & extracts concerning church history, Anonymous
16pp (20x30)
fols 450, 451 & 453, blank

456r-459v
Notes cont'd, church history, ca 1095-1106
5pp & 8 lines (20x31) Latin/French
fol 459, blank

460r-465v
Notes cont'd, church history
10pp (20x31)
fol 465, blank

466r-474v
Notes & extracts cont'd, church history, 11-12th centuries
18pp (20x31) French/Latin

475r-483v	Notes & chronology, church history, 11-12th centuries 17pp (20x31) Latin/French	
484r-487v	Treatise, Anonymous 6¾pp (20x31) "Traicté de la Clause Regnaus nostre/ Seigneur Jesus Christ."	undated
488r-495v	Notes & chronology, church history, 260-1200 AD 12pp (20x31) Latin fols 494 & 495, blank	
496r-501v	Extracts; catalogue of ecclesiastical works 12pp (20x31) Latin	
502r-505v	List, authors & books 8pp (20x33) Latin	
506r-513v	Notes & chronology, church history, 11-12th centuries 16pp (20x31) Latin fol 510, blank	
514r-516v	Fragment 6pp (20x31) Latin	undated

▼ ▼ ▼

COLLECTION BOULLIAU, VOLUME XXXVII, FONDS FRANÇAIS 13055

1r-25r	Bibliographic notes in B's hand, alphabetical 27pp (17x22) Greek/Latin fols 2-25, bibliographic notes fols 1,21,&23, blank
25v-26v	Extract in B's hand 3pp (17x22) Latin "Ex libris in Vitam Constantini"
27r-28v	Notes in B's hand 4pp (20x31) Latin/Greek "Socratis Scholastici Chronologia" [yrs. 306-439]
29r-34v	Notes in B's hand 11pp (20x31) Latin/Greek "Ex eodem Socrate" [notes from Bks. I-VII, pp 171-276]
35r-36v	Notes in B's hand 4pp (20x31) Greek/Latin "Theodoreti indiculus" [Bks. I-V]
37rv	Fragment in B's hand From Theophilo Alexandriae Archiẽpo 5 lines (4x17)
38rv	Fragment in B's hand 11 lines (7x12) Greek/Latin "In Episcorporum Illyricorum Synodo decretum/In Epistola ad Asiae Episcopos"
39r	Notes in B's hand ⅓p (20x31) Greek/Latin "Eugenius Tyrannus"
39v-42v	Notes in B's hand 7pp (20x31) Greek/Latin "Sozomeni Indiculus" [Bks. I-IX]
43rv	Fragment in B's hand 19 lines (11x14) Greek/Latin "Epistola Juliani...ad Arascium"

44rv	Notes in B's hand 17 lines (11x14) Greek/Latin "Soc. lib. 2·C·17, 3·C·9," etc.
45rv	Extracts in B's hand 16 lines (18x25) Greek/Latin "Ex Socratatis Hist. Eccles. Lib 7·C·21· p 271·Editionis Stephani"
46r-52v	Brief extracts in B's hand $7\frac{1}{4}$pp (17x22) Latin/Greek [from Socrates the Scholastic and from Eusebius], etc. fols 51 & 52, blank
53r-54v	Autograph dissertation $2\frac{1}{3}$pp (20x31) Latin/Greek Dissertation on the date of Christ's birth; dated: 29 November 1669; signed: Ismaël Bullialdus
55r-64v	Extracts & notes in B's hand 17pp (20x31) Latin/Greek "Tertulliani--Apolegetico;" "De Oratione;" "Alia ejusdem sententia varia circa Christanos militiam exercentes" "Adversus Hermogenem" "Adversus Valentinianos" "De Anima" "De Carne Christi;" "De Resurrectione Carnis" "Adversus Marcionem" [Lib. I-V] etc. "Ex Athenagorae Legatione ad Impp. M. Aurelium Antoninum, et Lucium Aurel/Commodum pro Christianis" "Aeonum Genealogia" fol 64, blank
65r-83v	Extracts & notes in B's hand 38pp (18x23) Greek/Latin "Ex Origene contra Celsum" [Bks. I-VIII] [fol 79, "S. Basilius Homilia in psalm 44" (15x18)]
84r-85v	Notes in B's hand 3pp & 2 lines (12x16) Latin Notes from "Collatio epistolae Cypriani ad Donatum" (pp 5,6,8)

86r-110v	Index of letters* of St. Jerome 7pp (20x31) Latin "S. Hieronymi epistolarum indiculus/ Ordines Ecclesiastici et quae ad Minesteria clericorum pertinent." "Consaetudines circa Sacramentorum usum" (p 4) "Ceremoniae Ecclesiasticae, et quae ad Liturgiam pertinent" (p 8) "Mores Christianorum, Politeia Eccelesiastica, et Concilia" (p 11) "Consuetudines Ethnicorum" (p 17) "Philosophi et Opiniones Philosophorum" (p 19) "Haeretici" (p 22) "Grammatica" (p 23) *[fols 86-110, originally numbered 1-24 by B] fols 87,88,90-92,94,95,98-102,104,106- 107, blank
111r-120v	Notes in B's hand 14pp & 3 lines (15x20) Latin "Ex Epistolis Diui Hieronymi" [notes from Jerome's Epistles 103,104,108, 109, etc.] fols 119 & 120, blank
121r-122v	Notes & extracts in B's hand 3pp (20x31) Latin/French [from Theodoretus] "Appendix prior ad postremam D Augustini editionem"
123r-126v	Extracts in B's hand 3pp (17x22) Latin "Ex Facundo hermianensi Episcopo Afro Ad Imperat. Justinianum" fols 125 & 126, blank
127v-134v	Copy, Greek text w/Latin translation in B's hand 14pp (19x26) Latin/Greek "Sanctissimi Patriarchae Hierosolymitani Domini/Gregorii Confutationum, ac Responsionum/adversus Becci dogmata, et scripta/C. XXXIII" [interleaved with Greek text]: "Ex MS Bibl. Regiae N° 1727"

135r-159v	Extracts in B's hand 37⅔pp & 5 lines (19x26) Greek/Latin "Chronico Bibliotheca Regiae N° 1349," [list of Popes to Honorius II] (fol 135) "Theophanes ad annum mundi 6036" Extract from MS N° 2216 (fol 136) Extract from MS N° 1027: "Ex Chronico Hebraico de Cain" (fol 138) "Ex MS Bibl. Reg. 1050" (fol 140) etc. "Ex MS Bibl. Reg. 1219" (fol 147v) "Patriarchus Aegypti" (fol 151) "Ex Cod. MS 5 Bibl. Reg." (fol 153) "Nicolai Cabasilae in Stolam sacram Sacredotalum considerationes" (fol 156v) etc. "Particula Panopliae Dogmaticae Evthimii Zigabeni à Frid. Sylburgio editus 1595 ex Bibliotheca Palatina" (fol 159) fols 139,149,&155, blank
160rv	Extract 1½pp (15x20) Latin "Ex Anonymo"
160v-174v	Extracts in B's hand 20⅓pp & 7 lines (15x20) Greek/Latin "Theophanes Monachus" (fol 160v) "In MS 1219" (fol 163) etc. "Ex Epistola Zonarae" (fol 171) "MS Bibl. Regiae 944" [and 1662, 1769 & 1067] fols 162&172, blank
175r-182v	Extract & notes in B's hand 3½pp (14x18) Latin "Epistola Decretali Clementis Papae I," etc. fols 177-182, blank
183r-184v	Extract in B's hand 2½pp (17x22) Greek "Ex MS Bibl. Regiae in quo/Theophanes et Syncellus continentur"
185r-186v	Extract in B's hand 1½pp (20x31) Latin "Ex Chronico Bibl. Reg. No. 79" "fol 15 - Nomina librorum sacre scripturae enumerat cum appellationibus Hebraïcis, Graecis et Latinis hoc pacto" fol 186, blank

187r-188v
Extract in B's hand
2pp (17x22) Greek/Latin
"Epitome Sacrorum et divinorum
canonum condita a pansebasto/
Nonophylace et Thessalonicae Braetore
Dnõ. Constantino Armenopulo"
fol 188, blank

189r-190v
Fragment in B's hand
10 lines (10x15) Latin
"Lanfrancus in epistolam Pauli Apostoli
...Ex MS Bibl. Puteanorum"
fol 190, blank

191r-194v
Extract in B's hand
5$\frac{2}{3}$pp (17x22) Latin
"Incipit Prologus in vita S$\overline{ci}$. Ac
Beatissimi/Iuliani Confessoris
Cenomanensis Episcopi"
fol 194, blank

195r-209v
Extract [in the hand of Rigault]
28pp & 2 lines (17x22) Latin
"Passio sanctae Agathae, ex veterib.
libris emandata"

210r-215v
Extracts & notes in B's hand
4$\frac{1}{2}$pp (17x22) Greek/Latin
"Ex Theodoro Balsamone Antiocheno
Patriarcha"
fols 213-215, blank

216r-223v
Extract in B's hand
10$\frac{1}{2}$pp (17x22) Latin
"Ex Petro Rhemensi quondam Cantoris/
Parisiensis, de Sacramentis, et casibus/
Scilicet de animae consiliis"
fols 222&223, blank

224r-225v
Copy in B's hand
3pp & 4 lines (20x31) Greek/Latin
Pope John VII to [Photius]

226r-244v
Extracts & notes in B's hand
37$\frac{1}{2}$pp (20x31) Latin/Greek
"Compendium Historiae Concilii Florentini
Oecumenici Scriptae/a quodam Ecclesiasticum
officium..."

245r-247v
Extract in B's hand
3pp (17x22) Greek
"Ex MS·BR·Historia Synodi Florentinae.
No. 1247"
fol 247, blank

248r-249v
Copy of a letter in B's hand
of Nephon, Archibishop of Constantinople,
1497.
2pp (17x22) Latin
fol 249, blank

250r-257v
Notes in B's hand [ca 1649]
12⅓pp (20x31) Latin/Greek
"Notes sur l'Histoire de Ducas"
fol 257, blank

258r-259v
List & notes in B's hand
3⅓pp (20x31) Latin
[list of Avignon popes and their Roman
counterparts; reflections on schism
and Gallicanism]

260r-269v
Extract in B's hand
19pp (20x31) Greek/Latin
"Ex MS Bibliothecae Regiae 916"

270rv
Copy of a letter of Edwardus Barton undated
½p & 4 lines (20x31) Greek

271r-272v
Copy of a letter [31 Oct 1599]
From Meletius, Patriarch of Alexandria
To Georges Dousa
1p (20x31) Greek/Latin
fol 272, blank

273r-275v
Copies of texts in B's hand
4½pp (20x31) Greek
of Cyrille Lucar, patriarch of
Constantinople, 1624

276r-277v
Extract of a letter in B's hand
2pp (17x22) Greek/French
"Extraict d'une lettre de Constantinople
par le Sr Nathanael Secretaire du Patriarche
Cyrille"
fol 277, blank

278r-280v	Catalogue $5\frac{1}{4}$pp (20x31) Latin "Catalogus librorum Socinianorum quos vidi" See MC, X, p 746, n 1	
281r-286v	Mémoire, Anonymous 12pp (18x27) Italian [concerns doctrines and acts condemned by the Italian Inquisition]	undated
287r-295v	Extracts in B's hand 9pp (17x22) Latin "Ex Cod. Theod. lib. XII...de Decurionib.·1·49·Constantii et constantis." [from the Theodosian Code and others] fols 290&294, blank	
296r-298v	Extract in B's hand 2pp (15x20) Latin "Ecclesiae Gallicanae in Schismate Natus" fol 298, blank	
299r-300v	Extract in B's hand 1p (17x22) Latin "Tertull. in Apologetico" fol 300, blank	
301rv	Extract in B's hand $\frac{2}{3}$pp (17x22) Latin "Ex Athenagore Legatione haec transsumpta sunt"	
302r-304v	Extracts in B's hand $1\frac{2}{3}$pp (13x18) Latin "Baronius Annal. Ecclesiast." and others fol 304, blank	
305rv	Extract, Council Canons 1p (20x21) Latin "MS 113 Bibl. Thuanae quil faut voir"	

306r-336v	Extracts and notes in B's hand 61pp & 6 lines (17x22) Latin/Greek "Rabi Moses...Directoris dubitantium" [from the Guide of the Perplexed of Maimonides]	
337r-338v	Short extracts in B's hand 2pp & 7 lines (17x22) Latin/Greek Extracts, Origen, Tertullian, Theodoretus, Eusebius, etc.	
339rv	Extracts in B's hand 1p (17x22) Latin/Greek Extracts, Cicero; Dion. Cassius, Josephus	
340r-341v	Short extracts 1p + ad (11x19) Greek/Latin Extracts, Eusebius & Tertullian, Dion. Cassius, etc. fol 341, blank	
342rv	Extracts & notes in B's hand 2pp (17x22) Latin/Greek "Angli cuiusdam Medici Propositiones de cultu Dei/Nomen illius est Worsley" [w/refutations by Boulliau]	
343r-344v	Extracts in B's hand 4pp (17x22) Latin/Greek "Tertullianus de Idololatria paulopost initium" "Theologia Phoenicum" "Clemens Alexandr."	
345r-346v	Treatise 2½pp (20x31) "Raisons et conjectures qui preuvent la/possession...du procès verbal des excorcismes de l'Official d'Auxonne" [1660]	
347r-350v	Treatise 6pp (20x31) "Raisons contre la Possession" fol 350, blank	
351r-352v	From Anonymous at Dijon 2pp (20x31) fol 352, blank	22 Dec 1660

353r-356v	Printed text 8pp (18x26) Latin "Instructio/Proformandis processibus in causis/Strigum, Sortilegiorum, et/ maleficiorum" [Rome 1657]	
357rv	To Cromhausen at Danzig ¾p (17x22) Latin	12 Dec 1662
358rv	Copy of a mémoire 2pp (17x22) "Pretendue possession des Ursulines de Loudun/anno 1634 tiree de l'histoire de l'esdit de Nantes" [Tome 2, fol 538]	
359r-360v	Extract in B's hand 2⅓pp (20x31) Latin "Ex Tractu Dnī...Joly cononici Eccles.../ de Assumptione B. Mariae Virginis"	
361r-364v	Extract in B's hand, copied in 1664 5⅔pp (20x31) Latin "Sociniani cujusdam Haeretici operis, cui Titulus, Imago Prodromi Veritatis Primaevae/Capita seu These" [Tract by Stanislas Lubienietzki] fol 364, blank	
365r-366v	Extract in B's hand 2½pp (17x22) Latin "Ex Codex No. 106 In MS Thuano est 52 epistola"	
367r-372v	Signed Mémoire by B 11pp (17x22) Latin/Greek "Canon concilij $\overline{IV}$ Toletani XXXVe. De Conventu Territorii et Basilicis ad quem/Episcopum pertinent"	
373r-382v	Extracts & notes in B's hand 12pp (17x22) Latin [notes on various councils] fols 379-382, blank	
383rv	Extract in B's hand ¾p (17x22) Latin "Ex Canone XV°, Concilii Emeritensis..."	

384r-386v	Extract in B's hand ¾p & 12 lines (17x22) Latin "Concilo Toletano IV," etc. fol 385, blank	
387r-389v	Extract in B's hand 3pp & 7 lines (17x22)* Latin "Concilium Turonense II celebratum anno Christi 567·can III," etc. *fol 389 (5x17)	
390r-391v	Extract in B's hand 3½pp (17x22) Latin "Ad Canonem III concili Turonensis II Nota/Canon III De compositione Corporis Domini in altari."	
392r-398v	Assorted extracts & fragments in B's hand 14pp (17x22) Latin/Greek [various councils]	
399rv	Copy of a letter in B's hand From Gregory I [Pope] To Emperor Maurice [of Byzantium] 1¼pp (20x23) Latin	undated
400r-401v	Copy in B's hand ⅔p (17x22) Latin/French "Affranchissement donné par le/Roy Charles de Bel à un de ses hommes de/ condition serve...1322 au mois de Septembre" fol 401, blank	
402r-403v	Extract in B's hand ¾p (17x22) Latin "Opus Thegani Chorepi Treuirens de Gestis Ludovici pü Imperatores" fol 403, blank	
404r-408v	Extracts in B's hand 3¼pp + ad (17x22) Latin [extracts from Platina concerning Zosimus, Anastasius II, Eugenius, Léon IV, etc.] fols 407&408, blank	

409r-410v	Extract in B's hand 13 lines (14x20) Greek/Latin "Canon 19·Synodi Nicaenae I" fol 410, blank	
411rv	Extract of a letter in B's hand 10 lines (17x22) Latin "Petri Blesensis [Pierre de Blois] epist. 45 ad [the Archbishop of Padua]"	undated
412rv	Notes in B's hand ¾p (17x22) Latin/French "Cap. 28 seu ultimo Act. Apostolor."	
413r-415v	Notes in B's hand ¾p (17x22) Latin [concerns canons of the councils of Laodicea and Toledo] fols 414&415, blank	
416rv	Notes in B's hand ½p (17x22) Latin [concerns canons of the councils of Venice (465), Rome (813), etc.]	
417rv	Extracts in B's hand 1½pp (17x22) Latin "Gregor. Turon. Histor. Lib. 9 c. 39-lib. 10. c.16"	

▼ ▼ ▼

COLLECTION BOULLIAU, VOLUME XXXVIII, FONDS FRANÇAIS 13056

1r-4v	Signed mémoire in B's hand 6¼pp (20x29) Latin/French "La naissance du Jansenisme et Molinisme" [B writes (4r) "En 1645 estant a Venise avec M^{r}. de Gremonville Amb^{r}. du Roy/il me pria d'escrire quelque discours touchant ce que je croyois des/Jansenistes & des questions du temps d'alors entre eux & les Jesuites/ Ismael Boulliau"]	
5r-6v	Copy of a letter From Pope Urban VIII at Rome 1p & 5 lines (20x29) Latin [concerning Jansenism] fol 6, blank	[24 Oct 1643]
7r-8v	Extracts of letters To Pope Innocent X 3pp & 2 lines (20x31) Latin/Italian [concerning Jansenism]	
9r-10v	Printed text 4pp (19x27) Latin "Decretum/Sanctissimi D.N./D. Innocentii/ .../Papae X./...Parisiis...1647..." [24 Jan 1647]	
11r-12v	From De St. Amour at Rome To M. Fleury 3pp + ad (20x31)	17 Aug 1652
13r-14v	Extract 2pp (17x22) "Extrait d'une lettre de Rome/Du 27 Juin 1653" [concerns Jansenism]	
15rv	Printed text, folding 1p (26x32) "Constitution de...Pape Innocent X" [condemns five propositions of Jansenism]	

16rv	Printed text 2pp (21x33) "Mandement de M^{sgr} L'Evesque D'Angers pour la Publication et l'Observance de la Constitution Cy-Dessus" [Angers, 14 Aug 1653, signed Henry, Evesque d'Angers, Musard]
17r-20v	Printed letter 7½pp (18x26) "Response du Provincial/aux deux primiers lettres de son ami du 2 fevrier 1656" [Pascal]
21r-24v	Printed letter 8pp (20x26) "Lettre Pastorale/de Monseigneur l'Evesque et Comte de Beauvais.../sur le suject des entreprises faites par le Chapitre de son/ Eglise Cathedrale contre sa Jurisdiction"
25r-28v	Printed letters 8pp (20x26) "ADVIS/De Messieurs les Curez de Paris/ ...sur le subject des mauvaises maximes de quelques nouveaux Casuistes" [Paris, 13 Sep 1656 & Rouen, 28 Aug 1656]
29r-30v	Printed text 4pp (18x26) "Table/des propositions Contenues/... de quelques-unes des plus dangereuses propositions de la Morale..."
31r-38v	Printed text 16pp (18x25) "Continuation/des impostures/Que les Jansenistes Publient/dans leur Lettres contre les Jesuites" [undated]
39r-42v	Printed letter 8pp (18x27) "Trezieme Lettre/escrite par l'Autheur des Lettres au Provincial/aux Reverends Peres Jesuites/du 30 Septembre 1656"

43r-48v	Printed letter 10½pp (18x26) "Lettre/escrite/a une personne/de condition,/ sur la conformité des reproches/et des calomnies que les Jansenistes publient/ contra les Peres de la Compagnie de Jesus/ Avec Celles/Que le Ministre de Moulin a publiées...à Geneve en l'annee 1632"	
49r-54v	Printed text 11pp (18x24) "Bulle/de N.S. Pere le Pape/Alexandre VII/ Leuës, publiées, et registrée en Parlement, le Roy y seant,/le 19 Decembre 1657... a Paris...1657" [concerns the 5 propositions]	
55r-56v	Text, untitled 3½pp (20x31) Latin	undated
57r-66v	Printed declaration 16½pp (19x26) French/Latin "Declaration/du Roy/sur la Bulle de N.S.P. le Pape Alexandre VII/...au su-/ ject des cinq Propositions extraites du Livre/de Jansenius, intitulé *Augustinus*. Leuë, publiée et registrée en Parlement, le Roy y seant,/le 29 Avril 1665...a Paris...1665" fol 66, blank	
67rv	Printed list 1p (19x29) "La Liste De Seigneurs/qui composent L'Assemblée Générale/du Clergé de France..."	
68r-69v	Copy of a letter, Anonymous To Pope Alexander VII 2pp (20x31) Latin fol 69, blank	29 Jul 1663
70r-71v	Extracts of letters in B's hand From Clement IX at Rome To [Various French Bishops, etc.] 1½pp (20x31) Latin fol 71, blank	19 Jan 1669

72rv	Printed text 1p (19x26) Declaration of Antoine Arnauld, at Paris [concerns Frequente Communion]	[14 Mar 1644]
73r-76v	Printed text 5pp & 6 lines (18x25) "Ordonnance/de Monseigneur/...Hardoüin de Perefixe/Archevesque de Paris... a Paris...1667."	
77r-78v	Printed text 4pp (19x25) "Arrest/du Conseil/D'Estat/Contre la Traduction du Nou-/veau Testament imprimée/à Mons/Du 22 Novembre 1667/" [Paris, 1667]	
79rv	List 1p (14x18) [names & cities]	undated
80r-83v	Extracts of letters in B's hand 8pp (17x22) Italian/Latin Extracts from Vienna, 15 Feb 1661, and 6 Feb 1661 Copy: "Epistola Patris Valeriani Magni ad P. Ludonicum a Sale ejusdem ordinis Viennae...18 Feb 1661" See MC, XIII, p 481, fn 1	
84r-97v	Various extracts and copies in B's hand (17x22) Italian/Latin From P. Valeriani Magni To the Theological Faculty of Paris, Papal Nuncios, etc.	10 Jul 1656 28 Apr 1656 5 May 1656 5 Jul 1655 10 Feb 1631 1 Apr 1633 26 Jun 1656
98r-103v	Copy of a letter in B's hand 8½pp (17x22) Latin/French From Clergy of France To Pope [Innocent XI] [B notes "Le Vendredi XI Juin 1677. cette lettre fut leüe dans le conseil du Roy environ la moitié & trouvee fort belle...] fol 103, blank	

104r-111v	Printed text 14⅓pp (10x18) "Sentiment/des/Jesuites/touchant/le Peché philosophique/...a Paris... 1690"	
112r-113v	Fragment 9 lines (17x22) fol 113, blank	undated
114r-117v	Printed text 8pp (16x20) Latin "Rationes/Quod non expediat novam Academiam Leopoli: PP. So-/ciet: JESU erigere"	
118r-119v	Printed text 4pp (16x20) Latin "Rationes/Quod non tantùm expediat, sed etiam necesse sit Academiam/ Leopoli in Collegio Patrum Societatis JESU erigere" [Warsaw, 1661]	
120r-131v	Copy of a letter 20pp & 2 lines (17x22) "Lettre de M. l'Archev. d'Ambrun/ a/M. l'Archev. de Paris 22 Juin 1686" fol 131, blank	
132r-137v	Printed text 8½pp (19x26) "Arrest/du Parlement/sur le Decret de l'Inquisition,/Contre plusieurs Propositions, tirées de differends/ Casuistes/A Paris...1679" fol 137, blank	
138r-145v	Copy of a letter 6pp & 5 lines [B notes: "Lettre sans le nom de la Dame qui l'a escrite, selon l'aduis des plus/clairs voyans elle est partie de la main d'un scavant et addroit personage/pour faire voir l'inconstance du Siecle"] fol 145, blank	

146r-149v	Printed letter 7pp & 3 lines (19x26) Latin "Epistola/.../D.D. Jacobi Boonen/ Archiepiscopi Mechliniensis/ad Eminentissimos Cardinales Inquisitio-/nis Romanae..." [Brussels, 17 Juillet 1654]	
150r-165v	Discourse, untitled, Anonymous 24pp (20x31) Italian [contra Society of Jesus; Fr. Paolo?] fols 163-165, blank	undated
166r-167v	Copy in B's hand 1½pp (20x31) "Dialogue d'entre le P. Arnoux et Mad. la Comtesse de Soissons" fol 167, blank	dated Mar 1619
168r-171v	Notes in B's hand 3pp (20x31) [concerning "le P. Guiot" dating from 8 Apr 1625 and "L'Affaire du P. Guyot"] fols 170&171, blank	
172r-175v	Copy in B's hand 6pp (17x22) Italian/French "Relation de ce qui se passa a Pignerol au mois de/Juillet 1658. entre Mad. la Marquise de Dade fille de/ Mr le Prince Thomas de Savoie, et deux P.P. Jesuites/le Pere Scarducci et le Pere Bosco Predicateur/Milanez. faict par lad. dame Marquise" fol 175, blank	
176r-181v	Copy of a letter From Anonymous at The Hague 10¾pp (20x31)	3 Feb 1659
182r-183v	Copy of a letter [Lettre Supposée] From Johannes Paulus Oliva [Jean Paul Oliva] To P. la Chaise et P. Mainbourg ½p (14x20) Latin fol 183, blank	8 Nov 1680

184r-191v	Copy in B's hand 14½pp (17x22) "Discourse sur le different d'entre le Roy/et Monseigneur le Cardinal de Retz/Archevesque de Paris"	
192r-193v	Fragment 4pp (18x27)	undated
194r-200v	Extracts & notes in B's hand 14pp (17x22) "l'Affaire de Mr le Cardinal de Raiz"	
201r-202v	Copy of a letter in B's hand From de Retz at Marena To [Alexander VII] at Rome 1p (20x31) fol 202, blank	5 Aug 1656
203r-204v	Copy of a letter From de Retz at Paris To the Pope at Rome 1p (20x31) Latin fol 204, blank	undated
205r-209v	Printed letters 8¼pp (19x24) "Lettres de Mr/le Duc de Crequy, au Roy/envoyee par sa majeste/a M. le Mareschal d'Aumont/Gouverner de la Ville de Paris...à Paris...1662"	
210r-214v	Printed letter 7pp & 3 lines (18x25) "Lettre/du Roy/Envoyee à Monsieur le Mareschal/d'Aumont...à Paris 1662"	
215r	Copy of a letter in B's hand From Louis XIV at St-Germain-en-Laye To Pope Alexander VII at Rome 1p (17x22)	30 Aug 1662
215v-216v	Copy of a letter in B's hand 1p (17x22) "Louis XIV aux Cardinaux...30 Aug 1662" fol 216, blank	

217r-218v	Copy of a letter in B's hand From Pope Alexander VII at Rome To Louis XIV 1¾pp (17x22) Latin fol 218, blank	28 Aug 1662
219r-220v	Copy in B's hand 2pp & 7 lines (17x22) Latin "Discours du P. Alexandre VII faict aux Cardinaux dans le Consistoire du/4 Septembre 1662"	
221r-222v	Copy in B's hand 1p & 7 lines (17x22) "Conseil du Cardinal Sacchetti au Pape Alexandre VII/de ne point entreprendre la guerre sur le different qu'il eut avec S.M.T. en 1662·63 en 64" fol 222, blank	
223rv	Copy of a letter in B's hand From Alexandre VII at Rome To Louis XIV 2pp (17x22) Latin	12 Sep 1662
224rv	Copy of a letter in B's hand From Louis XIV To Mr le Duc Cesarini ½p (17x22)	15 Sep 1662
225r-230v	Copy of a letter in B's hand 10pp (17x22) "Copie d'une lettre de Mr de Lionne a la Reine Christine de Suede sur les affaires de Rome du 15e Septembre 1662 a St. Germain" fol 230, blank	
231r-232v	Printed news 4pp (18x25) "Nouvelles Ordinaires," no 102 [2 Sep 1662]	
233r-236v	Printed news 8pp (18x24) "Gazette," no 103	

237r-238v	Printed news "Nouvelles Ordinaires," no 118 [7 Oct 1662] 4pp (18x24)	
239r-242v	Printed news "Gazette," no 119 8pp (18x24)	
243r-244v	Copy of a letter in B's hand 4 pp (17x22) "Lettre du Roy a la Reine Christine de Suede"	21 Dec 1662
245r-246v	Copy of a letter in B's hand 1p & 5 lines (17x22) Latin "Lettre a Mr de Creqùi Ambassadeur a Rome, escrite par le Sacré College des Cardinaux..." fol 246, blank	23 Nov 1663
247r-248v	Copy of a letter in B's hand 2pp & 9 lines (17x22) Latin/French "Lettre du Sacré College des Cardinaux au Roy du 23 Nov. 1662"	
249rv	Copy of a letter in B's hand From Alexandre VII at Rome To Louis XIV at Paris 2pp (17x22) Latin	22 Oct 1662
250rv	Copy of a letter in B's hand ¾p (17x22) Italian "dal Sig.^{r} Card. [Flavio] Chigi al Ré di Francia," 22 Oct 1662	
251r-258v	Mémoire 16pp (20x31) Italian "Compendiosa relatione del fatto sequit tia la famiglia dell/ Ambasciatore francese, e li corsi della militia del Papa" [Rome: 26 Aug; 2 and 9 Sep 1662]	

259r-262v	Mémoire 7pp (20x31) "Second manifeste touchant l'insulte fait a Rome a Monsr. de Créqui, Ambassadeur extraordinaire/du Roi au mois d'Aoust de l'annee 1662"	Aug 1662
263r-264v	Newsletter, Anonymous 4pp (20x31) Italian Rome	21 Jul 1663
265r-266v	Newsletter, Anonymous 4pp (20x31) Italian Rome	28 Jul 1663
267r-268v	Newsletter, Anonymous 4pp (20x31) Italian Rome	18 Aug 1663
269r-292v	Printed news 48pp (19x24) "Nouvelles Ordinaires" and "Gazette," nos 38,39,35,36; 22,23,20,21	Jan, Feb, Mar 1663
293r-294v	Copy of a letter From Alexandre VII at Rome To Louis XIV at Paris 3pp (20x31) Italian	27 Jan 1663
295r-296v	Printed text 4pp (17x25) "Lettre/d'un gentil-/homme Génois, contenant/ce qui s'est passé en la sortie/du Cardinal Impériale, de la/ville de Génes, par l'ordre/du Sénat," Genoa, 6 Feb 1663 Printed, 24 Feb 1663	
297r-298v	Note in B's hand 3pp (17x22) [concerns Créqui affair etc.]	undated
299r-301v	Copy of verse, anonymous 4pp (20x23)* Italian [satirical verse regarding the Imperial Cardinal] *fols 300&301 (20x31)	undated

302r-303v	Copy in B's hand 2⅓pp (17x22) "Escrit donné par le Roy à Mes. les Amb.rs. du Roy Catholique,/ et de la République de Venisi"	[4 Jan 1664]
304r-305v	Copy of a letter in B's hand 2⅔pp (17x22) "Lettre du Roy a Mr le Cardinal de Medicis du 8/Janvier 1664"	
306r-307v	Note in B's hand ½p (17x22) "Ce qui se passa entre le Sr. Rasponi et Mr de Crequi avant la Traicté de Pise" fol 307, blank	undated
308r-319v	Printed text 22¼pp (18x23) "Traité/de Pise/Entre/nostre Tres- Saint Pere le Pape/Alexandre VII/ et/Tres-Haut...Louis XIV...Du 12 Fevrier 1664/a Paris...1664"	
320r-331v	Printed text 21pp & 6 lines (19x25) Italian "Trattato/di Pisa,/Tra/il Santissimo nostro Padre/Alessandro VII/...Luigi XIV...in Parigi...1664"	
332r-333v	Copy of an account in B's hand Rome 1p & 2 lines (17x22) Italian [concerns ratification of the Treaty of Pisa] fol 333, blank	 15 Mar 1664
334r-337v	Printed verse by Claudius Guilletus 5½pp (19x28) Latin "Ad/Alexandrum VII/Pontificem Opt. Max./Pro Pace/Inter Reges Conciliianda/ Carmen Protrepticon/" fol 337, blank	undated

338r-349v | Mémoire | undated
18½pp (18x28)
"Memoire/pour monstrer que l'omission/du titre de *Roy de Navarre* dans/le pouvoir donné par le pape au/sieur Raspony son plenipoten-/tiare pour le traicté de Pise,/n'est pas de l'importance que/quelques-uns sè sont Imaginé"
fols 348&349, blank

350rv | Extracts in B's hand
¾p (17x22) French/Italian
Concerning "Claude Frederic Gillot a Francia Notaire Apostolique" undated;
From Anonymous at Rome, 9 Jun 1668

▼ ▼ ▼

COLLECTION BOULLIAU, VOLUME XXXIX, FONDS FRANÇAIS 13057

1rv&3r-71v	Printed text by Charpentier of the Académie française 136pp (20x27) "Relation/de/l'Etablissment/de la Compagnie Francoise,/pour le commerce/ des Indes Orientales...A Paris,... 1665" fols 3-6, letter, Charpentier to Louis XIV fols 7-9, Table des Matieres fols 10-71, Text (pp 1-123)	
2rv	Note ⅔p (18x25)	[dated Oct 1716]
72r-81v	Manuscript copy, Anonymous 16pp (20x31) [concerns deliberation of 7 Nov 1664 determining costs of vessel armaments] fols 80&81, blank	25 Apr 1665
82r-83v	Copy, Anonymous 2pp & 2 lines (20x31) "Personnes necessaires a la conduite d'une Gallaire"	undated
84r-85v	Copy, Anonymous 1½pp (20x31) "Estat du coust et esquipages de 12 fregattes" fol 85, blank	undated
86r-91v	Extracts in B's hand 8½pp (17x22) "Extraict de l'Instruction donnee par Sr Caron (qui a servi/30 ans la Compagnie des Indes Orientales des Provences/Unies) a Mr Colbert pour la Compagnies Françoise aux mesme/Indes/pour la China" "Extrait de l'Instruction touchant le Japon" fol 91, blank	undated undated

92r-99v	Printed list 13½pp (19x25) "Liste des Interessez/en la compagnie/ des Indes Orientales, qui ont voix/ active et passive pour la nomina-/tion des Directeurs Caissiers et/Secretaire de la Compagnie" fol 99, blank	Mar 1665
100r-103v	Copy, Anonymous 6pp & 13 lines (22x31) [Statues of commerece of the Compagnie des Indes Orientales]	undated
104r-109v	List 10pp (21x31) [list of trade goods, arms, instru- ments, etc.] fol 109, blank	undated
110r-117v	Copy of a letter 12½pp (20x31) "Lettre escrite par Messieurs les Directeurs de la/Compagnie du commerce des Indes Orientales, a Messieurs/Les Officiers du Con.el. de l'isle Dauphine..." fol 117, blank	21 Jun 1665
118r-121v	Copy, Anonymous 6pp (20x31) "Division Generalle des Colleges et subdivision/particuliere des Employes... de la Compagnie des Indes Orientales..." fol 121, blank	[1665]
122r-131v	Copy, Anonymous 17pp (23x35) "Reglements pour...administration de la justice...a la Compagnie...Indes Orientales establie...Isle Dauphine..." fol 131, blank	undated
132r-133v	Mémoire, Anonymous 2⅓pp & 4 lines (20x31) "Memoire du Reglement qui sera necessaire d'observer/a l'énvoy de deux Houcres d'avance a Madagascar"	undated

134r-135v	Copy 3pp (23x43) "De Par le Roy/Et Messieurs les Directeurs de la Chambre generalle/ de la Compagnie du Commerce des indes orientales" [reglement]	undated
136r-137v	Extract 1p (20x31) "Extrait du livre des deliberations des/directeurs de la Chambre generale de la Compagnie/du commerce des indes orientales" fol 137, blank	27 Jun 1665
138rv	Printed avertissement 11 lines (23x24)	undated
139r-140v	Printed avertissement $2\frac{1}{3}$pp (18x26)	undated
141r-152v	Printed text $19\frac{2}{3}$pp (17x23) "Articles/et conditions/sur les-quelles les Marchands Ne-/gotians du Royaume supplient/tres-humblement le Roy de/leur accorder sa Declaration, et/les graces y continuës pour l'e-/tablissement d'une Compagnie/pour le commerce des Indes/Orientales/ A Paris/1664" fol 152, blank	
153rv	Printed price list 1p (9x29) [price list of trade goods, 17 Sep 1668]	
154r-155v	Printed, 2 identical price lists 2pp (12x35) [items arriving in France May, Jun 1672]	
156r-157v	Printed, 2 identical price lists 4pp (12x35) [items arriving in France 10 Sep 1670]	

158r-161v	Copy, Anonymous 3½pp (22x31) "Instruction pour les navires partant de France/pour aller a Madagascar et pendre leur route la plus assuree." fols 160&161, blank	undated
162r-169v	Copy 12½pp & 3 lines (20x31) "Copie du Contract fait avec Louis de Charbonier d'Alençon/pour conduit des Colons et establir deux Forges/ dans l'Isle Dauphine·19 du Aoust 1665 [169v]	
170r-185v	Copy (sent by Mr. Janot) 28⅔pp & 2 lines (23x35) "Instruction Des vents qui se/ Rencontrent.../ès passages des Pays-Bas en l'Isle de Java"	undated
186r-195v	Copies of letters Between M. J. de Bebben; M. Mariage and M. Du Pont at [Ispahan] And the Directors of the French East Indies Company 17½pp (20x31)	18 Jul 1665 8 Nov 1665
196r-201v	Extract 9½pp (20x31) "Extrait du Journal du Sr. de la Boullaye President/au conseil. Ledit S[r] de la Boullaye/Du 6 Septembre 1665 en Hispahaam" fol 201, blank	
202r-209v	Copy of a letter 11⅓pp (20x31) From M. Mariage "A Messieurs les Directeurs de la Compagnie des Indes Orientales" [Chiras, 17 Jun 1667] fols 208&209, blank	17 Jun 1667
210r-211v	List 1½pp (20x31) [publications concerning the French East India Company]	undated

212r-215v	List (same hand as fol 210) 2pp (20x31) [list of proper names] fols 213-215, blank	undated
216r-229v	Index and Notes, Anonymous 27pp (17x22) "Table de ce que contient le Livre Intitulé: la Loy/des vaisseaux, dedié au Roy d'angleterre" "La Carte au Commerce..."	undated
230r-231v	Trade information 2⅔pp (20x31)	9 Jun 1657
232r-233v	Trade information 1½pp (20x31) fol 233, blank	[2?] Jun 1657
234r-237v	Trade information 5pp (20x31) fol 237, blank	1 Jun 1657
238r-241v	Trade information 3pp & 3 lines (20x31) fols 240&241, blank	8 Jun 1657
242r-249v	Trade information 11pp (20x31) fols 248&249, blank	12 Jun 1657
250r-279v	Extracts, Anonymous 57pp (20x31) [extracts & information concerning the French East India Company; same hand & format] fol 279, blank	undated
280r-283v	Copy of a decree [Conseil du Roi] 7½pp (23x31)	20 Dec 1657
284r-285v	Printed letter From Don Carlos To the General Chamber at Madrid [Paris, 1669] 4pp (20x30) Spanish/French	1 Nov 1669

286r-297v	Printed text 24pp (19x25) "Status, Ordannances et/Reglemens que sa Majesté veut estre observez/ par tous les Marchands Maistres Teinturiers en/soye, laine et fils des Villes et Bourgs des son/Royaume" [internally numbered pp 53-76; see below]	[1669]
298r-323v	Printed text 51pp (19x25) "Reglemens/et Status/Generaux/pour les Longueurs, Largeurs,/et Qualities des Draps, Serges, et autres...à Paris 1669" [internally numbered pp 1-52] [fol 286r (another set of reglements; see above) continues this pagination, pp 53-76] fol 323v, dated 20 May 1669, St-Germain-en-Laye	20 May 1669
324r-327v	Printed text 7pp (18x25) "Arrest du Conseil/de Commerce,/Donné en faveur des Marchands, et de/tous autres qui seront cy-aprés bastir/ des Vaisseaux, ou trafiquerant sur/ mer/Du 5 Decembre 1664/a Paris/... 1664"	5 Dec 1664
328r-331v	Printed text 5pp (18x25) "Extrait des Registres/du Conseil d'Estat," Louis, 21 Jan 1665 fol 331, blank	
332r-335v	Printed text 5½pp (18x25) "Extrait des Registres/du Conseil d'Estat," César de Vendosme and M. Matharel, 30 Mar 1665 fol 335, blank	
336r-337v	Printed text 1½pp (18x25) "Extrait des Registres/de Parlement," Du Tillet, 24 Jul 1665 fol 337, blank	

338r-341v	Printed text 4½pp (18x25) "Arrest/Du Conseil/D'Estat/...A Paris/ 1666," includes 2 printed letters of De Lionne, 22 Mar 1666 fol 341, blank	
342r-343v	Copy, Anonymous 3½pp (20x31) "Extrait des Registres du Conseil d'Estat."	7 Dec 1666
344rv	Copy of a letter From Anonymous at The Hague To Louis XIV at Paris 1p (20x31)	20 Jul 1667
345r-348v	Printed text 5½pp (18x24) "Extrait des Registres/du Conseil d'Estat," includes 2 printed letters of Le Tellier fol 348, blank	21 Sep 1668
349r-356v	Printed text 16pp (14x20) "Relation/Succincte/de tout ce qui s'est passé de plus considerable en la guerre que la/*Compagnie Hollandoise* des *Indes Orientales* a euë contre le Roy et les au-/tres jusques à l'annee 1669" [printed in Holland]	24 Jul 1670
357r-360v	Printed text 4½pp (20x24) "Arrest/Du Conseil/D'Etat/Portant que les Passeports/pour les Vaisseaux François qui iront aux/Iles de l'Amerique, seront donnée par le/Roi" fol 360, blank	12 Jun 1669
361r-362v	Printed text 2½pp (20x24) "Arrest/Du Conseil/De Commerce/.../ Du 7 Dec 1669/A Paris/.../1669"	

363r-366v Printed text
5¼pp (20x24)
"Edict/Du Roy/Qui Declaire le Commerce/
de Mer ne point déroger à la Noblesse/
Du mois d'Aoust 1669/.../A Paris/.../
1669"

367r-368v Printed text 7 Jan 1671
2¾pp (20x24)
"Arrest/Du Conseil/D'Estat:/Qui defend
a tous les sujets de sa/Majesté de
charger aucunes Eaux de Vie sur les
Vaisseaux/Holandois dans toute l'étenduë
des costes et Ports Maritimes de son
Royaume.../A Paris/.../1671"

369r-370v Printed text 14 Apr 1672
3pp (20x24)
"Ordonnance/Du Roy:/Portant Permission
aux Hollandois/qui peuvent estre dans
le Royaume, d'y demeurer encore six
mois.../Du 14 Avril 1672/A Paris/.../
1672"

371r-374v Printed text
7pp (20x24)
"Edit/Du Roy:/Portant Revocation/de la
Compagnie/des Indes Orientales./Donné
à S. Germain en Laye au mois de Decembre
1674, Registré en Parlement et Chambre
des Comptes les 18 Janvier et/9 Février/
A Paris/.../1675"

375r-376v Printed text
2½pp (20x24)
"Ordonnance/Du Roy:/Pour l'Ouverture/
des Ports du Royaume, et la liberté/
du Commerce./du 7·Juin 1678/A Paris/.../
1678"

▼ ▼ ▼

COLLECTION BOULLIAU, VOLUME XL, FONDS FRANÇAIS 13058

1rv		Title page: 1p (20x29) "suppl. fr. 997bis/Observations/ Astronomiques/Par mr. Bouillaud et autres auteurs./Carton 430/LL./25 Cahiers Mss ou Imprimés"	
2r-8v		Tracts, Comet 1664-1665 9pp	
	2r	[Observations, Florence] 1p (19x26)	7 Apr 1665
	3r	Drawing, comet between Andromeda & Pegasus 1p (19x26)	undated
	4r	Drawing, "Observationes Cometae novissimi in/Pegaso/a die 5:[a] ad 9 Aprilis mane/in crepuscolo Roma" 1p (26x36)	
	5r	Drawing, "Via Cometae per fixas mense Decembri 1664 Romanis observationibus Bayeri Tabulae..." [Cometary path, 18-29 December] 1p (30x38)	
	6rv	Observations of ☄ 3 Jan 1665; 30&31 Dec 1664 2pp (16x20)	
	7rv	Observations of ☄ 9 Jan 1665 2pp (16x20)	
	8r	Observations of ☄ [Strasbourg, 25 Dec, OS] 1p (15x20) [fol 8r, 2 lines in B's hand:] "Ce papier et les Imprimés Allemands m'ont esté donnez par Mr Johan./ Christophor Wagenseil Gentilhomme de Nuremberg"	
9r-19v		Thermometric observations: in graph form giving date, hour, degrees and sky condi- tion (3 or 4 obs./day) [in Des Noyers' hand] 11pp + ad (15x20)	
	9r	Warsaw	19-26 Jul 1659
	10r	Warsaw	12-19 Jul 1659
	11r	Warsaw	2-11 Jul 1659

12r	Warsaw	25 Jun-2 Jul 1659
13r	Warsaw	11-18 Jun 1659
14r	Warsaw	14-21 May 1659
15r	Warsaw	23-30 Apr 1659
16r	Warsaw	9-16 Aug 1659
17v	Warsaw	30 Apr-7 May 1659
18v	Warsaw	26 Jul-2 Aug 1659
19v	Warsaw	2-9 Aug 1659

20r-24v Tract, "Annotazioni sopra... [Pisa, 23 Feb 1665]
della Cometa/di Monsıi Aisout."
8½pp (20x27) Italian
fol 23r, drawing

25rv Observations, "Observatio.../Solaris/
...Anno D. 1659/Novem. 14. PM./...Paris"
1¾pp (18x23) Latin

26rv Observations, "Eclipse de ☾ partiale
observee/a Paris En l'annee 1659/le
29.me Septembre. Par [Agarrat]"
2pp (18x23)

27r-40v Observations, "Observationes habitae/
Viziliae, vel Gratianopoli ubi/
altitudo..."
28pp (20x27) Latin

27r-30r	Saturn: Observations	Feb-Dec 1622
		Jan 1623
		Jan-Dec 1623
		Jan-Feb 1624
		Feb-Dec 1624
		Jan-Dec 1625
		Jan-Mar 1626
		Apr-Dec 1626
		Feb-Dec 1628
		Jan-Dec 1629
		Jan-Jul 1630
		Feb-May 1631
		May-Aug 1631
		Feb-Jul 1632
30r-33r	Jupiter: Observations	Sep 1621
		Feb-Dec 1622
		Jan-Apr 1623
		May-Dec 1623
		Jan-Mar 1624
		Mar-Dec 1624
		Jan-Jun 1625
		Dec 1625

		Jupiter: Observations, continued	Jan-Dec 1626
			Feb-Mar 1628
			May-Aug 1628
			Jun-Dec 1629
			Jun-Nov 1630
			Jun-Dec 1631
			Jan-Jul 1632
	33v-37v	Mars: Observations	Oct 1621
			Mar-Dec 1622
			Jan-May 1623
			Jun-Oct 1623
			Nov-Dec 1623
			Jan-Mar 1624
			Mar-Dec 1624
			Jan-Aug 1625
			Aug-Dec 1625
			Dec 1625
			Jan-Mar 1626
			Jan-Jun 1628
			Jul-Dec 1629
			Jan-Mar 1630
			Apr-Jul 1630
			Jul-Sep 1631
			Oct-Dec 1631
			Jan-Jul 1632
	38rv	Venus: Observations	Feb-Apr 1622
			Feb 1624
		Venus & Mercury: Observations	Jan-Aug 1625
			Sep-Dec 1625
			Dec 1626
			Mar-Dec 1628
	39r	Mercury: Observations	Mar 1629
	39r	Venus & Moon: Observations	Jan-Dec 1630
	39r	Moon: Observations	Jan 1632
	39v	"Observationes altitudinis/solis ex umbra Vizilia" [1623,1624,1625]	
	40r	"Anno 1627..."	
	40rv	Saturn: Observations	Feb-Dec 1627
		Jupiter: Observations	Feb-Aug 1627
		Mars: Observations	Jun-Jul 1627
			Aug-Dec 1627
		Venus: Observations	Apr-Aug 1627
41r-65v		Observations in B's hand 47pp (18x26) Latin "Observationes Astronomicae/Ismaelis Boullialdi"	

41rv	Excerpts from various astronomical works [lunar and solar cycles]	
42r	"Anno Aerae Vulgaris Christi 1623" [planetary observations, 1623-1625]	
42v	Observations	1626-1627
43r		1628-1630
43v		1630-1631
44r		1632-1634
44v-47r	w/diagrams & text -	1634
47v-49r	"Analysis huius Observationis"	
49v-51v	[Observations & discussion w/diagrams]	1634&1635
52r	"Petrus Gassendus eam observavit..."	
52v	"Constitutio Verna huis anni 1635"	
53rv	"Notatio Eclipsis ☾ quae Contigit/ Anno 1635. Augusti die 27 astronomicè"	
54r	"Observatio eclipseos missa Aquis Sextiis/ Accepta per D." and "Observatio eclipseos ☾ celebratae anno 1636/februarii die 20. Per Doctissimum Petrum Gassendum/Coelitus habitae Diniae Stylo nouo"	
54v	"Ad eclipsim ☾ Anni 1635 Mense Augusto die 27." etc.	
55r	"Quarumdam stellarum distantiae.../ Tychonicis observationibus"	
55v-57r	Observations and text	1636-1637
57rv	1638 - Petrus Gassendus [Digne and Paris]	
58rv	"Parallaxeos ☽ inquisitio"	
58v-61r	Observations, continued	1639
61rv	w/extract of a letter of M. de Saint-Légier (Avignon) to M. Valois, 24 Jul 1640	1640
61v-62v	Observations	Anno 1641
62v	Observations, continued	Anno 1642
63rv	"Observatio eclipsis ☾ quae contigit anno 1642. Aprilis 14/astronomicè. observatores fuerunt R.P. Fournier/Soc. Jesu presb. Petrus Gassendus, et Ismael Bullialdus"	
63v-64v	Observations	Anno 1643
65rv	Observations, continued	Anno 1644
65v	"Eclipsis Lunae Februarii die 10"	1645

66r-68v Extracts in B's hand
5pp & 6 lines (18x26) Latin
"Ex libro Machometis Tinen Ben-acharrani qui et/Albategni dicitur, alias dicitur Ben-Gebir Ben-crueni./Inscripto de Scientia stellarum"

Folio		Description	Date
69rv		Extracts in B's hand	
		2pp (18x26) Greek/Latin	
		"ex Codice MS Bibliothecae Regiae Numero CXIV. Tij/quo continentur libri sequentes"	
70rv		Observations in B's hand	
		2pp (18x26) Latin	
		Observations	(Oct-Dec) 1643
			1644
			1645
71r-73v		Observations in B's hand	
		6pp (18x26)	
		"Ex Petri Gassendi Observationibus"	
	71rv	Observations	(Sep-Dec) 1636
			1637
			1638
			1639
			1642
	72rv	"Anno 1643 Observationes circa Lunae discum"	
		[w/planetary observations, Jupiter and Saturn]	
	73rv	Fragment	(1652-1653)
74r-85r		Observations in B's hand	
		19½pp & 10 lines (21x30)*	
		"Ex Petri Gassendi observationibus"	
	74rv	Observations	1620-1630
			1630-1634
	75r-77v	Mercury: Observations	1634-1635
		Mercury: Observations, continued	1635-1636
		Mercury: Observations, continued	1636
	78r-81v	Mercury & Venus: Observations	1636
			1628
	82r	Scratch sheet	
	82v&		
	83v-85r	"Schema Positions Orbium Mercurii..."	
		*fols 82-85 (18x26)	
85v-87v		Observations in B's hand	
		5pp (18x26) Latin	
	85v-87r	Observations, B in Loudun	1634
			1635
	87v	Scratch sheet	
88r-89v		Astronomical tables in B's hand	
		3¼pp (18x26) Latin	
		"Loca 84 Stella ♂ a Tychone observata/	

Folio	Description	Date
	excerpta e commentariis in Stellam ♂ / Johannis Kepleri"	
90r-92v	Observations & text in B's hand 6pp (20x30) Latin	
90rv	"Anno 1645 Parisiis/Eclipsis ☾ Februarii die 10 observata"	
91r	"1646/Parisiis Januarii die 31. mane Gassendus observavit Lunae/Eclipsim prout sequitur"	
91rv	[Smyrna from 2 Dec 1646-20 Jan 1647 observations w/Johannes Faber]	1647
92r	[Constantinople, 5 Apr 1647-27 Apr 1647]	1647
92r	"Ad Castra Dardanella sive sestum et/ Abydum in Hellesponto" [8 May 1647-9 May]	
92v	[obs., Smyrna, 17 May-6 Sep]	1647
92v	Obs., Smyrna, Paris	1648-1649
93r-95v	Observations in B's hand 4pp & 13 lines (20x30) Latin	
93r	"Philolaicae/Calculus eclipsis ☽ Anno 1648 Novembris 29"	
93v	Observations, continued ☾ obs.	1649
94rv	Observations, continued (obs. of Faber in Smyrna and B in Paris)	1649
95rv	Observations, continued (B and Gassendi)	1649
96r-102v	Observations & text in B's hand w/diagrams 10½pp (15x20) Latin	
96r-97r	"Anno 1632 die 19 Augusti..." etc.	
97v	Calculations	
98r-99v	"Alia Notatio eiusdem observationis tempore facta [w/diagrams]	
101v-102v	Scratch sheet fol 100, blank	
103A&104A	Calculations & data in B's hand 3pp (18x24) Latin "Calculus Eclipsis Lunae celebrandae Septembris 27/anno 1643•H•7•'23•Medio tempore Uraniburgi-/Medii motus ☉ juxta Tabulas Philolai" etc.	
103B	Observations in B's hand [Observations at Loudun, Oct, 1634] 1p (18x24) Latin	

104Br-105v	Printed text w/diagram 4pp (20x24) Latin "Observatio Eclipsis Lunaris,/Quae Contigit Anno/D. 1653. Martii 13. P.M./ Facta Parisiis In Claro-/Montano Collegio Parisiensi Societatis Jesu/Clarissimo Viro D.I. Bullialdo. P. Bourdin D. S.P."	
106rv	Diagram annotated by B 1p (23x33) Latin "Septem Soles observati Gedani/Anno Aerae Christi 1661/Die ☉ Februar. 20/ stilo novo/Hor. $\overline{XI}$/a/Johanne Hevelio/"	
107r-108v	Folding diagram annotated by B 4pp (27x41) "Parhelii Gedani visi die 6. Aprilis/ Hora vespertina 5.½ Anno 1660/A Johanne Hevelio./"	
109r	Copy of a letter 1p (21x32) Latin "Iter Cometae anno 1665/a die 14 Decemb. [usque] ad 30 Romae/observati [per] Patrem Athan. Kircherum/Ad. R.P. Albertum Curtium. S.J./"	
109v-110v	Copy of a letter 2½pp (21x32) Latin "Tridento a P. Simone Mair/[haec] de Cometa Scribuntur" and "Observatio Lugdunensis" [fol 110v, 6 line note, 26 Jan 1665, Ratisbon, in hand of Lubienietzki]	
111rv	Copy of a letter From Jonas Tornelius at Uppsala To [N. Heinsius?] 2pp (21x32) Latin	14 Feb 1665
112rv	Observations [in hand of Agarrat] 2pp (17x23) "Eclipse de ☾ le 13me Mars 1653..."	

113rv Fragment
10 lines (10x18) Latin
Concerns lunar eclipse, Feb 1645

114rv Fragment
5 lines (6x18) Latin
Eclipse data, Sep 1643

115rv Fragment
1p (17x23)
2 crude sketches of spherical angles

116r-123v Data & text in B's hand
14pp (21x33) Latin

116r-117r "Observationes a Tychone factae locorum ♄ /ex Historia Coelesti excerptae/" etc. [years 1582-1586]

118rv "Loca ☉ et ♄ ex Tabulis Philolaicis computata; et cum observatis comparata"

119r Fragment (12x16)
Observations of Mercury 1674&1661

120r "Loca ☉ et ♄ computata ad diem 13 Aprilis H 10 Parisiis/Hora vero 10'40• Uraniburgi anno 1650/"

120v "Loca ☉ ♄ et ♀ computata ex Tabulis Philolaicis/ad Diem Novembris 14•H18•'0 ..." etc.

121r "Loca ☉ et ♄ supputata ad Meridianum Gedanen=/sem Martii 27•St.no. H•13•'31• Uraniburgi H•13'•3•/Anno 1661"

121r "Loca ☉ et ♄ supputata ad diem Septemb. 12 hora 8/Parisiis, Uraniburgi, Hor•8•' 40 Anno 1662"

121v "Calculus locorum ☉ ♄ et ♃ ad dies Octobris 20 et 21/in meridie anno 1663• Uraniburgi"

122r "Motus ♃ ad praedict. diem XII•Octob. h 7•'25•1663•/Ab Aequinoctio. Aphelii. ☊/" also

122rv "Calculus motuum ☉ et ♄ Diem Maii 20 Hor 14•'20/Parisiis Hor. 15 Uraniburg anno 1665./"

122v "Calculus ☉ ♄ et ♂ ad diem 3 Decembris/H 5•'30•Parisiis, Hora vero 6•'10 Uraniburgi/Anno 1668" etc.

123r " ♄ motus ad. praedict diem et Horam" etc.

124r-125v
Tract on comets in B's hand
2pp & 6 lines (18x25) Latin
[comets of: Jul 1596, Sep 1607,
Dec 1618, and Sep 1621]

126r-128v
Observations & text in B's hand, continued
6pp (21x33) Latin
126r "Ad diem 27 Febrarii H 8 ' Uraniburgi
Parisiis vero H 7•'20•1678"
126v-127r "Ad Hor. 9•'10 Uraniburgi Parisiis Hor.
8•'30• TM"
127r-128v "Juxta Tabulas itaque Philolaicas apparuit
hujus/Saturni a ☾ occultationis Parisiis."
[w/data, notes and diagrams]

129r-130v
Observations
1½pp (17x22)
"Eclipse de ☾ observee a Paris/En
l'année 1657 le 25.me du mois de Juin/
au soir par Agarrat"
[Antoine Agarrat]
fol 130, blank

131r-134v
Printed text
8pp (16x21) Latin/German
"Astronomica et Astrologica/Observatio/
COMETAE TERRIBILIS,/Mense Februario,
huius currentis Anni adhuc spectati;/
nec non/Tristis Eclipsis Stellae fixae
Regiae/...1665" [Johanne Placentio]

▼ ▼ ▼

COLLECTION BOULLIAU, VOLUME XLI, FONDS FRANÇAIS 13059

1r-4v Calendric conversion charts in B's hand
8pp (21x34) Latin
"Descripta series annorum Arabicorum ex Lib. Jo. Marianae S.J. cui titulus, Septem Tractatus"
Description:
Large sheets with three major columns; each column gives dates in four categories:
1) Aera Caesaris Curr. [660-1783]
2) Anni Christi Curr. [622-1745]
3) Neomenia Muharram [15 Juli-1 Feb]
4) Anni Arabum Inchoan. [cycle prin. Bl-1158]

5r-6v Calendric chart & text
$2\frac{1}{3}$pp (21x34) Latin
"Sethi Caluisii Methodus ad reperiendos annos Arabicos quam/secutus est David Origanus in suis Ephemeridibus"
fol 6, text regarding conversion of dates w/examples
fol 5r, empty grid of columns as used in fols 1-4

7rv Calendric chart in B's hand
1p (21x34) Latin
"Ex Nicolai Muleri Libro cui Titulus, Judaeorum Annus Lunae Solaris,/et Turc-Arabum Annus mere Lunaris."

8r-12v Calendric chart in B's hand
1p (21x34) Latin
"Ex Praedicti Nicol. Muleri Libro citato"/
fol 9r, chart [anni Christi Currentes, 1595-1675]
fols 10-12, blank

13r-14v Calendric charts & text in B's hand
4pp (21x34) Latin
"Methodus Computandi annos Judaicos"

15rv Calendric charts in B's hand
2pp (21x34) Latin
"Tabula exhibens in Annis Christi stylo novo Neomenias Mensis Tisri,/et Pacha Judaeorum; Pascha etiam Christianorum. &

alio characteres/Calendarii ab Anno
Judaïco 5397. qui incipit anno Christi
1636. estque/primus Cycli Judaici 285.
ad annum 5434. qui ultimus est cycli
286. anno Christi 1673."

16r-17v Calendric charts in B's hand
4pp (21x34) Latin/Greek
"Kalendarum Judaicum anni 5426.
Embolimaei Abundantis, qui cycli 286
est decimus/..."

18r-24v Notes in B's hand
3pp (21x34) Latin
"Collectio annorum Hebraeorum ex
Tabulis per totum cyclum 285."
fol 19r, calculation sheet [yrs 1664-1674]
fols 20-24, blank

25r-36v Calendric conversion charts (12)
January-December
12pp (21x34) Latin
Anni Christi Currentes (C 1655-F 1675)
Anni Hegirae Currentes (1065-1085 B)
Menses Arabici (Saphar-Sehewal)[cyclic]

37rv Fragment
1 line (3x15) Latin
"Ex Josephi Historia de Bello Judaïco
excerpta"

38r-52v Greek manuscript, extracts from the History of *The Jewish War* of Josephus, begins:
"'O δὲ Τίτος περαιωθεὶς..."*
12pp (21x34) Greek
fols 50-52, blank
*[cf. Loeb Classical Library edition,
Josephus, The Jewish War, Bk. III, 64,
Vol. II, p 594 et. seq.]

53r-80v Biblical extracts
$47\frac{2}{3}$pp & 6 lines (25x34) Greek
53r-56v Brief extracts, Jacob, Matth., Luc.,
etc. (25x34)
Brief extracts (18x23)
Brief extracts (20x25)
Extract
57r-59r Epist. Ad. Rom. (I-VII)
Epist. Ad. Rom. (VIII-XV)

	Epist. Ad. Rom. (XVI)
59r-62v	Corinth. I (I-VII)
	Corinth. I (VIII-XIV)
	Corinth. I (XV-XVI)
	Corinth. II (I-X)
	Corinth. II (VI-XIII)
63rv	Ad Galat. (I-VI)
64rv	Ad Ephes. (I-VI)
64r-65r	Ad Philipp. (I-II)
	Ad Philipp. (II-IV)
65rv	Coloss. (I-II)
	Coloss. (II-IV)
65v-66r	Ad Thessal. I (I-II)
	Ad Thessal. I (III-V)
	Thessal. II (I-IV)
66v-67v	Ad Timoth. I (I-V)
	Ad Timoth. I (V-VI)
	Ad Timoth. II (I-II)
	Ad Timoth. II (III-IV)
67v	Ad Titum (I-III)
68r	Ad Philem.
68r-69v	Hebr. (I-V)
	Hebr. (VI-VIII)
	Hebr. (IX-X)
	Hebr. (XI-XIII)
70rv	Jacob Ep. (I-II)
	Jacob Ep. (III-V)
71r-72v	I Petri Ep. I (I-III)
	I Petri Ep. I (IV-V)
	II Petri (I)
	II Petri (I-III)
	II Petri (III)
72v-73v	Joann. Ep. I (I-III)
	Joann. Ep. I (III-V)
	Joann. Epist. II (II)
	Joann. Epist. II (III)
73v	Judae Epist.
74r-80v	Apocal. Joann. (I-II)
	Apocal. Joann. (III-V)
	Apocal. Joann. (VI-IX)
	Apocal. Joann. (IX-XII)
	Apocal. Joann. (XII-XIV)
	Apocal. Joann. (XIV-XVII)
	Apocal. Joann. (XVIII-XX)
	Apocal. Joann. (XX-XXII)
	Apocal. Joann. (XXII)
	fols 79&80, blank

81r-86v

Printed text
11pp (24x41) Greek/Latin
"In Galilaeâ et Capernaum, Mense Adar, Februario, Ante II Pascha."*
Description: Biblical chronology in chart form treating period Const. 5539/ Jul. 4744 to the time of Christ. Horizontal divisions establish chronology, vertical divisions supply Greek biblical extracts of Matthew, Mark, Luke, John.
*[unidentified text, pp 20-21, 41-48)

▼ ▼ ▼

NAME INDEX

HAMLET: Thou dost lie in't,
to be in't and say it is thine.

Hamlet, Act V, Sc 1

A

B

C

D

E

F

G

H

H

I

J

K

M

N

O

R

T

U

V

W

Y

Z

SOURCES

If you are troubled with a pride of
Accuracy, and would have it completely
taken out of you, print a catalogue.

--epigraph in D.M. Morris
*A History of Cataloguing and
Cataloguing Methods, 1100 -
1850* (1939)

A. ORIGINAL SOURCES

1. Manuscript Sources of Ismaël Boulliau

AUSTRIA

Vienna

Österreichische Nationalbibliothek
Briefsammlung Cod. 7049: Contains letters exchanged between Boulliau and Gabriel Naudé, Vincent Fabricius, Peter Lambeck, Johann Rhodius, Johann Friedrich Gronovius, Jacques de Billy, Leopoldo of Tuscany, Johann Mochinger, and others
Briefsammlung Cod. 7050: Contains letters exchanged between Boulliau and Claude Saumaise, Nicolas Rigault, Isaac Vossius, Jacques de La Rivière, Johann Rhodius, Jacques-Auguste de Thou, Claude Sarrau, and others
Briefsammlung Cod. 9712: Letter, Boulliau to Peter Lambeck, 5 Apr 1649
Briefsammlung Cod. 10889: Copy of a letter, Boulliau to Wilhelm Schickard, XVII Kal Apr 1633
Signatur. Autogr. 78/26: Letter, Boulliau to Johannes Hevelius, 29 May 1665

DENMARK

Copenhagen

Det Kongelige Bibliotek
MS Thott 1259: Epistolae doctorum virorum ad J. G. Graevium; Letters, Boulliau to Johannes Georgius Graevius, 9 Dec 1672, undated, 5 Jul 1679, 16 Aug 1679, 22 Oct 1680, 2 Nov 1682, 9 Jan 1683
MS Gl. Kongelig Saml. 2134: Copy of a letter, Boulliau to Peter Lambeck, 5 Apr 1649

ENGLAND

London

The British Library
MSS Add. 4255: Latin copy of a letter, Boulliau to Samuel Hartlib, ca. 1659
MSS Add. 4365: Letters, Boulliau to Henry Oldenburg, 12 May 1676, 12 May 1677, 25 Jul 1676

MSS Add. 4400: Copies of letters, Boulliau to Henry Oldenburg, 4 May 1672, 25 Sep 1674

MSS Add. 4443: Letters, Boulliau to Henry Oldenburg, 4 May 1672, 24 Nov 1674

MSS Add. 16912: "Ismaelis Bullialdi adversaria literaria" (248 fols.)

MSS Add. Lat. 24212: Letters, Nicolaas Heinsius to Boulliau, 5 Dec 1657; J. G. Graevius to Boulliau, 17 Aug 1658

Royal Society

MS Cl P VIII (1): "Ismaelis Bullialdi observationes factae anno 1667/Circa Nebulosam in Cingulo Andromedae, et/stellam mirabilem in collo Ceti"; Astronomical data, 22 Apr 1671

MS EL B1: Letters, Boulliau to Henry Oldenburg, 16 May 1667, 16 May 1667 (the Royal Society), 6 Feb 1675, 17 Jul 1675

MS LB 2: Draft of a letter, Henry Oldenburg to Boulliau, 22 Apr 1667

MS LBC 2: Copies of letters, Henry Oldenburg to Boulliau, 22 Apr 1667; Boulliau to Henry Oldenburg, 16 May 1667; Boulliau to the Royal Society, 16 May 1667

MS LBC 7: Copies of letters, Boulliau and Henry Oldenburg; eclipse data

MS LBC 7: Copies of letters, Boulliau and Henry Oldenburg; astronomical data

MS LBC 8: Copy of a letter, Boulliau to Henry Oldenburg, 25 May 1678

MS LBC 9: Copy of a letter, Francis Aston to Boulliau, 10 Mar 1683/84

MS LBO 2: Letter, Henry Oldenburg to Boulliau, 22 Apr 1667; Boulliau to Henry Oldenburg, 16 May 1667 O.S.; Boulliau to the Royal Society, 16 May 1667 O.S.

MS LBO 8: Letter, Boulliau to the Royal Society, 25 May 1678

MS RBC 3: Copies of astronomical data and calculations

MS RBO 4: Original drafts of astronomical data and calculations

Oxford

Bodleian Library

MS Ashmole 436: Copy of astronomical calculations, 1663

MS D'Orville 239: "Ex Excerptis ismaelis Bouillaud 1730," 34 epigrams from Boulliau

MS D'Orville 243: Copy, "in anthologiam in editam Ad p. 5. Ep. 2. excerpta Ismaelis Bouillaud . 1730..."

MS D'Orville 468: Copy of a letter, Boulliau to Isaac Vossius, 9 May 1645

MS Rawlinson C. 364: Copy, annotated by Boulliau, "Discours sur la necessité qu il y a de conserver les biens que possede l'eqlise ..." (by Boulliau)

MS Smith 5: Letters, Boulliau to Edward Bernard, 12 Jan 1683; 12 Jan 1683 (different letter)

FRANCE

Aix-en-Provence

Bibliothèque Méjanes

Correspondance littéraire de Peiresc

MS 1020 (T. II): Copies of letters, Boulliau to Pierre Gassendi, 7 Sep 1634; Boulliau to N. C. Fabri de Peiresc, 17 Oct 1636; Boulliau to M. de Valavès, 21 Aug 1637

MS 1031 (T. XIII): Copy of a letter, Boulliau to N. C. Fabri de Peiresc, 17 Oct 1636

Musée d'Arbaud

Série A: Letter, Pierre Gassendi to Boulliau, 20 Mar 1653

Caen

Bibliothèques de la ville de Caen

Correspondance P-D Huet, 205 (MSS in 4°): Copy of a letter, Boulliau to Pierre-Daniel Huet, 1 Oct 1656, cat. Lavalley 241

Correspondance P-D Huet, 206 (MSS in 4°): Copies of letters, P-D Huet to Boulliau, Eid. Oct 1654; Boulliau to P-D Huet, 30 Dec 1654; astronomical observations

Carpentras

Bibliothèque d'Inguimbert

MS 1810: "Lettres françoises de plusiers sçavans personages. Lettres françoises meslées." Peiresc materials. Copy of a letter, Boulliau to Pierre Gassendi, 7 Sep 1634

MS 1832: Miscellany. Copy of a letter, Pierre Gassendi to Boulliau, 13 Aug 1633

Lille

Bibliothèque Municipale

MS 854: Letter, Boulliau to Johannes Hevelius, 31 Dec 1670

Loudun

Archives Communales

Série BB 30: Certified copy by Ismael Boulliau, Sr., Greffier de la Ville, 7 May 1632 (7pp); Copy by Ismael Boulliau, Sr., concerning the plague of Loudun dated 8 May 1632 (21pp); Copy of a decision of the Assembly of Loudun delivered by Ismael Boulliau, Sr., 31 Aug 1632 (2pp)

Série II, 1: 12 line document (33x54cm) of Gilbert du Puy du Fou, Gentilhomme ordinaire de la chambre du Roi, Seigneur Baron de la Jaille, dated 12 Jul 1606, for Ismael Boulliau, Sr.

Lyon

Bibliothèque de Ville

MS 1629: Copies of letters, Boulliau to Claude Saumaise, 6 Dec 1641; Boulliau to Pierre Gassendi, 7 Sep 1634

Paris

Archives du Ministère des Affaires Étrangères

Correspondance Politique-Pologne

MSS Pol. 14: Letters of Pierre Des Noyers to Boulliau, 1660-1665, with diverse letters, copies, and enclosures (552 fols.)

MSS Pol. 25: Letters of Pierre Des Noyers to Boulliau, 1666-1669, with diverse letters, copies, and enclosures (499 fols.)

MSS Pol. 37: Letters of Pierre Des Noyers to Boulliau, 1670-1673, with diverse letters, copies, and enclosures (349 fols.)

Archives Nationales

Chartrier de Tours

MSS 1 AP 646: Diverse letters of Ismael Boulliau, Sr., and Jean-Pierre Boulliau to the familly Trémoille

Bibliothèque de l'Arsenal

MS 4630 (839 E.H.F.)6°: "Catalogus bibliothecae manuscriptae Thuani"; 7°: "Catalogus librorum manuscriptorum qui habentur in musaeo cl. viri Ismaelis Bullialdi Juliodunensis, Parisiis, in aedibus Thuanis. Anno 1650"

Bibliothèque Interuniversitaire Sainte-Geneviève

MS 2289: Contains copies of astronomical observations by Boulliau

Bibliothèque Mazarine

MS XXIII (4551)2°b: Copies of a letter, Christiaan Huygens to Boulliau, 12 Feb 1660

Bibliothèque Nationale

Collection Dupuy

MS 16: Diverse letters

MS 18: "Lettres d'Ismaël Boulliau à Jacques Dupuy (1645-1651)" (220 fols.)

MS 541: Copies of letters in Boulliau's hand

MS 625: "Discours latin contre les cruautés du cardinal de Richelieu," by Boulliau

MS 630: "Mélanges literaires et historiques. (Notes sur les oeuvres de Théodoret et de saint Augustin, par Boulliau"; "Lettre d'Ismaël Boulliau sur les différences de la musique française et de la musique italienne"; "Indices sive tituli librorum, quos Gaspar Sciopplus, Claraevallis comes, editioni paratos habet", in Boulliau's hand

MS 631: "Avis de Paris, relatif au duc de Beaufort," 27 Aug [1642], in Boulliau's hand
MS 648: "Nouvelles de Venise," 19 May 1646 by Boulliau
MS 667: "Inscription grecque trouvée à Smyrne et envoyée de là aux frères Dupuy par Ismaël Boulliau," Apr 1647
MS 678: "Note d'Ismaël Boulliau sur les Jésuites de Venise"
MS 691: "Notes sur le partriarche Photius" by Boulliau
MS 777: "Notes sur le Roi Dagobert" by Boulliau
MS 835: "Recueil de pièces relatives à l'histoire de France pendant le XVIIe Siècle, (1616-1656)." Copies. Boulliau to Nicolaas Heinsius, 30 Jul 1660
MS 879-880: "Catalogue alphabétique de la Bibliothèque de président Jacques-Auguste de Thou," by Pierre Dupuy, Jacques Dupuy, and Boulliau
MS 886-891: "Catalogue méthodique de la Bibliothèque de président Jacques-Auguste de Thou, rédigé, sous la direction d'Ismaël Boulliau--Fiches originales montées, revue et complétées" by Boulliau, Louis Billaine, and Joseph Quesnel

Fonds français
MS 892: Brief note on the life of Boulliau
MS 3930: Letters of Boulliau, Pierre Bourdelot, Gabriel Naudé, Samuel Sorbière, and others to Claude Saumaise; letters, Boulliau to Saumaise, 6 Dec 1641, 21 Feb 1642, 25 May 1646, 13 Jul 1646, 14 Sep 1652, 14 Dec 1652
MS 9778: "Lettres de Dom. Christophe Dupuy, procureur général des Chartreux à Rome, à Ismaël Boulliau, sur les affaires de Rome (1641-1653)" (102 fols.)
MS 13019-13059: *Collection Boulliau*, Correspondance et papiers politiques et astronomiques d'Ismaël Boulliau (1605-1694)
I FF 13019: Letters, Pierre Des Noyers to Boulliau, Oct 1655-Dec 1657 (320 fols.)
II FF 13020: Letters, Pierre Des Noyers to Boulliau, Dec 1657-Nov 1659 (335 fols.)
III FF 13021: Letters, Pierre Des Noyers to Boulliau, Jan 1680-Dec 1683 (352 fols.)
IV FF 13022: Letters, Pierre Des Noyers to Boulliau, Jan 1684-Dec 1688 (385 fols.)
V FF 13023: Letters, Pierre Des Noyers to Boulliau, Jan 1689-Oct 1692 (ending f. 116); Letters of Vair to Boulliau, Jan 1655-Sep 1657 (305 fols.)
VI FF 13024: Letters, Emery Bigot to Boulliau, Oct 1657-Oct 1658 (60 fols.)
VII FF 13025: Letters, to Boulliau from Carlo Antonio de Gondi, M. De Boüillon, Pierre de Bonsy, Paul Parenti, Vincenzio Viviani, and others (329 fols.)
VIII FF 13026: Letters, drafts, and copies of Boulliau to Johannes Hevelius, J-A de Thou, Stanislas Lubienietzki, Jacques de La Rivière, Vincenzio Viviani, J. H. Boecler,

Wilhelm Schickard, J. G. Graevius, P. C. Forstner, Stephanus Gradius, Martin Hortensius, Mazarin, Colbert, and others (336 fols.)

IX FF 13027: Letters and drafts of letters, Boulliau to Leopoldo of Tuscany, Johannes Hevelius, Nicolaas Heinsius, Henry Oldenburg, George Caspar Kirchmayer, Johann Ravius, Vincent Fabricius, Johannes Ernest von Rautenstein, and others (262 fols.)

X FF 13028: Collection of horoscopes with biographical information, by Boulliau (371 fols.)

XI FF 13029: Letters and drafts of letters, Christiaan Huygens, Jacques de Valois, Nicolaas Heinsius, Isaac de Gruter, J. H. Boecler, P. C. Forstner, and others to Boulliau (222 fols.)

XII FF 13030: Letters, Jacques de Valois to Boulliau, Sep 1642-Sep 1654; three letters of M. Vair (212 fols.)

XIII FF 13031: Letters and newsletters, Stanislas Lubienietzki to Boulliau, 1662-1672, etc. (401 fols.)

XIV FF 13032: Letters, Pieter and Philip Pels to Boulliau, May 1661-Dec 1667 (308 fols.)

XV FF 13033: Letters, Pieter and Philip Pels to Boulliau, Jan 1668-Jan 1680 (interrupted Mar 1672)(339 fols.)

XVI FF 13034: Letters, to Boulliau from Johann Albrecht Portner, Vincent Marucelli, Henry Oldenburg, Emery Bigot, Johannes Ernest von Rautenstein, and others (241 fols.)

XVII FF 13035: Letters, Francesco Gratta to Boulliau, Nov 1660-Dec 1665 (294 fols.)

XVIII FF 13036: Letters, Francesco Gratta to Boulliau, Jan 1666-Dec 1669 (256 fols.)

XIX FF 13037: Diverse letters, to Boulliau, Wilhelm Schickard, Pierre Gassendi, Martin Hortensius, Galileo Galilei, Francesco Bardi, Michael van Langren, Antonio Santini, Jacob Golius, John Heydon, Wilhelm Langius, P-D Huet, and others (257 fols.)

XX FF 13038: Letters, Nicolas Bretel de Grémonville and Ludovico de Matharel to Boulliau, etc. (200 fols.)

XXI FF 13039: Letters, manuscripts, and copies of unpublished tracts by Italians: Letters of Leopoldo of Tuscany, Carlo Rinaldini, tracts by Vincenzio Viviani, Carlo Rinaldini, Paulo del Buono, and others (172 fols.)

XXII FF 13040: Diverse letters to Boulliau, Isaac de la Peyrère, Henri Justel, and others; copies of letters in Boulliau's hand (262 fols.)

XXIII FF 13041: Letters, from Johann Albrecht Portner and J. C. Lenz with drafts of Boulliau's responses, Dec 1652-Dec 1657, etc. (351 fols.)

XXIV FF 13042: Diverse letters to Boulliau from François Luillier, Herouard de La Boissière, Jacques de Valois, M. Chauvet, Abraham de Wicquefort, N. de Gersdorff, J. B.

Barducci, Pieter de Groot, J-A de Thou, Pierre Bourdelot, l'Empereur, Antoine d'Aubray, and others (272 fols.)

XXV FF 13043: Letters, to Johannes Hevelius with drafts of Boulliau's responses; letters of Albert Linneman, Christophe Storanus, Rostainy de Bury; copies of letters of Marin Mersenne, Johannes Hevelius, G. B. Riccioli, Maria Cunitz, and others (163 fols.)

XXVI FF 13044: Letters of Johannes Hevelius, Tito Livio Burratini, Vincenzio Viviani, Pierre Des Noyers, and others (310 fols.)

XXVII FF 13045: Letters, M. Bernart to Boulliau, Oct 1657-Dec 1662; diverse letters of François Dulaurens, M. Jannot, Abraham Wicquefort, and others (278 fols.)

XXVIII FF 13046: Letters, M. Bernart to Boulliau, Jan 1664-Dec 1666, and others (263 fols.)

XXIX FF 13047: Letters, M. Bernart to Boulliau, Jan 1667-Dec 1669, and others (298 fols.)

XXX FF 13048: Letters, M. Bernart to Boulliau, Jan 1670-Sep 1680, and others (358 fols.)

XXXI FF 13049: Letters, to Boulliau from M. Valeran, Leopoldo of Tuscany, Johannes Bleau, Robert Scott, Edward Sherburne, and others (201 fols.)

XXXII FF 13050: Letters, to Boulliau from J-A de Thou, Marie Picardet, Jacques de La Rivière, Francis Aston, J. F. Gronovius, Henry Smith, Gilles Ménage, Pierre Gassendi, Claude Soëfve, Seth Ward, and others (263 fols.)

XXXIII FF 13051: This volume is a record of scholarly and epistolary materials found in the estate of Boulliau (69 fols.)

XXXIV FF 13052: Extracts and copies of classical authors, in Boulliau's hand, including Lucan, Livius, Terence, Lucretius, Virgil, Catullus, and others, with biographical and philological notes (196 fols.)

XXXV FF 13053: Letters, newsletters, extracts, and copies of letters addressed to Boulliau, concerning Rome, Milan, Genoa, and other cities of Italy, 1659-1677 (443 fols.)

XXXVI FF 13054: Diverse materials and notes regarding ecclesiastical affairs of the 17th century (516 fols.)

XXXVII FF 13055: Extracts of classical authors and ecclesiastical writers in Boulliau's hand; Socrates the Scholastic, Eusebius, Tertullian, St. Jerome, and others. Notes and extracts on various ecclesiastical councils, notes on the History of Ducas, and so forth (417 fols.)

XXXVIII FF 13056: Diverse manuscripts and printed tracts on Jansenism, Molinism and ecclesiastical affairs of Rome (350 fols.)

XXXIX FF 13057: Manuscripts and printed tracts concerning French commerce, particularly the East India Company (376 fols.)

XL FF 13058: Astronomical observations, calculations, and geometrical demonstrations of Boulliau, with observations of Pierre Gassendi, Johannes Hevelius, and others (134 fols.)

XLI FF 13059: Diverse manuscripts of Boulliau concerning the Arabic and Jewish calendar, and problems of chronology; a 16th-century manuscript extract of the "Historia de Bello Judaico" of Josephus (86 fols.)

MS 15189: "Correspondance littéraire et privée d'une partie du siècle de Louis XIV, ou recueil de lettres adressées à Huet, évêque d'Avanches." Copy of a letter, P-D Huet to Boulliau, 1 Oct 1656

MS 15506: "Fragment orig. d'Ismaël Boulliau sur l'échéance du jour de Pâques"

MS 15713: "Recueil de pièces, manuscrits et imprimées, relative à différentes matières ecclésiastiques." Contains "Diatriba de S. Benigno, ab Ismaele Bullialdo"

MS 16793: "Recueil de pièces, pour la plupart imprimées, parmi lesquelles on remarque un grand nombre d'éloges, panegyriques," etc. Contains letters of Boulliau to Jacques Dupuy; epitaph of J. Dupuy by Boulliau

Nouvelles acquisitions françaises

MS 1096: "Recueil de lettres originales de XVIe, XVIIe et XVIIIe siècles, provenant de la collection Rathery." Letter, Pierre Gassendi to Boulliau, [5 Apr?] 1639

MS 5158: Letter, Pierre Bourdelot to Boulliau, 12 Apr 1653

MS 5163: "Papiers de Théodore Godefroy." Contains letters of Boulliau

MS 5214: "Recueil de Lettres autographes de divers personnages des XVIe, XVIIe et XVIIIe siècles." Letters, Boulliau to Johannes Hevelius, 31 Mar 1672, 9 Jul 1676

MS 5856: "Recueil de Lettres, autographes et copies, d'astronomes et savants du XVIIe siècle, provenant en grande partie de la correspondance d'Hévélius." Letters, Boulliau to Johannes Hevelius, III Id. Dec 1648, 25 Oct 1652, 3 Oct 1653

MS 6197: "Observations et correspondances astronomiques des XVIIe et XVIIIe siècle." Letters, Boulliau to Pierre Des Noyers, 28 May 1671; Seth Ward to Boulliau, 14 Feb 1656/7 OS

MS 6202: Correspondance de Huet. Contains three letters of Boulliau

MS 6204, 6205: "Recueil, formé par le frère Hilarion de Coste, de lettres originales adressées au P. Marin Mersenne." Letters, Boulliau to Marin Mersenne, 17 Feb 1645 (6204), Boulliau to Mersenne, 16 Dec 1644 (6205)

MS 22763: "Mélanges scientifiques, provenant en partie d'Ismaël Boulliau"

Fonds Latin
- MS 8000: Contains letters of Boulliau
- MS 10347-10349: Copies of letters, Johannes Hevelius to Boulliau
- MS 17866: Contains scientific discourses, autograph of Boulliau's *Philolaus*; treatise on astrology

Nouvelles acquisitions Latines
- MS 1637: Contains letters of Boulliau to Pierre Gassendi
- MS 1640: Contains letters/minutes of Boulliau
- MS 1642: Contains letters and minutes of Boulliau and Johannes Hevelius

Bibliothèque de l'Observatoire
- MS B3 11: Contains diverse materials of Boulliau, Eleazar Feronce, and le P. Maire
  Boulliau's astronomical observations are in two parts:
  1. Observations in five groups of the planets (Saturn, Jupiter, Mars, Venus, Mercury)
  2. Two groups of observations, some are copies from BN FF 13058, and others
- MS B5 11: Astronomical observations of Boulliau. Data, calculations, and notes; extracts from the writings of Pierre Gassendi, Longomontanus, Johannes Kepler, C. Rothmann, F. Viète, P. Cruger, Galileo, and others; several printed pamphlets (686pp)
- MS B5 12: Astronomical observations of Boulliau. Data, notes, letters, and copies of letters; Ismael Boulliau, Sr., Pierre Gassendi, Johannes Hevelius, Pierre Des Noyers, François Luillier, François Vincent, Vincent Wing, Emery Bigot, Nicolas Mercator, and others. Loose folios and sewn signatures (135 fols.) followed by bound manuscripts (652pp)
- MS C1 1-16: Epistolae Hevelium. Letters exchanged between Boulliau and Johannes Hevelius, Volumes 2-15

Bibliothèque du Protestantisme Français
- MS 811: Letter, Boulliau to Christophe-Daniel Findekeller, 7 Dec 1678

Poiters

Bibliothèque Municipale
- MS C 4, 454: Letters, Boulliau to Johannes Hevelius, 5/15 Oct 1651, 9 Mar 1663

GERMANY

Berlin

Staatsbibliothek Preussischer Kulturbesitz

Slg. Darmst. J. 1638: Letters, Boulliau to Johannes Hevelius, 26 Mar 1680; Boulliau to [J-A de Thou], 13 Nov 1655

Slg. Darmst. A. 1641: Letter, Johannes Hevelius to Boulliau, Apr 1654

Hamburg

Staats- und Universitätsbibliothek

MS Sup. Ep. (4°) 1: Letters, Boulliau to J. H. Boecler, 19 Mar 1658, 12 Apr 1659

MS Sup. Ep. (4°) 36: Letters, Boulliau to J. H. Boecler, 28 Jan 1654, 11 Sep 1654 (eclipse data, 12 Aug 1654), 20 Jul 1657, 23 Jul 1657, 5 Feb 1655, 10 Jul 1658, 17 Jul 1654

MS Sup. Ep. (4°) 36: Book list

MS Sup. Ep. (4°) 36: Copy of a letter, Boulliau to J. A. Portner, 3 Dec 1655

MS Sup. Ep. (2°) 48: Eclipse chart and text, 12 Aug 1654

Munich

Universitätsbibliothek

MS (2° Cod.) 610: Letters, Boulliau to J. F. Gronovius, 7 Mar 1643, 8 Sep 1651, 1 Feb 1662

Stuttgart

Wurttembergische Landesbibliothek

MS Cod. hist. 2°, 563ab: Letter, Boulliau to Wilhelm Schickard, 19 Jun 1634

MS Q201A: Contains copies of letters of Wilhelm Schickard

HOLLAND

Amsterdam

Universiteits-Bibliotheek

MS Rem. Kerk. (III E. 8^{72}): Letter, Boulliau to Isaac Vossius, 9 May 1645

The Hague

Koninklijke Bibliotheek

MS 130 E: Album Amicorum of J. F. Gronovius, inscription by Boulliau, Paris, XIV Kal Dec 1641

MS 131 E2: Album Amicorum of S. Tennulius, inscription by Boulliau, Paris, 26 Nov 1659

Leiden

Bibliotheek der Rijksuniversiteit

Department of Occidental Manuscripts

Burm. F. 9: Contains letters to Nicolaas Heinsius; Boulliau, Claude Sarrau, the brothers Valois, and others

Burm. Q. 18: Contains letters, Boulliau to Nicolaas Heinsius

BPL 885: Letter, Christiaan Huygens to Boulliau, 9 Jan 1659

BPL 1923: Copy of a letter, Boulliau to Jean Chapelain [19 May 1667]

Hugenius 45: Contains fifteen letters, Christiaan Huygens to Boulliau

Pap. 7: Contains letters, Boulliau to Claude Saumaise

Pap. 15: Contains letters, Boulliau to Stanislas Lubienietzki

ITALY

Florence

Archivio di Stato

Archivio Mediceo

MS Series 4783: Contains letters, Boulliau to C. A. de Gondi

Biblioteca Mediceo-Laurenziana

Collezione Ashburham-Libri

MS Carteggio Huet C. 8, 21, 23: Contains correspondence between P-D Huet and Boulliau

Biblioteca Nazionale Centrale

MS C.V.: Letter, Boulliau to Isaac Vossius, Jun 1643

MS Gal. 87: Contains copies of letters in Boulliau's hand, Galileo to Boulliau, Kal Jan 1638, 30 Dec 1639; Galileo to Diodati, 11 Dec 1639

MS Gal. 94: Contains letters, Boulliau to Galileo

MS Gal. 256: Letters, Boulliau to [M. Charlier], 5 Aug 1677; Boulliau to Vincenzio Viviani, 16 Mar 1678, 7 Jun 1683, 20 Sep 1683

MS Gal. 272: Copy of a letter, Adrien Auzout to [Boulliau], 30 Apr 1665

MS Gal. 274: Letter, Boulliau to Leopoldo, 17 Jan 1653

MS Gal. 275: Letter, Boulliau to Leopoldo, 19 Dec 1659
MS Gal. 276: Letters, Boulliau to Leopoldo, 30 Jul 1660, 5 Apr 1661, 5 Apr 1661, 15 Dec 1661, 30 Aug 1662, 18 May 1663, 10 Aug 1663
MS Gal. 277: Letters, Boulliau to Leopoldo, 6 Jun 1664 (with extract of a letter, Johannes Hevelius to Boulliau, 28 Mar 1664); 29 Aug 1664 (with astronomical data), 24 Oct 1664, 30 Jan 1665 (with data on Comet of 1664-1665), 12 Feb 1665 (with data), 27 Mar 1665 (with data), 24 Apr 1665, 10 Jul 1665, 4 Sep 1665
MS Gal. 278: Letters, Boulliau to Leopoldo, 7 Jan 1667, 14 Jan 1667, 10 Apr 1669
MS Gal. 279: Letter, Boulliau to Leopoldo, 12 Aug 1672; Latin tract by Boulliau on ecclesiastic affairs; Boulliau to Leopoldo, 14 Oct 1672; Latin tract, ecclesiastic history
MS Gal. 285: Letter, Boulliau to the Bishop fo Beziers (Pierre de Bonsy) 11 Nov 1660
MS Gal. 286: Letter, Boulliau to Leopoldo, 1 Apr 1670
MS Gal. 312: Letters, Boulliau to Leopoldo, 25 Apr 1652 to 29 Aug 1664 with enclosed copies of letters, astronomical data etc. Copies of letters, Boulliau to Pierre de Bonsy, 11 Nov 1660; Leopoldo to Boulliau, undated

UNITED STATES

Cambridge, Massachusetts

Harvard University, Houghton Library
Letter, Boulliau to Stanislas Lubienietzki, 5 Dec 1664

New York, New York

Columbia University Library
Special Manuscript Collection, David Eugene Smith Collection
Letter, Boulliau to [J-A de Thou], 30 Oct 1655
Letter, Boulliau to Johannes Hevelius, 15 Jan 1666
Manuscript (in Boulliau's hand),
"Pour/Monseigneur le Tellier Ministre & Secretaire/d'Estat./.../Ismael Boulliau," entitled
"Discours/Sur la necessité qu'il y a de conserver les biens que possede/l'Eglise, pour maintenir la Religion Christienne,/Catholique, Apostolique & Romaine." Undated

New York Public Library
Manuscript and Archives Division
Samuel Ward Autograph Collection
No. 7: Letter, Boulliau [to Francesco Gratta], 4 May 1668

2. Printed Works of Ismaël Boulliau

Boulliau, Ismaël. *De Natura lucis, authore Ismaele Bullialdo...*, Parisiis, apud de Heuqueville, 1638 [8°, xviii + 156pp].

-----. *Philolai, sive Dissertationis de vero systemate mundi libri IV*, Amsterdami, Apud Gvil. & Iohannem Blaev. 1639 [4°, 159pp].

-----. *ΘΕΩΝΟΣ ΣΜΤΡΝΑΙΟΤ ΠΛΑΤΩΝΙΚΟΥ*, τῶν κατὰ μαθηματικὴν χρησίμων εἰς τὴν τοῦ Πλάτωνος ἀνάγνωσιν. *Theonis Smyrnaei Platonici, Eorum, quae in Mathematicis ad Platonis lectionem vtilia sunt, Expositio. E Bibliotheca Thvana. Opus nunc primum editum, Latina versione, ac Notis illustratum ab Ismaele Bvllialdo Iuliodunensi.* Lvtetiae Parisiorvm, Apud Lvdovicvm de Hevqveville...M.DC.XLIV [4°, 308pp].

-----. *Ismaelis Bvllialdi astronomia philolaica. Opus novvm, in quo motus Planetarum per novam ac veram Hypothesim demonstrantur. Mediique motus, aliquot observationum authoritate, ex Manuscripto Bibliothecae Regiae quae hactenus omnibus Astronomis ignotae fuerunt, stabiliuntur. Superque illa Hypothesi Tabvlae constructae omnium quotquot hacenus editae sunt, facillimae. Addita est nova methodus cvivs ope Eclipses Solares, absque vlla solutione triangulorum Sphaericorum ad Parallaxes investigandas, expeditissimè computantur. Historia ortvs et progressvs astronomiae in Prolegomenis discribitur, & breuiter recensentur ea, quae in hoc opere nunc primum prodeunt.* Parisiis, Sumptibus Simeonis Piget ...M.DC.XLV [2 vols in 1; folio, 469 + 232pp].

-----. "Ismaelis Bvllialdi Epistola de solis defectu, cujus Livius lib. XXXVII. meminit." In *Johannes Fredericus Gronovius, Ad T. Livii Patavini libros svperstites Notae.* Lugd. Batav, cIↄ Iↄ c XLV [24°, pp 787-808].

-----. *Ducae, Michaelis Ducae Nepotis, Historia Byzantina. . .a Joanne Palaeologo I. ad Mehemetem II. Accessit Chronicon Breve [A.D. 1421-1523] quo Graecorum, Venetorum, et Turcorum aliquot gesta continentur. . .Nunc primùm. . .edita, versione Latina & Notis illustrata. Studio & opera Ismalis Bullialdi.* Parisiis, M.DC.XLIX [folio, ix + 288].

-----. *Ismaelis Bullialdi observatio secundi deliqvii lunaris anno 1652. mense Septembri facti, vna cvm calcvlo illivs, et fvtvri alius Lunae defectus mense Martio 1653. ex Tabulis Philolaicis. Et observationes circa cometam qui mense Decembri 1652. fulsit, tam ab ipso, quàm ab aliis factae.* Parisiis, E Typographia Edmvndi Martini...M.DC.LIII [4°, 22pp].

-----. "Animadversiones." In *Marci Manilii Astronomicon. A Iosepho Scaligero ex vetusto codice gemblacensi infinitis mendis repurgatum. Ejusdem Iosephi Scaligeri notae, quibus auctoris prisca astrologia explicatur, castigationum caussae redduntur, portentosae transpositiones in eo auctore antiquitus commissae indicantur. Nunc primum ex codice Scaligeri, quem sua manu ad tertiam editionem praeparauerat, plurimisque accessionibus suarum curarum locupletauerat, post longas moras latebrasque, in lucem publicam cum auctario tam nobili, restitutae. Accesserunt quaedam. . . Thomae Renesii et Ismaelis Bvllialdi animadversiones.* Argentorati, sumptibus Ioannis Ioachimi Bockenhofferi, 1655 [4°; pp 1-18 (after p 462)].

-----. *Ismaelis Bvllialdi de popvlis fundis epistola ad Amplissimum Eruditissimumque Virum Nicolavm Rigaltivm.* Parisiis, ex officina Cramosiana. M.DC.LV. [4°, 20pp].

-----. "Ismaelis Bvllialdi Epistolae de supremis Gassendi" (Ismael Bvllialdvs Ioanni Alberto Portnero S.P.D.). In *Lessus mortualis illustri viro Petro Gassendo praeposito Diniensi regioque matheseos in academia Parisiensi professori celeberrimo canatvs a Ioanne Alberto Portnero. . .Accessit Praefatio eiusdem ad Virum Cl. Ioannem de Lavnoy Theologum Parisiensem; accesserunt simul ipsius Lavnoii, & Cl. itidem Viri Ismaelis Bvllialdi Epistolae de supremis Gassendi.* Parisiis, Ex Typographia Edmvndi Martini...M.DC.LVI. [4°].

-----. *Ismaelis Bullialdi pro ecclesiis Lvsitanicis ad clerum Gallicanum libelli dvo; accessit ejusdem ad ampliss. virum Nicolavm Rigaltium in suprema Metensi curia Senatorem dissertatio de populis fundis.* Argyropoli, Sumt. Io. Ioach. Bockenhoferi...CIↃ IↃ CLVI [16°, 60pp].

-----. *Nicolai Rigaltii, Ismaelis Bvllialdi et Henrici Valesii observationes de popvlis fvndis.* Divione, Apud Philibertvm Chavance, M.DC.LVI. [8°, xvi + 56pp].

-----. *Diatriba de Sancto Benigno, ab Ismaele Bullialdo Scripta anno M.DC.XL.* Parisiis, Apud Sebastianvm Cramoisy...Et Gabrielem Cramoisy...M.DC.LVII. [8°, 27pp].

-----. *Ismaelis Bvllialdi exercitationes geometricae tres. I. Circa demonstrationes per inscriptas & circumscriptas figuras. II. Circa conicarum sectionum quasdam propositiones. III. De Porismatibus. Astronomiae philolaicae fundamenta clariùs explicata, & asserta Aduersùs Clariss. Viri Sethi Wardi, Oxoniensis Professoris impugnationem.* Parisiis, Apud Sebastianum Cramoisy...Et Gabrielem Cramoisy...M.DC.LVII. [4°, vi + 56pp].

-----. *Ismaelis Bvllialdi de lineis spiralibus demonstrationes novae.* Parisiis, Apud Sebastianum Cramoisy... Et Gabrielem Cramoisy... M.DC.LVII. [4°, viii + 138pp].

-----. Κλαυδιου Πτολεμαιου Περὶ κριτηρίου και ἡγεμονικοῦ. *Claudii Ptolemaei tractatus de judicandi facultate et animi principatu. Accesserunt de illius patria & astronomiae sedibus Olympiodori & Theodori Meliteniotae testimonia. Et Inscriptio Canobi. . . repertorum ab eo in Astronomia summa capita continens. . .* Edidit, Latine reddidit & commentarius illustravit I. Bullialdus. Apud Sebastianum Cramoisy...Parisiis, 1663 [4°, ix ℓ. + 248pp].

-----. *Ismaelis Bullialdi ad astronomos monita dvo. Primum, De Stella Noua, quae in Collo Ceti ante annos aliquot visa est. Alterum, De Nebulosa in Andromedae cinguli parte Borea ante biennium iterum orta.* Parisiis, apud Sebastianvm Mabre-Cramoisy...M.DC.LXVII. [4°, 17pp].

-----. *Catalogus bibliothecae Thvanae, à clariss. vv. Petro et Iacobo Pvteanis ordine alphabetico primùm distributus; Tvm secundum scientias & artes à clariss. viro Ismaele Bvllialdo digestus. Nvnc vero editus à Iosepho Qvesnel,...* Parisiis, Impensis directionis, 1679 [2 vols., 8°].

-----. *Ismaelis Bullialdi opvs novvm ad arithmeticam infinitorum Libris sex comprehensum, In quo plura à nullis hactenus edita demonstrantur.* Lutetiae Parisiorum, Sumptibus viduae Joannis Pocquet, M.DC.LXXXII. [folio, iv + 426pp].

3. Contemporary Printed Works

Aubin, Nicolas. *Histoire des diables de Loudun, ou de la possession des religieuses Ursulines, et de la condemnation et du supplice d'Urbain Grandier, curé de la même ville.* 3rd ed. Amsterdam, 1716.

Boecler, Johann Heinrich. *Marci Manilii Astronomicon. A Josepho Scaligero ex vetusto codice gemblacensi infinitis mendis repurgatum. Eiusdem Josephi Scaligeri notae, quibus auctoris prisca astrologia explicatur, castigationum caussae redduntur, portentosae transpositiones in eo auctore antiquitus commissae indicantur. Nunc primum ex codice Scaligeri, quem sua manu ad tertiam editionem praeparaverat, plurimisque accesionibus suarum curarum locupletaverat, post longas moras latebrasque, in lucem publicam cum auctario tam nobili, restitutae. Accesserunt quaedam...Thomae Renesi et Ismaelis Bullialdi animadversiones.* Strassburg, 1655.

Borelli, Giovani Alfonso. *Theoricae mediceorum planetarum ex causis Physicis deductae.* Florence, 1666.

Bossuet, Jacques-Bénigne de. *Correspondance de Bossuet.* Edited by Charles Urbain and Eugène Levesque. 15 vols. Paris, 1909-1925.

Brahe, Tycho. *Tychonis Brahe Dani Opera Omnia.* Edited by J. L. E. Dreyer. 15 vols. Copenhagen, 1913-1929.

Burattini, Tito Livio. *Intorno alla vita ed ai lavori di Tito Livio Burattini.* Edited by Antonio Favaro. Memorie del Reale Istituto Veneto de Scienze, Lettere ed Arti, vol. 25, no. 8. Venice, 1896.

Burman, Pieter, ed. *Sylloges Epistolarum a viris illustribus scriptarum . . .* 2d ed. 5 vols. Leyden, 1727.

Caspar, Max, ed. *Bibliographia Kepleriana, ein Führer durch das gedruckte Schrifttum von Johannes Kepler.* 2d ed. by Martha List. Munich, 1968.

Chapelain, Jean. *Lettére inedíte á corrispondénti Italíani.* Ed. Petre Ciureanu. Genoa, 1964.

-----. *Lettres de Jean Chapelain.* Ed. Philippe Tamizey de Larroque. 2 vols. Paris, 1880-1883.

Chevreau, Urbain. *Chevraeana.* 2 vols. Paris, 1697-1700.

Chiaramonti, Scipione. *Antiphilolaus. . .in quo Philolao rediuiuo de motu terrae & solis, ac fixarum quiete repugnatur rationesq; eius, quas ipse pro demonstrationibus affert fallaces detegunter. Insuper positio eadem de re Copernici confutator, & Galilaei defensiones reijeiuntur. . .* Caesenae, 1643.

Cousin, Louis. "Eloge de Monsieur Boulliau." *Journal des Sçavans* (14 Feb 1695):74-84.

Cunitia, Maria. *Urania propitia, sive Tabulae astronomicae mire faciles, vim hypothesium physicarum a Kepplero proditarum complexae; facillimo calculandi compendio, sine ulla logarithmorum mentione, phaenomenis satisfacientes. . .* Silesia, 1650.

Descartes, René. *Correspondance.* Edited by Charles Adam and Gaston Milhaud. 8 vols. Paris, 1936-1963.

-----. *Oeuvres de Descartes.* Edited by Charles Adam and Paul Tannery. 13 vols. Paris, 1897-1913.

Des Noyers, Pierre. *Lettres de Pierre Des Noyers, secrétaire de la reine de Pologne, Marie-Louise de Gonzague, princesse de Montaue et de Nevers, pour servir à l'histoire de Pologne et de Suède de 1655 à 1659 (Lettres de Pierre Des Noyers à Ismael Boulliaud.).* Librairie de B. Behr (E. Bock), Berlin, 1859.

Dupuy, Pierre. *Traittez concernant l'Histoire de France, sçavoir la condamnation des Templiers, avec quelques actes; l'histoire du schisme, les papes tenans le siège en Avignon, et quelques procez criminels. . .* Paris, 1654.

-----. *Traitez touchant les Droits du Roy très-Chrestien sur plusieurs estats et seigneuries possedées par divers princes voisins et pour prouver qu'il tient à juste titre plusiers provinces contestées par les princes estrangers. . .* Paris, 1655.

Eichstadius, Laurentius. *Pars prima ephemeridum novarum et motuum coelestium quinquennalis.* Stettin, 1634.

Fabroni, Angelo, ed. *Lettre inedite d'uomini illustri.* 2 vols. Florence, 1773 and 1776.

Faugère, A. P. *Journal du voyage de deux jeunes hollandais à Paris en 1656-1658.* Paris, 1899.

Fermat, Pierre. *Oeuvres de Fermat.* Edited by Paul Tannery, Charles Henry, and Cornélis de Waard. 5 vols. Paris, 1891-1922.

Flamsteed, John. *The "Gresham Lectures" of John Flamsteed.* Edited and Introduced by Eric G. Forbes. London, 1975.

Galilei, Galileo. *Le Opere di Galileo Galilei, Edizione Nazionale.* Edited by Antonio Favaro. 20 vols. Florence, 1890-1939.

Gassendi, Pierre. *Institutio astronomia.* Paris, 1647.

-----. *Lettres familières à François Luillier pendant l'hiver 1632-1633.* Edited by Bernard Rochot. Paris, 1944.

-----. *Petri Gassendi. . .Opera Omnia. . .hactenus edita auctor ante obitum recensuit. . .posthuma vero. . .in lucem nunc primum prodeunt. . .* Edited by H. L. Habert de Montmor and F. Henry. 6 vols. Lyon, 1658-1675.

-----. *The Selected Works of Pierre Gassendi.* Edited and translated by Craig B. Brush. The Hague, 1972.

Gaultier, Joseph. *Les Correspondants de Peiresc. IV. Joseph Gaultier, prieur de la Valette, Lettres inédites écrites d'Aix à Peiresc, de 1609 à 1632, publiées et annotées par Ph. Tamizey de Larroque.* Aix, 1881.

Hevelius, Johann. *Illustribus viris. . .Gassendo et Is. Bullialdo.* Danzig, 1652.

-----. *Machina coelestis.* 2 pts. Danzig, 1673-1679.

-----. *Mercurius in Sole visus Gedano anno. . .1661. . .cum aliis quibusdam. . .observationibus. . .cui annexa est, Venus in Sole pariter visa, anno 1639. . .Liverpoliae a Jeremia Horroxio, nunc primum edita. . .* Danzig, 1662.

Horrox, Jeremiah. *Opera postuma, viz Astronomia Kepleriana, defensa et promota. . .* London, 1673.

Huet, Pierre-Daniel. *Memoirs of the Life of Peter Daniel Huet, Bishop of Avranches.* Translated by John Aikin. 2 vols. London, 1810.

-----. *P.-D. Huetii, Episcopi Abricensis, Commentarius de rebus ad eum pertinentibus.* Amsterdam, 1718.

Huygens, Christiaan. *Oeuvres Complètes de Christiaan Huygens, publiées par la Societé Hollandaise des Sciences.* 22 vols. The Hague, 1888-1950.

Journal des Sçavans. Paris, 1665-

Kepler, Johannes. *Gesammelte Werke*. Edited by Walther van Dyck, Max Caspar, and Franz Hammer. Munich, 1937-

-----. *Joannis Kepleri astronomi opera omnia*. Edited by Christian Frisch. 8 vols. Frankfurt, 1858-1871.

-----. *Tabulae Rudolphinae*. Ulm, 1627.

Lackman, A. H. *Epistolae diversi argumenti maximam partem a variis ad Clarissimum multorumque meritorum virum Lucam Lossium illustris Lycei apud Luneburgenses dum viveret Prorectorem et post eum a Duraeo, Langwedelio, Boeclero, Portnero, Berneggero, Freinshemio aliisque ad alios exaratae*. . . Hamburg, 1728.

Leeuwenhoek, Antoni van. *The Collected Letters of Antoni van Leeuwenhoek*. Amsterdam, 1939-

Longomontanus, Christianus Severinus. *Astronomia Danica*. Amsterdam, 1622.

Lubienietzki, Stanislas. *Theatri Cometici pars prior Communicationes de Cometis 1664 et 1665 cum Viris per Europam Cl. habitis. . .Auctore Stanislao de Lubienietz Lubieniecio Rolitsio*. Amsterdam, 1667.

Luillier, François. "Lettres de Luillier, père de Chapelle." In *Les Historiettes des Tallemant des Réaux*, edited by Louis-J.N. de Monmerqué and Paulin Paris, vol. 4, pp. 489-516. 3d ed. Paris, 1850-1856.

Mazarin, Cardinal. *Lettres du Cardinal Mazarin pendant son ministère (1642-1661)*. Edited by A. Chéruel and G. D'Avenel. 9 vols. Paris, 1872-1896.

Mercator, Nicholas. *Hypothesis astronomica nova, et consensus eius cum observationibus*. London, 1664.

-----. *Nicolai Mercatoris. . .Institutionum Astronomicarum Libri Duo*. London, 1676.

-----. "Some considerations of Mr. Nic. Mercator, concerning the geometrick and direct method of Signior Cassini for finding the apogees, eccentricities, and anomalies of the planets. . ." *Philosophical Transactions* 5(1670):1168-1175.

Mersenne, P. Marin. *Correspondance du P. Marin Mersenne*. Edited by Paul Tannery, Cornélis de Waard, and Armand Beaulieu. 14 vols. Paris, 1932- .

-----. *Harmonie universelle*. Paris, 1636.

Morin, Jean-Baptiste. *J. B. Morinus. . .ab I. Bullialdi convitiis iniquissimis. . .vindicatus.* 2 pt. [Paris?, 1640?].

Nabielak, Ludwik. "Listy Piotra des Noyers, sekretarza królowej Maryi Kazimiry z lat 1680-1683 rzeczy polskick dotyczace." *Biblioteca Ossolińskich* X(1868):1-79.

Naudé, Gabriel. *Judgement de tout ce qui à este imprime contra le Cardinal Mazarin depuis le septième janvier jusques à la déclaration du premier. . .* Paris, 1650.

Newton, Isaac. *The Correspondence of Isaac Newton.* Edited by Herbert W. Turnbull, J. F. Scott, and A. Rupert Hall. 7 vols. Cambridge, 1959-1977.

-----. *The Mathematical Papers of Isaac Newton.* Edited by Derek T. Whiteside (Vol. 4, edited by J. F. Scott). To be complete in 8 vols. Cambridge, 1967- .

Newton, John. *Astronomia Britannica, exhibiting the doctrine of the sphere, and theory of the planets. . .according to the Copernican system as it is illustrated by Bullialdus and the easie way of calculating lately published by Doctor Ward.* London, 1657.

Oldenburg, Henry. *The Correspondence of Henry Oldenburg.* Translated and edited by A. Rupert Hall and Marie Boas Hall. 9 vols. Madison, 1965-1973; since vol. 10, London, 1975- .

Olhoff, J. E. *Excerpta ex literis illustrium et Clarissimorum Virorum ad Nobilissimum, Ampliss. et Consultiss. Dn. Johannem Hevelium. . ., Studio ac opera Johannis Erici Olhofii secretarii.* Danzig, 1683.

Omont, Henri. "Lettres d'Émeric Bigot à Gilles Ménage et à Ismaël Bouillaud au cours de son voyage en Allemagne lors de l'élection de l'empereur Léopold Ier (1657-1658)." *Annuaire bulletin du Société de l'Histoire de France* 23(1886):227-255.

-----. "Trois lettres de Boulliau à Dupuy (1645-1651)." *Archives Historiques, Artistiques, et Littéraires* 2(Paris, 1890-1891): 134-138.

Le Opere dei Discepoli di Galileo Galilei. Vol. I (1642-1648). Edited by Paolo Galluzzi and Maurizio Torrini. Florence, 1975- .

Pagan, Blaise François de. *La théorie des planètes du Comte de Pagan, où tous les orbes célestes sont geométriquement ordonnez, contre le sentiment des astronomes.* Paris, 1657.

-----. *Les tables astronomiques du Comte de Pagan. . .* Paris, 1658.

Pascal, Blaise. *Oeuvres de Blaise Pascal*. Edited by Léon Brunschvicg, Pierre Boutroux, and Félix Gazier. 12 vols. Paris, 1908-1914.

Peiresc, Nicolas-Claude Fabri de. *Les correspondants de Peiresc, Lettres inédites*. 2 vols. Reimpression, Geneva, 1972.

-----. *Lettres de Peiresc*. Edited by Philippe Tamizey de Larroque. 7 vols. Paris, 1888-1898.

Philolaus [pseud.]. *A Character of Popery and Arbitrary Government, With a Timely Caveat and Advice to all the Freeholders, Citizens and Burgesses, How They May Prevent the Same, By Choosing Good Members to Serve in this New Parliament*. London, 1681.

Philosophical Transactions: giving some Accompt of the present understandings, Studies and Labours of the Ingenious in many considerable parts of the World. Edited by Henry Oldenburg. London and Oxford, 1665-1677.

Portner, Johann Albrecht. *Lessus mortualis illustri viro Petro Gassendo praeposito Diniensi regioque matheseos in academia Parisiensi professori celeberrimo cantatus a Joanne Alberto Portnero*. Regensburg, 1655.

Quesnel, Joseph. *Catalogus bibliothecae Thuani, a Petro et Jacobo Puteanis ordine alphabetico primum distributus, tum secundum scientias et artes a. . .Ismaele Bullialdo digestus, nunc vero editus a Josepho Quesnel*. 2 vols. Paris, 1679.

Riccioli, Giovanni Battista. *Almagestum novum. . .* Bologna, 1651.

-----. *Astronomia reformata*. Bologna, 1665.

Rigaud, Stephen P., and Stephen J. Rigaud, eds. *Correspondence of Scientific Men of the Seventeenth Century. . .in the Collection of. . . the Earl of Macclesfield*. 2 vols. Oxford, 1851.

Schickard, Wilhelm. *Pars responsi ad epistolas P. Gassendi. . .de Mercurio sub Sole viso, & alijs novitatibus uranicis*. Tübingen, 1632.

Shakerley, Jeremy. *Anatomy of Urania Practica*. London, 1649.

-----. *Synopsis Compendiaria*. London, 1651.

-----. *Tabulae Britannicae. . .Calculated for the meridian of London from the hypothesis of Bullialdus and the observations of Mr. Horrox*. London, 1641.

Sherburne, Sir Edward. *The Sphere of M. Manilius made an English Poem: with annotations and an Astronomical Appendix*. London, 1675.

Streete, Thomas. *Astronomia Carolina. A new theorie of the coelestial motions. . .* London 1661.

Tamizey de Larroque, Philippe. "Document relatif à Urbain Grandier." *Cabinet Historique* 3(1879):1-14.

-----. "Quelques Lettres inédites d'Isaac de la Peyrère à Boulliau" *Plaquettes Gontaudaises*, No. 2 (1878) Paris.

Torricelli, Evangelista. *Opere di Evangelista Torricelli*. Edited by Gino Loria and Giuseppe Vassura. 4 vols. in 5 pts. Faenza, 1919-1944.

Wallis, John. *Arithmetica infinitorum*. Oxford, 1656 [1655].

Ward, Seth. *Astronomia geometrica: ubi methodus proponitur qua primariorum planetarum astronomia sive elliptica sive circularis possit geometrice absolvi*. Oxford, 1656.

-----. *De cometis, ubi de cometarum natura disseritur, nova cometarum theoria, et novissimi cometae historia proponitur*. Oxford, 1653.

-----. *Idea trigonometriae demonstratrae. . .Item praelectio de cometis et inquisitio in Bullialdi Astronomiae philolaicae fundamenta*. Oxford, 1654.

-----. *In Ismaelis Bullialdi astronomia philolaicae fundamenta inquisitio brevis*. Oxford, 1653.

Wing, Vincent. *Astronomia Britannica*. London, 1669.

-----. *Astronomia instaurata: Or a New and Comprehensive Restauration of Astronomy*. London, 1656.

-----. *Harmonicon coeleste: Or the Coelestial Harmony of the Visible World*. London, 1651.

-----. *Urania practica*. London, 1649.

B. LATER AND SECONDARY SOURCES

Abraham, Claude Kurt. *Gaston d'Orléans et sa cour.* Chapel Hill, 1965.

Adam, Antoine. "Gassendi: L'Influence Post-hume." In *Pierre Gassendi (1592-1655), sa vie et son oeuvre*, pp. 157-182. Paris, 1955.

-----. *Histoire de la littérature française au XVIIe siècle.* 5 vols. Paris, 1948-1956.

Adam, Charles. "Le P. Mersenne et ses correspondants en France." *Bulletin Historique et Philologique du Comité des travaux hist. et scient.* (1896):486-500.

Aiton, E. J. *The Vortex Theory of Planetary Motions.* New York, 1972.

Allen, E. J. B. *Post and Courier Service in the Diplomacy of Early Modern Europe.* The Hague, 1972.

Applebaum, Wilbur. "Kepler in England: The Reception of Keplerian Astronomy in England, 1599-1687." Ph.D. Dissertation, State University of New York, Buffalo, 1969.

Armitage, Angus. "'Borell's Hypothesis' and the Rise of Celestial Mechanics." *Annals of Science* 6(1950):268-282.

Aubin, N. *Histoire des diables de Loudun, ou de la possession des religieuses ursulines.* 2d ed. Amsterdam, 1716.

Auger, Leon. "Les Idées de Roberval sur le système du monde." *Revue d'Histoire des Sciences*, July-September 1957, pp. 226-234.

-----. *Un Savant Méconnu: Gilles Personne de Roberval (1602-1675): Son Activité intellectuelle dans les domaines mathématique, physique, mécanique et philosophique.* Paris, 1962.

Barker, Thomas M. *Double Eagle and Cresent, Vienna's Second Turkish Siege and its Historical Setting.* Albany, New York, 1967.

Bayle, Pierre. *Dictionnaire historique et critique.* 4 vols. Rotterdam, 1720.

Beauchet-Filleau, H. and P. Beauchet-Filleau (avec le concours de RR. PP. H. et G. Beauchet-Filleau, de plusiers membres des sociétés savantes de la province et la collaboration pour la partie héraldique de M. Maurice de Gouttepagnon). *Dictionnaire historique et généalogique des familles du Poitou.* 2d ed. 4 vols. Poitiers, 1891-1915.

Belloc, Alexis. *Les postes français.* Paris, 1886.

Bertrand, Joseph. *L'Académie des Sciences et les académiciens de 1666 à 1793*. Paris, 1869.

Bigourdan, Guillaume. *Histoire de l'astronomie d'observation et des observatoires en France*. Paris, 1918.

-----. "Inventaire général et sommaire des manuscrits de la Bibliothèque de l'observatoire de Paris." In *Annales de l'Observatoire de Paris, Memoires*, vol. 21, pp. F1-60. Paris, 1895.

-----. *Les Premières sociétés savantes de Paris et les origines de l'Académie des Sciences*. Paris, 1919.

Bots, Hans. *Correspondance de Jacques Dupuy et de Nicolas Heinsius (1646-1656)*. The Hague, 1971.

-----. "Ismaël Boulliau, ses voyages en Hollande." *Lias* 1(1974):29-62.

Boyer, Carl B. "Note on Epicycles & the Ellipse from Copernicus to Lahire." *Isis* 38(1947):54-56.

Brown, Harcourt. *L'Académie de Physique de Caen (1666-1675) d'après les lettres d'André de Graindorge*. Caen, 1938.

-----. *Science and the Human Comedy, Natural Philosophy in French Literature from Rabelais to Maupertuis*. Toronto and Buffalo, 1976.

-----. *Scientific Organizations in Seventeenth-Century France (1620-1680)*. Baltimore, 1934.

Brugmans, Henri L. *Le Séjour de Christian Huygens à Paris et ses relations avec les milieux scientifiques français, suivi de son Journal de Voyage à Paris et à Londres*. Paris, 1935.

Busson, H. *La Pensée religieuse française de Charron à Pascal*. Paris, 1933.

Cabeen, D. C., ed. *Seventeenth Century. A Critical Bibliography of French Literature*, vol. 3. Syracuse, 1961.

Caspar, Max. *Kepler*. Edited and translated by C. Doris Hellman. London and New York, 1959.

Chapin, Seymour L. "The Astronomical Activities of Nicolas Claude Fabri de Peiresc." *Isis* 48(1957):13-29.

Chaplain, Jean. *Jean Chaplain: 77 lettres inédites à N. Heinsius (1649-1658)*. Edited by Bernard Bray. The Hague, 1966.

Cheney, C. R. *Handbook of Dates for Students of English History*. London, 1970.

Chéruel, A. "Biographie normande, Nicolas Brétel, seigneur de Grémonville, ambassadeur de France à Rome et à Venise, 1644-1648." *Extrait du Précis analytique des Travaux de l'Académie Royal des Sciences, Belles-Lettres et Arts de Rouen*. Rouen, 1847.

Clark, Joseph T. "Pierre Gassendi and the Physics of Galileo." *Isis* 54 (1963):352-370.

Clarke, Jack A. "A Book Buying Tour in 1645: A Note on Ismael Boulliau in Italy." *The Journal of Library History* 4(1969):330-336.

Clément, Pierre. *Histoire de Colbert et de son administration*. 3rd Ed. 2 vols. Paris, 1892.

Cohen, Gustave. *Ecrivains français en Hollande dans la première moitié du XVIIe siècle*. Paris, 1920.

Cohen, I. Bernard. "The Dynamics of the Galileo--'Plato Problem'--its Relation to Newton's *Principia*." In *Actes du IXe congrès international d'histoire des sciences*, pp. 187-196. Barcelona, 1959.

-----. *Introduction to Newton's 'Principia.'* Cambridge, 1971.

-----. *The Newtonian Revolution, with Illustrations of the Transformation of Scientific Ideas*. Cambridge, 1980.

Cole, C. W. *Colbert and a Century of French Mercantilism*. 2 vols. New York, 1939.

Coornaert, Emile. *Les corporations en France avant 1789*. Paris, 1941.

Delambre, Jean B. J. *Histoire de l'astronomie moderne*. 2 vols. Paris, 1821.

Delaroche, Pierre. *Une épidémie de peste à Loudun en 1632*. Bordeaux, 1936.

Delattre, Pierre. *Les Etablissements des jésuites en France depuis quatre siècles*. 5 vols. Wetteren, Belgium, 1949-1957.

Delisle, Léopold. *Le Cabinet des manuscrits de la Bibliothèque imperiale*. 4 vols. Paris, 1868-1881.

-----. *Catalogue des Manuscrits des Fonds Libri et Barrois*. Paris, 1888.

Denis, Jacques. *Sceptiques ou Libertins de la première moitié du XVIIe siècle: Gassendi, Gabriel Naudé, Guy Patin, La Mothe le Vayer, Cyrano de Bergerac*. Caen, 1884.

Dethan, Georges. *Gaston d'Orléans: Conspirateur et prince charmant.* Paris, 1959.

Dibon, Paul. *Inventaire de la correspondance d'André Rivet (1595-1650).* The Hague, 1971.

Dibon, Paul, Hans Bots, and E. Bots-Estourgie. *Inventaire de la correspondance de Johannes Fredericus Gronovius (1631-1671).* The Hague, 1974.

Dictionary of National Biography. 2d ed. 22 vols. London, 1908-1937.

Dictionnaire de Biographie Française. Paris, 1933- .

Dictionnaire des lettres françaises: le dix-septième siècle. Under the direction of Cardinal Georges Grente. Paris, 1954.

Doucette, Leonard E. *Emery Bigot, Seventeenth-Century French Humanist.* Toronto, 1970.

Drake, Stillman. "The Accademia dei Lincei." *Science* 151(11 March 1966): 1194-1200.

Dreux Du Radier, J.-F. *Bibliothèque historique et critique du Poitou.* 3 vols. Niort, 1842-1849.

Dumoustier de la Fond. *Essais sur l'histoire de la ville de Loudun.* Poitiers, 1778.

Fauré-Fremiet, E. "Les Origines de L'Académie des Sciences de Paris." *Notes and Records of the Royal Society of London* 21(1966):20-31.

Feret, L'Abbe P. *La Faculté de théologie de Paris et ses docteurs les plus célèbres.* Vols. 3-5. Paris, 1904.

Fermi, Stefano. *Lorenzo Magalotti, scienziato e letterato, 1637-1712.* Piacenza, 1903.

Fouillée, Alfred. "Notice sur la vie et les travaux de M. Adolphe Franck." *Séances et Travaux de l'Académie des Sciences Morales et Politiques, Compte Rendu* 43(1895):177-197.

Franklin, Alfred. *Historie de la Bibliothèque Mazarine et du Palais de l'Institut.* Paris, 1901.

-----. *Les Anciennes Bibliothèques de Paris.* 3 vols. Paris, 1867-1873.

-----. *Précis de l'histoire de la Bibliothèque de roi, aujourd'hui Bibliothèque nationale.* 2nd ed. Paris, 1875.

Gasztowtt, A.-M. *Une mission diplomatique en Pologne au XVIIe siècle: Pierre de Bonzi à Varsovie (1665-1668)*. Paris, 1916.

Gauja, Pierre. "L'Académie Royale des Sciences (1666-1793)." *Revue d' Histoire des Sciences*, 2, No. 4 (1949):293-310.

Gaythorpe, Sidney B. "Jeremiah Horrocks and His 'New Theory of the Moon.'" *Journal of the British Astronomical Association* 67(1957): 134-144.

-----. "On Horrock's Treatment of the Evection and Equation of the Centre, with a Note on the Elliptic Hypothesis of Albert Curtz and its Correction by Boulliau and Newton." *Monthly Notices of the Royal Astronomical Society* 85(1925):858-865.

George, Albert J. "The Genesis of the Académie des Sciences." *Annals of Science* 3(1938):372-401.

Graesse, J. G. Th. *Orbis latinus oder Verzeichnis der wichtigsten lateinischen Orts-und Ländernamen*. Berlin, 1909.

Guhrauer, G. E. *Joachim Jungius und sein Zeitalter*. Stuttgart and Tübingen, 1850.

Haag, Eugène, and Emile Haag. *La France protestante, ou Vies des protestants français*. 10 vols. Paris and Geneva, 1846-1859.

Hahn, Roger. *The Anatomy of a Scientific Institution: The Paris Academy of Sciences, 1666-1803*. Berkeley, 1971.

Hall, A. Rupert. "Wren's Problem." *Notes and Records of the Royal Society of London* 20(1965):140-144.

Hanson, Norwood R. "The Copernican Disturbance and the Keplerian Revolution." *Journal of the History of Ideas* 22(1961):169-184.

Harrisse, H. *Le Président de Thou et ses descendants, leur célèbre bibliothèque, leurs armoiries et les traductions françaises de J. A. Thuani Historiaram sui Temporis*. Paris, 1905.

Henry, Charles. *Huygens et Roberval*. Leyden, 1879.

Herivel, John W. *The Background to Newton's 'Principia': A Study of Newton's Dynamical Researches in the Years 1664-1684*. Oxford, 1965.

Hetherington, N. S. "The Hevelius-Auzout Controversy." *Notes and Records of the Royal Society of London* 27(1972):103-106.

Hirschfield, John Milton. *The Académie Royale des Sciences, 1666-1683.* New York, 1981.

The History of the Devils of Loudun. Edited and translated by Edmund Goldsmid. Edinburgh, 1887.

Holton, Gerald. "Johannes Kepler's Universe: Its Physics and Metaphysics." *American Journal of Physics* 24(1956):340-351.

Hoppen, K. Theodore. *The Common Scientist in the Seventeenth Century, A Study of the Dublin Philosophical Society, 1683-1708.* London, 1970.

Hoskin, Michael A. "Novae and Variables from Tycho to Bullialdus." *Sudhoffs Archiv* 61(1977):195-204.

Hoskin, Michael A. and Christine (Schofield) Jones. "Problems in Late Renaissance Astronomy." In *Le Soleil à la Renaissance*, pp. 19-31. Paris, 1965.

Humbert, Pierre. *Un Amateur: Peiresc, 1580-1637.* Paris, 1933.

-----. *L'Astronomie en France au dix-septième siècle.* Paris, 1952.

-----. "Lieux astronomiques à Paris de 1610 à 1667." *XVIIe Siècle* (1953-1954):1-5.

-----. "Mersenne et les astronomes de son temps." *Revue d'Histoire des Sciences* 2(1948):29-32.

-----. "L'Oeuvre Astronomique de Gassendi." *Actualitiés Scientifiques et Industrielles*, No. 378, *Exposés d'Histoire et Philosophie des Sciences.* Paris, 1936.

Huxley, Aldous. *The Devils of Loudun.* New York, 1965.

Jammer, Max. *Concepts of Force: A Study in the Foundation of Dynamics.* Cambridge, 1957.

Jourdain, Charles M. G. *Histoire de l'Université de Paris au XVIIe et XVIIIe siècles.* Paris, 1862-1866.

Kerviler, R. *Henri-Louis Habert de Montmor, de l'Académie française et bibliophile (1600-1679).* Bibliophile français, 1872.

King, James E. *Science and Rationalism in the Government of Louis XIV, 1661-1683.* Baltimore, 1949.

Kinser, Samuel. "An Unknown Manuscript Catalogue of the Library of J. A. De Thou." *The Book Collector* 17 (1968):168-176.

-----. *The Works of Jacques-Auguste de Thou.* The Hague, 1966.

Koyré, Alexandre. "A Documentary History of the Problem of Fall from Kepler to Newton." *Transactions of the American Philosophical Society* 45(1955):329-395.

-----. "La Gravitation universelle de Kepler à Newton." *Archives Internationales d'Histoire des Sciences* 30(1951):638-653.

-----. *La Révolution Astronomique: Copernic, Kepler, Borelli.* Paris, 1961.

Labrousse, Elisabeth. *Inventaire critique de la correspondance de Pierre Bayle.* Paris, 1961.

Lachevre, Frédéric. *Le Libertinage au XVIIe siècle.* 11 vols. Paris, 1904-1924.

Legué, Gabriel. *Documents pour servir à l'histoire médicale des possédées de Loudun.* Paris, 1874.

-----. *Urbain Grandier et les possédées de Loudun. Documents inédits de M. Charles Barbier.* Paris, 1880.

Lenoble, Robert. *Mersenne, ou la naissance du Mécanisme.* 2d ed. Paris, 1971.

Le Prince, Nicholas Thomas. *Essai historique sur la Bibliothèque du Roi.* Paris, 1856.

Lerosey, Auguste-Louis. *Loudun, histoire civile et religieuse.* Loudun, 1908.

Libri, Guillaume. *Histoire des sciences mathématiques en Italie.* 4 vols. Paris, 1838-1841.

Lindberg, S. G. "Christina and the Scholars." In *Christina Queen of Sweden, a Personality of European Civilisation,* pp. 44-53. Stockholm, 1966.

Lister, Martin. *A Journey to Paris in 1698.* London, 1698. Reprint, New York, 1971.

McClaughlin, Trevor. "Sur les rapports entre la Compagnie de Thévenot et l'Académie Royale des Sciences." *Revue d'Histoire des Sciences* 28 (1975):235-242.

McKeon, Robert M. "Etablissement de l'astronomie de précision et oeuvre d'Adrien Auzout." Doctoral dissertation, University of Paris, 1965.

McPike, Eugene F. *Hevelius, Flamsteed and Halley: Three Contemporary Astronomers and their Mutual Relations*. London, 1937.

Maddison, R. E. W. "A Tentative Index of the Correspondence of the Honourable Robert Boyle, F.R.S." *Notes and Records of the Royal Society of London* 13(1958):128-197.

Maeyama, Yasukatsu. *Hypothesen zur Planetentheorie des 17. Jahrhunderts*. Frankfurt on the Main, 1971.

Mandrou, Robert. *Des humanistes aux hommes de science: XVIe et XVIIe siècles*. Paris, 1973.

-----. *Magistrates et sorciers en France au XVIIe siècle: Une analyse de psychologie historique*. Paris, 1968.

Michaud, J. F., ed. *Biographie Universelle, ancienne et moderne*. 45 vols. Paris, n.d.

Middleton, W. E. Knowles. *The Experimenters, a Study of the Accademia del Cimento*. Baltimore and London, 1971.

-----. *A History of the Thermometer and Its Use in Meteorology*. Baltimore, 1966.

-----. "Paolo del Buono on the Elasticity of Air." *Archives for the History of the Exact Sciences* 6(1969):1-28.

Moréri, Louis. *Le grand dictionnaire historique ou le Mélange curieux de l'histoire sacrée et profane*. 18th ed. 8 vols. Amsterdam, 1740.

Murris, R. *La Hollande et les hollandais au XVIIe et au XVIIIe siècles, vus par les français*. Paris, 1925.

Nellen, H. J. M. "Ismael Boulliau (1605-1694), nieuwsjager en correspondent." Doctoral dissertation (in Dutch), Nijmegen, 1980.

Nicéron, Jean Pierre. *Mémoire pour servir à l'histoire des hommes illustres dans la République des Lettres, avec un catalogue raisonné de leurs ouvrages*. 43 vols. Paris, 1727-1745.

Olmsted, John W. "Recherches sur la biographie d'un astronome et géodésian méconnu: Jean Picard (1620-1682)." *Revue d'Histoire des Sciences* 29(1976):213-222.

Ornstein, Martha. *The Rôle of Scientific Societies in the Seventeenth Century*. Chicago, 1928.

Pelseneer, Jean. "Gilbert, Bacon, Galilée, Kepler, Harvey et Descartes; leurs relations." *Isis* 17(1932):171-208.

Perrault, Charles. *Les Hommes illustres qui ont paru en France pendant le XVIIe siècle*. 2 vols. Paris, 1701.

Perrens, François-Tommy. *Les Libertins en France au XVIIe siècle*. 2d ed. Paris, 1899.

Picolet, Guy. "La correspondance de Jean Picard avec Hevelius (1671-1679), Edition et traduction française." *Revue d'Histoire des Sciences* 31(1978):3-42.

Pingré, Alexandre-Guy. *Annales célestes du Dix-Septième Siècle*. Edited by G. Bigourdan. Paris, 1901.

Pintard, René. *Le libertinage érudit dans la première moitié du XVIIe siècle*. 2 vols. Paris, 1943.

-----. "Modernisme, humanisme, libertinage." *Revue d'histoire littéraire* 48(1948):1-52.

-----. "Les problèmes de l'histoire du libertinage, notes et réflexions." *XVIIe Siècle* 32(April/June 1980):131-161.

Piobert, G. "Relations des savants entre eux avant la création de l'Académie des Sciences en 1666." *Comptes-rendus of the Académie des Sciences* 54(1862):703-715.

Popkin, R. H. *The History of Scepticism from Erasmus to Spinoza*. Berkeley, Los Angeles, London, 1979.

Pottinger, David T. *The French Book Trade in the Ancien Régime 1500-1791*. Cambridge, Massachusetts, 1958.

"Le Procès de Urbain Grandier." In *Urbain Grandier/Sade*, pp. 9-156. Le Grands Procès de l'Histoire de France, vol. 18. Paris, 1968.

Rice, James V. *Gabriel Naudé, 1600-1653*. Baltimore and London, 1939.

Russell, John L. "Kepler's Laws of Planetary Motion: 1609-1666." *British Journal for the History of Science* 2(1964):1-24.

Salmon, J. H. M. *Cardinal de Retz, the Anatomy of a Conspirator*. London, 1970.

Seck, Friedrich. *Wilhelm Schickard 1592-1635: Astronom, Geograph, Orientalist, Erfinder der Rechenmaschine*. Tübingen, 1978.

Secret, François. "Astrologie et alchimie au XVIIe siècle. Un ami oublié d'Ismaël Boulliau: Pierre des Noyers, secrétaire de Marie Louise de Gonzague, reine de Pologne." *Studi Francesi (Nuova Serie)*, Settembre-Dicembre 1976, pp. 463-479.

Small, Robert. Foreword to *An Account of the Astronomical Discoveries of Kepler* by William D. Stahlman. 1804. Reprint. Madison, 1963.

Smith, A. Mark. "Saving the Appearances of the Appearances: The Foundations of Classical Geometrical Opticks." *Archive for the History of Exact Sciences* 24(1981):73-100.

Smith, George. "Gabriël Naudé, a Librarian of the Seventeenth Century." *Library Association Records* (1899):423-431; 483-493.

Solente, S. "Les Manuscrits des Dupuy à la Bibliothèque Nationale." *Bibliothèque de l'Ecole des Chartes* 88(1927):177-250.

Solomon, Howard M. *Public Welfare, Science, and Propaganda in Seventeenth Century France, the Innovations of Théophraste Renaudot*. Princeton, 1972.

Spink, J. S. *French Free-Thought from Gassendi to Voltaire*. London, 1960.

Stevens, Henry. *Thomas Hariot: The Mathematician, the Philosopher, and the Scholar*. London, 1900.

Stolpe, Sven. *Christina of Sweden*. London, 1966.

Stratton, F. J. M. "Horrox and the Transit of Venus." *Occasional Notes of the Royal Astronomical Society* 7(1939):89-95.

Szanser, Adam J. "Johannes Hevelius (1611-1687), Astronomer of Polish Kings." *Quarterly Journal of the Royal Astronomical Society* 17 (1976):488-498.

Tapié, Victor L. *La France de Louis XIII et de Richelieu*. Paris, 1967.

Targosz, Karolina. "Correspondance scientifique de Pierre Des Noyers et d'Ismael Boulliau, Fragmente d'histoire des relations scientific franco-polonais[es] au XVIIe Siècle." In *Etudes de l'histoire de la science et de la technique*, Section 2, pp 19-29. Warsaw 1971.

-----. "Johann Hevelius et ses démarches pour trouver des mécènes en France." *Revue d'Histoire des Sciences* 30(1977):25-41.

Taton, René. *Les origines de l'Académie Royale des Sciences.* Paris, 1966.

Thoren, Victor E. "Kepler's Second Law in England." *The British Journal for the History of Science* 7(1974):243-256.

Thorndike, Lynn. *A History of Magic and Experimental Science.* 8 vols. New York, 1923-1958.

Tolmer, Léon. *Pierre-Daniel Huet (1630-1721), Humaniste-Physicien.* Bayeux, 1949.

Tranquille (Le Père). *Véritable relation des justes procédures observées au fait de la possession des Ursulines de Loudun et au procès de Grandier.* Archives Curieuses de l'Histoire de France, vol. 5. 1634. Reprint. Paris, 1838.

Uri, I. *Un cercle savant au XVII*[e] *siècle: François Guyet (1575-1655): d'après des documents inédits.* Paris, 1886.

Vaillé, E. *Histoire générale des Postes françaises*, Vol. 3. *De la réforme de Louis XIII à la nomination de Louvois, à la surintendance générale des postes, 1630-1668.* Paris, 1950.

Van Helden, Albert. "The Invention of the Telescope." *Transactions of the American Philosophical Society* 67, Pt. 4 (1977).

Volgraff, J. A. "Deux Lettres de Christiaan Huygens." *Archives internationales d'Histoire des Sciences* 4(1951):634-637.

Vuilhorgne, L. *Gui Patin (1601-1672). Sa vie, ses ancêtres, ses enfants, ses relations dans le monde des medecins et des littérateurs.* 2d ed. Bois-Colombes, 1898.

Waard, Cornélis de. "A la recherche de la correspondance de Mersenne." *Revue d'Histoire des Sciences* 2(1948):13-28.

Wade, Ira O. *The Intellectual Origins of the French Enlightenment.* Princeton, 1971.

Westfall, Richard S. *Force in Newton's Physics.* New York, 1971.

-----. *Never at Rest: A Biography of Isaac Newton.* Cambridge, 1980.

Whiteside, Derek T. "Before the *Principia*: The Maturing of Newton's Thoughts on Dynamical Astronomy, 1664-1684." *Journal for the History of Astronomy* 1(1970):5-19.

-----. "Newton's Early Thoughts on Planetary Motion: A Fresh Look." *British Journal for the History of Science* 2(1964):117-137.

-----. "Newton's Lunar Theory: From High Hope to Disenchantment." *Vistas in Astronomy* 19(1975-1976):317-328.

-----. "Newton's Marvellous Year: 1666 and All That." *Notes and Records of the Royal Society* 21(1966):32-41.

Wickelgren, F. *La Mothe le Vayer, sa vie et son oeuvre*. Paris, 1934.

Wilson, Curtis. "The Error in Kepler's Acronychal Data for Mars." *Centaurus* 13(1969):263-268.

-----. "From Kepler's Laws, So-Called, to Universal Gravitation: Empirical Factors." *Archive for the History of the Exact Sciences* 6 (1970):89-170.

-----. "Horrox, Harmonies, and the Exactitude of Kepler's Third Law." In *Science and History: Studies in Honor of Edward Rosen* (Studia Copernicana, Vol. XVI), pp. 235-259. Edited by Erna Hilfstein, Pawel Czartoryski, and Frank D. Grande. Warsaw, 1978.

-----. "Kepler's Derivation of the Elliptical Path." *Isis* 59(1968):5-25.

-----. "Newton and Some Philosophers on Kepler's 'Laws.'" *Journal of the History of Ideas* 35(1974):231-258.

Wolf, Charles Joseph Etienne. *Histoire de l'Observatoire de Paris de sa fondation à 1793*. Paris, 1902.

Wrangel, F. U. *Liste des diplomates français en Suède, 1541-1891*. Stockholm, 1891.

Yates, Frances A. *The French Academies of the Sixteenth Century*. London, 1947.

▼ ▼ ▼

<< at extra fastidium ac taedium nulli absolvendum >>

--from a letter of Boulliau to
Storanus on completion of the
Catalogus Bibliothecae Thuanae,
11 Jul 1653, CB XXV: 147r

www.ingramcontent.com/pod-product-compliance
Lightning Source LLC
LaVergne TN
LVHW091521200926
848483LV00002B/274
9780871699848